Lecture Notes in Computer S

Edited by G. Goos, J. Hartmanis, and J.

Springer
Berlin
Heidelberg
New York
Barcelona
Hong Kong
London
Milan
Paris
Tokyo

Joan Daemen   Vincent Rijmen (Eds.)

Fast
Software Encryption

9th International Workshop, FSE 2002
Leuven, Belgium, February 4-6, 2002
Revised Papers

Springer

Series Editors

Gerhard Goos, Karlsruhe University, Germany
Juris Hartmanis, Cornell University, NY, USA
Jan van Leeuwen, Utrecht University, The Netherlands

Volume Editors

Joan Daemen
Proton World
Zweefvliegtuigstraat 10
1130 Brussel, Belgium
E-mail: joan.daemen@protonworld.com

Vincent Rijmen
Cryptomathic
Lei 8A
3000 Leuven, Belgium
E-mail: vincent.rijmen@cryptomathic.com

Cataloging-in-Publication Data applied for

Die Deutsche Bibliothek - CIP-Einheitsaufnahme

Fast software encryption : 9th international workshop ; revised papers / FSE
2002, Leuven, Belgium, February 2002. Joan Daemen ; Vincent Rijmen (ed.). -
Berlin ; Heidelberg ; New York ; Barcelona ; Hong Kong ; London ; Milan ;
Paris ; Tokyo : Springer, 2002
    (Lecture notes in computer science ; Vol. 2365)
    ISBN 3-540-44009-7

CR Subject Classification (1998): E.3, F.2.1, E.4, G.4

ISSN 0302-9743
ISBN 3-540-44009-7 Springer-Verlag Berlin Heidelberg New York

Springer-Verlag Berlin Heidelberg New York,
a member of BertelsmannSpringer Science+Business Media GmbH

http://www.springer.de

© Springer-Verlag Berlin Heidelberg 2002
Printed in Germany

Typesetting: Camera-ready by author, data conversion by PTP-Berlin, Stefan Sossna e.K.
Printed on acid-free paper   SPIN: 10870318   06/3142   5 4 3 2 1 0

Preface

This Fast Software Encryption workshop was the ninth in a series of workshops started in Cambridge in December 1993. The previous workshop took place in Yokohama in April 2001. It concentrated on all aspects of fast primitives for symmetric cryptography: secret key ciphers, the design and cryptanalysis of block and stream ciphers, as well as hash functions and message authentication codes (MACs).

The ninth Fast Software Encryption workshop was held in February 2002 in Leuven, Belgium and was organized by General Chair Matt Landrock (Cryptomathic Belgium), in cooperation with the research group COSIC of K.U. Leuven. This year there were 70 submissions, of which 21 were selected for presentation and publication in this volume.

We would like to thank the following people. First of all the submitting authors and the program committee for their work. Then Markku-Juhani O. Saarinen, Orr Dunkelman, Fredrik Jönsson, Helger Lipmaa, Greg Rose, Alex Biryukov, and Christophe De Canniere, who provided reviews at the request of program committee members. Bart Preneel for letting us use COSIC's Web-review software in the review process and Wim Moreau for all his support. Finally we would like to thank Krista Geens of Cryptomathic for her help in the registration and the practical organization.

May 2002   Joan Daemen and Vincent Rijmen

Fast Software Encryption 2002

February 4–6, 2002, Leuven, Belgium

Sponsored by the
International Association for Cryptologic Research

General Chair

Matt Landrock, Cryptomathic, Belgium

Program Co-chairs

Joan Daemen, Proton World, Belgium
Vincent Rijmen, Cryptomathic, Belgium

Program Committee

Ross Anderson .... Cambridge University, UK
Eli Biham .... Technion, IL
Don Coppersmith .... IBM, USA
Cunshen Ding .... Hong Kong University of Science and Technology, HK
Thomas Johansson .... Lund University, SE
Mitsuru Matsui .... Mitsubishi Electric, JP
Willi Meier .... Fachhochschule Aargau, CH
Kaisa Nyberg .... Nokia, FI
Bart Preneel .... Katholieke Universiteit Leuven, BE

Block Cipher Theory

Stream Cipher Design

Stream Cipher Cryptanalysis

Odds and Ends

Table of Contents

Block Cipher Cryptanalysis

Integral Cryptanalysis

New Results on Boomerang and Rectangle Attacks*

Eli Biham[1], Orr Dunkelman[1], and Nathan Keller[2]

[1] Computer Science Department, Technion.
Haifa 32000, Israel
{biham,orrd}@cs.technion.ac.il
[2] Mathematics Department, Technion.
Haifa 32000, Israel
nkeller@tx.technion.ac.il

Abstract. The boomerang attack is a new and very powerful crypt-analytic technique. However, due to the adaptive chosen plaintext and ciphertext nature of the attack, boomerang key recovery attacks that retrieve key material on both sides of the boomerang distinguisher are hard to mount. We also present a method for using a boomerang distinguisher, which enables retrieving subkey bits on both sides of the boomerang distinguisher. The rectangle attack evolved from the boomerang attack.In this paper we present a new algorithm which improves the results of the rectangle attack.

Using these improvements we can attack 3.5-round SC2000 with 2^{67} adaptive chosen plaintexts and ciphertexts, and 10-round Serpent with time complexity of $2^{173.8}$ memory accesses (which are equivalent to $2^{165.3}$ Serpent encryptions) with data complexity of $2^{126.3}$ chosen plaintexts.

1 Introduction

Differential cryptanalysis [3] is based on studying the propagation of differences through an encryption function. Since its introduction many techniques based on it were introduced. Some of these techniques, like the truncated differentials [11] and the higher order differentials [2,11], are generalizations of the differential attack. Some other techniques like differential-linear attack [14] and the boomerang attack [18] use the differential attack as a building block.

The boomerang attack is an adaptive chosen plaintext and ciphertext attack. It is based on a pair of short differential characteristics used in a specially built quartet. In the attack a pair of plaintexts with a given input difference are encrypted. Their ciphertexts are used to compute two other ciphertexts according to some other difference, these new ciphertexts are then decrypted, and the difference after the decryption is compared to some (fixed known) value.

---

* The work described in this paper has been supported by the European Commission through the IST Programme under Contract IST-1999-12324.

J. Daemen and V. Rijmen (Eds.): FSE 2002, LNCS 2365, pp. 1–16, 2002.

The boomerang attack was further developed in [10] into a chosen plaintext attack called the amplified boomerang attack. Later, the amplified boomerang attack was further developed into the rectangle attack [7].

In the transition from the boomerang attack to the rectangle attack the probability of the distinguisher is reduced (in exchange for easing the requirements from adaptive chosen plaintext and ciphertext attack to a chosen plaintext attack). The reduction in the distinguisher's probability results in higher data complexity requirements. For example, the data requirements for distinguishing a 2.5-round SC2000 [17] from a random permutation using a rectangle distinguisher is $2^{84.6}$ chosen plaintext blocks, whereas only $2^{39.2}$ adaptive chosen plaintext and ciphertext blocks are required for the boomerang distinguisher.

In this paper we present a method to retrieve more subkey bits in key recovery boomerang attacks. We also present a better algorithm to perform rectangle attacks. These improvements result in better key recovery attacks which require less data or time (or both) and are more effective. The improvement to the generic rectangle attack reduces the time complexity of attacking 10-round Serpent from 2^{217} memory accesses[1] to $2^{173.8}$ memory accesses which are equivalent to about $2^{166.3}$ 10-round Serpent encryptions. We also prove that these key recovery attacks succeed (with very high probability) assuming that the distinguishers are successful.

The paper is organized as follows: In Section 2 we briefly describe the boomerang and the rectangle attacks. In Section 3 we present our new optimized generic rectangle attack and analyze its application to generic ciphers and to SC2000 and Serpent. In Section 4 we present our optimized generic boomerang attack and analyze its application to both a generic cipher and real blockciphers like SC2000 and Serpent. Section 5 describes a new technique to transform a boomerang distinguisher into a key recovery attack that retrieves more subkey material. We summarize this paper and our new results in Section 6.

2 Introduction to Boomerang and Rectangle Attacks

2.1 The Boomerang Attack

The boomerang attack was introduced in [18]. The main idea behind the boomerang attack is to use two short differentials with high probabilities instead of one differential of more rounds with low probability. The motivation for such an attack is quite apparent, as it is easier to find short differentials with a high probability than finding a long one with a high enough probability.

We assume that a block cipher $E : \{0,1\}^n \times \{0,1\}^k \to \{0,1\}^n$ can be described as a cascade $E = E_1 \circ E_0$, such that for E_0 there exists a differential $\alpha \to \beta$ with probability p, and for E_1 there exists a differential $\gamma \to \delta$ with probability q. The boomerang attack uses the first characteristic $(\alpha \to \beta)$ for E_0 with respect to the pairs (P_1, P_2) and (P_3, P_4), and uses the second characteristic

---

[1] In [7] it was claimed to be 2^{205} due to an error that occurred in the analysis.

$(\gamma \to \delta)$ for E_1 with respect to the pairs (C_1, C_3) and (C_2, C_4). The attack is based on the following boomerang process:

- Ask for the encryption of a pair of plaintexts (P_1, P_2) such that $P_1 \oplus P_2 = \alpha$ and denote the corresponding ciphertexts by (C_1, C_2).
- Calculate $C_3 = C_1 \oplus \delta$ and $C_4 = C_2 \oplus \delta$, and ask for the decryption of the pair (C_3, C_4). Denote the corresponding plaintexts by (P_3, P_4).
- Check whether $P_3 \oplus P_4 = \alpha$.

We call these steps (encryption, XOR by δ and then decryption) a δ–shift.

For a random permutation the probability that the last condition is satisfied is 2^{-n}. For E, however, the probability that the pair (P_1, P_2) is a right pair with respect to the first differential $(\alpha \to \beta)$ is p. The probability that both pairs (C_1, C_3) and (C_2, C_4) are right pairs with respect to the second differential is q^2. If all these are right pairs, then they satisfy $E_1^{-1}(C_3) \oplus E_1^{-1}(C_4) = \beta = E_0(P_3) \oplus E_0(P_4)$, and thus, with probability p also $P_3 \oplus P_4 = \alpha$. Therefore, the probability of this quartet of plaintexts and ciphertexts to satisfy the boomerang conditions is $(pq)^2$. Therefore, $pq > 2^{-n/2}$ must hold for the boomerang distinguisher (and the boomerang key recovery attacks) to work.

The attack can be mounted for all possible β's and γ's simultaneously (as long as $\beta \neq \gamma$), thus, a right quartet for E is built with probability $(\hat{p}\hat{q})^2$, where:

$$\hat{p} = \sqrt{\sum_{\substack{\beta \\ \alpha \to \beta}} \mathrm{Pr}^2[\alpha \to \beta]} \quad \text{and} \quad \hat{q} = \sqrt{\sum_{\substack{\gamma \\ \gamma \to \delta}} \mathrm{Pr}^2[\gamma \to \delta]}.$$

We refer the reader to [18,7] for the complete description and the analysis.

2.2   The Rectangle Attack

Converting adaptive chosen plaintext and ciphertext distinguishers into key recovery attacks pose several difficulties. Unlike the regular known plaintext, chosen plaintext, or chosen ciphertext distinguishers, using the regular methods of [3,15,11,4,5,14] to use adaptive chosen plaintext and ciphertext distinguishers in key recovery attacks fail, as these techniques require the ability to directly control either the input or the output of the encryption function.

In [10] the amplified boomerang attack is presented. This is a method for eliminating the need of adaptive chosen plaintexts and ciphertexts. The amplified boomerang attack achieves this goal by encrypting many pairs with input difference α, and looking for a quartet (pair of pairs) for which, $C_1 \oplus C_3 = C_2 \oplus C_4 = \delta$ when $P_1 \oplus P_2 = P_3 \oplus P_4 = \alpha$. Given the same decomposition of E as before, and the same basic differentials $\alpha \to \beta, \gamma \to \delta$, the analysis shows that the probability of a quartet to be a right quartet is $2^{-(n+1)/2}pq$.

The reason for the lower probability is that no one can guarantee that the γ difference (in the middle of the encryption; needed for the quartet to be a right boomerang quartet) is achieved. The lower probability makes the additional problem (already mentioned earlier) of finding and identifying the right quartets even more difficult.

The rectangle attack [7] shows that it is possible to count over all the possible β's and γ's, and presents additional improvements over the amplified boomerang attack. The improvements presented in the rectangle attack improve the probability of a quartet to be a right rectangle quartet to $2^{-n/2}\hat{p}\hat{q}$.[2]

3   Improving the Rectangle Attack

The main problem dealt in previous works is the large number of possible quartets. Unlike in the boomerang attack, in which the identification of possible quartets is relatively simple, it is hard to find the right quartets in the rectangle attacks since the attacker encrypts a large number of pairs (or structures) and then has to find the right quartets through analysis of the ciphertexts. As the number of possible quartets is quadratic in the number of pairs[3], and as the attacker has to test all the quartets, it is evident that the time complexity of the attack is very large.

In this section we present an algorithm which solves the above problem by exploiting the properties of a right quartet, and which tests only a small part of the possible quartets. The new algorithm is presented on a generic cipher with the following parameters: Let E be a cipher which can be described as a cascade $E = E_f \circ E_1 \circ E_0 \circ E_b$, and assume that E_0 and E_1 satisfy the properties of E_0 and E_1 presented in Section 2 (i.e., there exist differentials $\alpha \to \beta$ with probability p of E_0 and $\gamma \to \delta$ with probability q of E_1). An outline of such an E is presented in Figure 1. We can treat this E as composed of $E' = E_1 \circ E_0$ (for which we have a distinguisher) surrounded by the additional rounds of E_b and E_f. As mentioned in Section 2, for sufficiently high probabilities $\hat{p}, \hat{q}$, we can distinguish $E_1 \circ E_0$ from a random permutation using either a boomerang or a rectangle distinguisher. However, we also like to mount key recovery attacks on the full E.

Recall that the rectangle distinguisher parameters are α, δ, $\hat{p}$, and $\hat{q}$. Given these parameters, the rectangle distinguisher of the cipher $E' = E_1 \circ E_0$ can easily be constructed.

Before we continue we introduce some additional notations: Let X_b be the set of all plaintext differences that may cause a difference α after E_b. Let V_b be the space spanned by the values in X_b. Note that usually $n - r_b$ bits are set to 0 for all the values in V_b. Let $r_b = \log_2 |V_b|$ and $t_b = \log_2 |X_b|$ (r_b and t_b are not necessarily integers). Let m_b be the number of subkey bits which enter E_b and affect the difference of the plaintexts by decrypting pairs whose difference after E_b is α, or formally

---

[2] This is a lower bound for the probability. For further analysis see [7].

[3] In the rectangle attack the quartet $[(x, y), (z, w)]$ differs from the quartet $[(x, y), (w, z)]$.

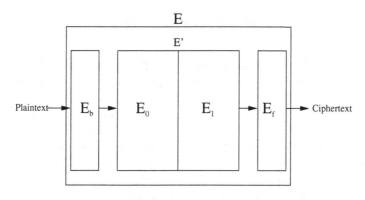

Fig. 1. Outline of E

$$m_b = \left| \left\{ K' \middle| w(K') = 1 \text{ and } \exists K, x : \begin{array}{c} E_{b_K}^{-1}(x) \oplus E_{b_K}^{-1}(x \oplus \alpha) \neq \\ E_{b_{K \oplus K'}}^{-1}(x) \oplus E_{b_{K \oplus K'}}^{-1}(x \oplus \alpha) \end{array} \right\} \right|$$

where $w(x)$ denotes the hamming weight of x.

Similarly, let X_f is the set of all ciphertext differences that a difference δ before E_f may cause. Let V_f denote the space spanned by the values of X_f and denote $r_f = \log_2 |V_f|$. Let $t_f = \log_2 |X_f|$. Let m_f be the number of subkey bits which enter E_f and affect the difference when encrypting a pair with difference δ or formally

$$m_f = \left| \left\{ K' \middle| w(K') = 1 \text{ and } \exists K, x : \begin{array}{c} E_{f_K}(x) \oplus E_{f_K}(x \oplus \alpha) \neq \\ E_{f_{K \oplus K'}}(x) \oplus E_{f_{K \oplus K'}}(x \oplus \alpha) \end{array} \right\} \right|.$$

We outline all these notations in Figure 2.

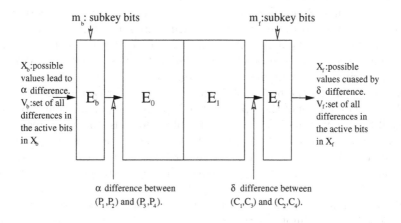

Fig. 2. The Notations Used in This Paper

Our new algorithm for using rectangle distinguisher in a key recovery attack is as follows:

1. Create $Y = \lceil 2^{n/2+2-r_b}/\hat{p}\hat{q} \rceil$ structures of 2^{r_b} plaintexts each. In each structure choose P_0 randomly and let $L = P_0 \oplus V_b$ be the set of plaintexts in the structure.

2. Initialize an array of $2^{m_b+m_f}$ counters. Each counter corresponds to a different guess of the m_b subkey bits of E_b and the m_f subkey bits of E_f.

3. Insert the $N = Y \cdot 2^{r_b}$ ciphertexts into a hash table according to the $n - r_f$ ciphertext bits that are set to 0 in V_f. If a pair agrees on these $n - r_f$ bits, check whether the ciphertext difference is in X_f.

4. For each collision (C_1, C_2) which remains, denote C_i's structure by S_{C_i} and attach to C_1 the index of S_{C_2} and vice versa.

5. In each structure S we search for two ciphertexts C_1 and C_2 which are attached to some other S'. When we find such a pair we check that the $P_1 \oplus P_2$ (the corresponding plaintexts) is in X_b, and check the same for the plaintexts which P_1 and P_2 are related to.

6. For all the quartets which passed the last test denote by (P_1, P_2, P_3, P_4) the plaintexts of a quartet and by (C_1, C_2, C_3, C_4) the corresponding ciphertexts. Increment the counters which correspond to all subkeys K_b, K_f (actually their bits which affect the α and δ differences, respectively) for which $E_{b_{K_b}}(P_1) \oplus E_{b_{K_b}}(P_2) = E_{b_{K_b}}(P_3) \oplus E_{b_{K_b}}(P_4) = \alpha$ and $E_{f_{K_f}}^{-1}(C_1) \oplus E_{f_{K_f}}^{-1}(C_3) = E_{f_{K_f}}^{-1}(C_2) \oplus E_{f_{K_f}}^{-1}(C_4) = \delta$.

7. Output the subkey with maximal number of hits.

The data complexity of the attack is $N = 2^{r_b}Y = 2^{r_b}\lceil 2^{n/2+2-r_b}/\hat{p}\hat{q} \rceil$ chosen plaintexts. The time complexity of Step 1 (the data collection step) is N encryptions. The time complexity of Step 2 is $2^{m_b+m_f}$ memory accesses in a trivial implementation and only one memory access using a more suitable data structures (like B-trees).

Step 3 requires N memory accesses for the insertion of the ciphertexts into a hash table (indexed by the $n - r_f$ bits which are set to 0 in V_f). The number of colliding pairs is about $N^2 \cdot 2^{r_f-n}/2$ as there are N plaintexts divided into 2^{n-r_f} bins (each bin correspond to a value of the $n - r_f$ bits). Note that we not necessarily use all the bins due to large memory requirements (i.e., we can hash only according the first 30 bits set to 0 in V_f). For each collision we check whether the difference of the ciphertexts of the colliding pair belongs to X_f. We keep all the 2^{t_f} values of X_f in a hash table, and thus, the check requires one memory access for each colliding pair. Out of the 2^{r_f} possible differences for a colliding pair, only 2^{t_f} differences are in X_f (i.e., can occur in a right quartet), and thus, about $N^2 \cdot 2^{t_f-n-1}$ pairs remain. The time complexity of this step is $N + N^2 \cdot 2^{r_f-n-1}$ memory accesses on average.

Step 4 requires one memory access for each pair which passes the filtering of Step 3. In a real implementation it is wiser to implement Step 4 as part of Step 3, but we separate these steps for the sake of simpler analysis. As there are

$N^2 \cdot 2^{t_f - n - 1}$ such pairs, the time complexity of this step is on average $N^2 \cdot 2^{t_f - n - 1}$ memory accesses.

Step 5 implements a search for possible quartets. In a right quartet both (P_1, P_2) and (P_3, P_4) must satisfy that $E_b(P_1) \oplus E_b(P_2) = E_b(P_3) \oplus E_b(P_4) = \alpha$, and thus any right quartet must be combined from some $P_1, P_2 \in S$ and $P_3, P_4 \in \tilde{S}$ where S and $\tilde{S}$ are two (not necessarily distinct) structures. Moreover, a right quartet satisfies that $E_f^{-1}(C_1) \oplus E_f^{-1}(C_3) = E_f^{-1}(C_2) \oplus E_f^{-1}(C_4) = \delta$, and thus C_1 is attached to S_{C_3} and C_2 is attached to S_{C_4} and as P_3, P_4 are from the same structure then $-$ $S_{C_3} = S_{C_4}$. Therefore, in each structure S we search for colliding attachments, i.e., pairs of ciphertexts in S which are attached to the same (other) structure $\tilde{S}$. The $N^2 \cdot 2^{t_f - n - 1}$ attachments (colliding pairs) are distributed over Y structures, and we get that approximately $(N \cdot 2^{t_f + r_b - n - 1})^2 / Y$ possible quartets are suggested in each structure (where a quartet corresponds to a pair of plaintexts from some structure attached to the same structure). We implement the test in the same manner as in Step 3, i.e., keeping a hash table H_S for each structure S and inserting each ciphertext C to H_{S_C} according to the index of the structure attached to C. Denoting the plaintexts of the suggested quartet by (P_1, P_2, P_3, P_4) and their corresponding ciphertexts by (C_1, C_2, C_3, C_4), we first check that $P_1 \oplus P_2 \in X_b$. This test requires one memory access for each possible quartet. The probability that the value $P_1 \oplus P_2$ is in X_b is $2^{t_b - r_b}$. A quartet which fails this test can be discarded immediately. Therefore, out of the $N^2 \cdot 2^{2t_f + 2r_b - 2n - 2}$ possible quartets only $N^2 \cdot 2^{2t_f + r_b + t_b - 2n - 2}$ quartets remain. As stated before, this filtering requires one memory accesses for each candidate quartet, thus the algorithm requires $N^2 \cdot 2^{2t_f + 2r_b - 2n - 2}$ memory accesses. We can discard more quartets by testing whether $P_3 \oplus P_4 \in X_b$. In total this step requires $N^2 \cdot 2^{2t_f + 2r_b - 2n - 2} \cdot (1 + 2^{t_b - r_b})$ memory accesses and about $N^2 \cdot 2^{2t_f + 2t_b - 2n - 2}$ quartets remain after this step.

In Step 6 we try to deduce the right subkey from the remaining quartets. Recall that a right quartet satisfies $E_b(P_1) \oplus E_b(P_2) = \alpha = E_b(P_3) \oplus E_b(P_4)$. Both pairs are encrypted by the same subkey, hence, a right quartet must agree on K_b (the m_b subkey bits which enter E_b and affect the output difference α). There are 2^{t_b} possible input differences that lead to α difference after E_b, therefore, $2^{m_b - t_b}$ subkeys on average take one of these values into the difference α. As each pair suggests $2^{m_b - t_b}$ subkeys, they agree on average on $(2^{m_b - t_b})^2 / 2(2^{m_b}) = 2^{m_b - 2t_b - 1}$ subkeys for E_b. We can find these options by keeping in a precomputed table either the possible values for any pair on its own, or for the whole quartet. Repeating the analysis for E_f, (C_1, C_3), and (C_2, C_4) we get about $2^{m_f - 2t_f - 1}$ subkeys suggestions from each quartet. Thus, each of the remaining quartets suggests $2^{m_b + m_f - 2t_f - 2t_b - 2}$ possible subkeys. There are $2^{m_b + m_f}$ possible subkeys and $N^2 \cdot 2^{2t_f + 2t_b - 2n - 2} \cdot 2^{m_b + m_f - 2t_f - 2t_b - 2} = N^2 \cdot 2^{m_b + m_f - 2n - 4}$ hits. The expected number of hits for a (wrong) subkey is about $N^2 \cdot 2^{-2n - 4}$. Since $N \leq 2^n$ is the number of plaintexts the expected number of hits per wrong subkey is less than $2^{-4} = 1/16$, and we can conclude that the attack almost always succeeds in recovering subkey bits (since the number of expected hits for the right subkey is 4), or at least reduces the number of candidates for the right subkey. We can insert the $2^{m_b-t_b}$ subkeys suggested by (P_1, P_2) into a hash table, and for each subkey suggested by the pair (P_3, P_4) we can check whether it was already suggested. Doing the same for E_f we get that for each quartet we need $3 \cdot 2^{m_b-t_b} + 3 \cdot 2^{m_f-t_f}$ memory accesses. We can optimize this a little bit by storing in advance a table and a list for each difference in X_b and save the time of building the hash table. This method saves $1/3$ of the number of memory accesses. Using this method this step requires about $N^2 \cdot 2^{2t_f+2t_b-2n-3} \cdot (2^{m_b-2t_b} - 2^{m_f-2t_f}) = N^2 \cdot 2^{-2n-3} \cdot (2^{m_b+2t_f} + 2^{m_f+2t_b})$ memory accesses for the entire attack.

Step 7 requires $2^{m_b+m_f}$ memory accesses using a trivial implementation, which can be reduced to 1–4 memory accesses using a more efficient data structure (e.g., B-trees or dynamic hash tables).

Overall, this algorithm requires $N = 2^{r_b}Y = 2^{r_b}\lceil 2^{n/2+2-r_b}/\hat{p}\hat{q}\rceil$ chosen plaintexts, and time complexity of $N + N^2(2^{r_f-n-1} + 2^{t_f-n} + 2^{2t_f+2r_b-2n-2} + 2^{m_b+t_b+2t_f-2n-1} + 2^{m_f+2t_b+t_f-2n-1})$ memory accesses. The memory complexity is $N + 2^{t_b} + 2^{t_f} + 2^{m_b+m_f}$.

Table 1 summarizes the time complexity of each step and the number of plaintexts / pairs / quartets that remain after each step of the algorithm.

Table 1. The Rectangle Attack Steps and their Effect

Step No.	Short Description	Time Complexity	# Remaining Texts/ Pairs/Quartets
1	Data generation	N encryptions	N plaintexts
2	Subkey counters' init.	1 MA	No change
3	First filtering	$N + N^2 2^{r_f-n-1}$ MA	$N^2 \cdot 2^{t_f-n-1}$ pairs
4	Suggesting quartets	$N^2 2^{t_f-n-1}$ MA	No change
5	Eliminating quartets	$N^2 2^{2t_f+2r_b-2n-2}$ MA	$N^2 2^{2t_f+2t_b-2n-2}$ qts.
6	Subkey detection	$N^2 2^{t_b+t_f-2n-1}(2^{m_b+t_f}+2^{m_f+t_b})$ MA	No change
7	Printing subkey	1–4 MA	No change

MA - Memory Accesses

Using this algorithm we can break 3.5-round SC2000, using the following decomposition: E_b consists of the first S4 layer, the following 1.25 rounds are E_0, the next 1.25 rounds are E_1, and the final S4 layer is E_f. For this decomposition the following properties were presented in [8]: $r_b = r_f = m_b = m_f = 40, t_b = 27, t_f = 27.9, n = 128, \hat{p} = 2^{-8.96}, \hat{q} = 2^{-9.16}$. Thus, we conclude that the data complexity is $N = 2^{84.6}$ chosen plaintexts and that the time complexity is $2^{84.6}$ memory accesses, which slightly improves the results in [8].

We can also break 10-round Serpent using the following decomposition: E_b consists of round 0. The following 4 rounds are E_0, the next 4 rounds are E_1, and round 9 is E_f. For this decomposition the following properties are presented in [7]:[4] $r_b = m_b = 76, r_f = m_f = 20, t_b = 48.85, t_f = 13.6, n = 128, \hat{p} =$

---

[4] As stated earlier, in [7] it was mistakenly claimed that $t_b = 64$. We use the correct value of 76, and derive the correct time complexity.

$2^{-25.4}, \hat{q} = 2^{-34.9}$. Thus, we conclude that the data complexity is $N = 2^{126.3}$ chosen plaintexts and that the time complexity is $2^{173.8}$ memory accesses. Note that the time complexity of the attack presented in [7] is 2^{217} memory accesses using 2^{196} memory cells (or $2^{219.4}$ memory accesses with $2^{131.8}$ memory cells).

Note that we can use the above algorithm in several other scenarios. One of them is a chosen ciphertext scenario, where the above algorithm is applied on $E^{-1} = E_b^{-1} \circ E_0^{-1} \circ E_1^{-1} \circ E_f^{-1}$. The analysis of this case is the same as before as long as we replace in the above equations all the sub-scripts b by f and vice versa.

Sometimes, we might want to attack a cipher E with additional rounds only at one side, i.e., to divide E to $E = E_1 \circ E_0 \circ E_b$ (or $E = E_f \circ E_1 \circ E_b$). In this case our analysis still holds with $m_f = r_f = t_f = 0$.

4 A Key Recovery Attack Based on Boomerang Distinguisher

In this section we apply our ideas from the previous section to the boomerang attack. We generalized the results of [18,8]. Like the rectangle attack, we have found that whenever the boomerang distinguisher succeeds then the key recovery attack also succeeds.

There are various standard techniques to use distinguishers for a key recovery attack [3,15,11,4,5]. The basic idea is to try all subkeys which affect the difference (or the approximation) before and after the distinguishers (i.e., in E_b and E_f), and to deduce that the correct subkey is the one for which the statistical distinguisher has the best results. However, this basic idea can be very expensive in terms of time and memory complexities. Moreover, due to the adaptive chosen plaintext and ciphertext requirement, using a boomerang distinguisher in a key recovery attack can be done if either E_b or E_f is present but not when both exist.

As both boomerang and rectangle distinguishers exploit the same α and δ, we use the same notations of $E = E_f \circ E_1 \circ E_0 \circ E_b$, m_b, r_b, t_b, m_f, r_f, and t_f as in the earlier sections.

The generic boomerang attack on $E = E_1 \circ E_0 \circ E_b$ is as follows:

1. Initialize an array of 2^{m_b} counters. Each counter corresponds to a different guess of the m_b subkey bits of E_b.
2. Generate a structure F of plaintexts, choose P_0 randomly and let $F = P_0 \oplus V_b$ be the set of plaintexts in the structure.
3. Ask for the encryption of F and denote the set of ciphertexts by G.
4. For each ciphertext $c \in G$ compute $c' = c \oplus \delta$, and define the set $H = \{c \oplus \delta | c \in G\}$.
5. Ask for the decryption of H, and denote the plaintexts set by I.
6. Insert all the plaintexts in I into a hash table according to the $n - r_b$ plaintext bits which are set to 0 in V_b.
7. In case of a collision of plaintexts in the hash table:

a) Denote the plaintexts which collide in the hash table by (P_3, P_4), and test whether $P_3 \oplus P_4 \in X_b$. If this condition is satisfied denote the plaintexts from F which correspond to (P_3, P_4) by (P_1, P_2). Test whether $P_1 \oplus P_2 \in X_b$. If any of the tests fails, discard this quartet.

b) For a quartet (P_1, P_2, P_3, P_4) which passes the above filtering we check all possible K_b which enter E_b (actually its bits which affect the α) and increment the counters which correspond to the subkeys for which $E_{b_{K_b}}(P_1) \oplus E_{b_{K_b}}(P_2) = E_{b_{K_b}}(P_3) \oplus E_{b_{K_b}}(P_4) = \alpha$.

8. Repeat Steps 2–7 until a subkey is suggested 4 times.

Steps 2–5 perform a δ-shift on structures (and not on the pairs directly).

From the analysis in [18] and the properties of the algorithm it is evident that the data complexity of the attack is about $8(\hat{p}\hat{q})^{-2}$. However, we generate at least 2^{r_b+1} plaintexts and ciphertexts, and thus the data complexity of the attack is $N = \max\{2^{r_b+1}, 8(\hat{p}\hat{q})^{-2}\}$.

The time complexity of Step 1 is equivalent to 2^{m_b} memory accesses. However, we can keep the counters in more efficient data structures (like B-trees, or dynamic hash tables) for which Step 1 takes only one memory access.

In steps 2–5 we encrypt $N/2$ plaintexts, compute $N/2$ XOR operations and decrypt $N/2$ ciphertexts. Thus, the total time complexity of Steps 2–5 for the whole attack is N encryptions/decryptions.

Repeating the analysis from the previous section, we restrict our attention to time complexity analysis for each F independently of other F's, as each execution of Steps 6–7 for a given structure is independent of the execution of these steps for other structures.

As we insert the plaintexts into a hash tables, each plaintext in I requires one memory access. Thus, the time complexity of Step 6 is 2^{r_b} memory accesses per structure, and therefore $N/2$ memory accesses for the entire attack.

Repeating the analysis from the previous section (with the relevant minor changes), we get that Step 7 for each structure I (and therefore, for each F) requires about 2^{3r_b-n-1} memory accesses. For the entire $N/2^{r_b+1}$ structures this step requires $2^{3r_b-n-1} \cdot N/2^{r_b+1} = N \cdot 2^{2r_b-n-1}$ memory accesses.

Using the same arguing about right quartets (as in right quartet both (P_1, P_2) and (P_3, P_4) have an α difference after E_b), we get that each pair suggests $2^{m_b-t_b}$ subkeys, and both pairs agree on $(2^{m_b-t_b})^2/2(2^{m_b}) = 2^{m_b-2t_b-1}$ subkeys for E_b on average. Hence, the expected number of memory accesses in Step 7(b) is $2^{t_b+m_b+r_b-n}$ for each structure. We conclude that the total time complexity of Step 7(b) is expected to be $N \cdot 2^{t_b+m_b-n-1}$ memory accesses for the whole attack.

Each structure F induces about $2^{2t_b+r_b-n-1} \cdot 2^{m_b-2t_b-1} = 2^{r_b+m_b-n-2}$ subkey hits. As there are $N/2^{r_b+1}$ structures, the total number of subkey hits is expected to be $N \cdot 2^{m_b-n-3}$, which are distributed over 2^{m_b} subkeys. Thus, the expected number of hits for each subkey is $N \cdot 2^{-n-3}$. As $N \leq 2^n$, the expected number of hits for a wrong subkey is less then $1/8$ while the right subkey is expected to get 4 hits. This is sufficient for either recovering the right subkey, or to reduce the subkey candidates space by a very large factor.

In Step 8 we check whether one of the counters has the value of 4 (or more). This has to be done whenever we finish Step 7(b) for some F. We can implement this step as part of Step 7(b). Whenever a counter is increased we check that it has not exceeded 4. However, this method results in enlarging the time of Step 7(b). Using more appropriate data structures, we can perform the check once whenever we replace the F structure we work with. This yields a time complexity of $N/2^{r_b+1}$ memory accesses.

We conclude that the attack requires $N = \max\{2^{r_b+1}, 8(\hat{p}\hat{q})^{-2}\}$ adaptive chosen plaintexts and ciphertexts and time complexity of about $N(1+2^{2r_b-n-2}+2^{t_b+m_b-n-1})$ memory accesses.

Table 2 summarizes the time complexity of each step and the number of plaintexts / pairs / quartets that remain after each step of the algorithm.

Table 2. The Basic Boomerang Attack Steps and their Effect

Step No.	Short Description	Time Complexity	# of Remaining Plaintexts/ Pairs/ Quartets
1	Subkey counters' init.	1 MA	—
2+3	Data generation	$N/2$ encryptions	$N/2$ plaintexts
4	Data generation	$N/2$ MA	No change
5	Data generation	$N/2$ decryptions	N plaintexts
6	Finding possible quartets	$N/2$ MA	$N \cdot 2^{2r_b-n-2}$ quartets
7(a)	Eliminating quartets	$N \cdot 2^{2r_b-n-2}$ MA	$N \cdot 2^{2t_b-n-2}$ quartets
7(b)	Subkey detection	$N \cdot 2^{r_b+m_b-n-1}$ MA	No change
8	Printing subkey	$N/2^{r_b+1}$ MA	No change

MA - Memory Accesses

Using this algorithm, we can break 3-round SC2000, using the following decomposition: E_b consists of the first S4 layer, the following 1.25 rounds are E_0 and the next 1.25 rounds are E_1. For this decomposition the following properties were presented in [8]: $r_b = m_b = 40, t_b = 27, n = 128, \hat{p} = 2^{-8.96}, \hat{q} = 2^{-9.16}$. Thus, we conclude that the data complexity of the attack is $N = 2^{41}$ adaptive chosen plaintexts and ciphertexts. The time complexity of the attack is about 2^{41} memory accesses.

We attack 9-round Serpent, using this algorithm and the following decomposition: E_b consists of round 0, the following 4 rounds are E_0, and the next 4 rounds are E_1. For this decomposition the following properties were presented in [7]: $r_b = m_b = 76, t_b = 48.85, n = 128, \hat{p} = 2^{-25.4}, \hat{q} = 2^{-34.9}$. The data complexity of the attack is $N = 2^{123.6}$ adaptive chosen plaintexts and ciphertexts, with time complexity of $2^{147.2}$ memory accesses. We can also attack rounds 1–9 of Serpent using the decomposition used in the previous section. In this attack we are attacking rounds 9–1 (i.e., the attack is on $E_0^{-1} \circ E_1^{-1} \circ E_f^{-1}$). For this decomposition we get: $r_f = m_f = 20, t_f = 13.6, n = 128, \hat{p} = 2^{-25.4}, \hat{q} = 2^{-34.9}$. Obviously the data complexity does not change, as we use the same underlying distinguisher. However, the time complexity of this attack drops to $2^{123.6}$ memory accesses (instead of $2^{147.2}$).

5   Enhancing the Boomerang Attack

In this section we present a new method to use the boomerang attack with both E_b and E_f. Recall that the main step of the boomerang attack is the δ-shift (encryption of each plaintext, XORing of the corresponding ciphertext with δ, and decryption of the outcome). Our method uses a generalization of the δ-shift.

The new algorithm to attack $E_f \circ E_1 \circ E_0 \circ E_b$ is as follows:

1. Initialize an array of $2^{m_b+m_f}$ counters. Each counter corresponds to a different guess of the m_b subkey bits of E_b and the m_f bits of E_f.
2. Generate a structure F of plaintexts, choose P_0 randomly and let $F = P_0 \oplus V_b$ be the set of plaintexts in the structure.
3. Ask for the encryption of F and denote the set of ciphertexts by G.
4. For each $c \in G$ and $\epsilon \in X_f$ compute $c' = c \oplus \epsilon$, and define the set $H = \{c \oplus \epsilon | \ c \in G \text{ and } \epsilon \in X_f\}$.
5. Ask for the decryption of H, and denote the plaintexts set by I.
6. Insert all the plaintexts in I into a hash table according to the $n - r_b$ plaintext bits which are set to 0 in V_b.
7. In case of a collision of plaintexts in the hash table:
   a) Denote the plaintexts which collide in the hash table by (P_3, P_4), and test whether $P_3 \oplus P_4 \in X_b$. If this condition is satisfied denote the plaintexts from F which correspond to (P_3, P_4) by (P_1, P_2). Test whether $P_1 \oplus P_2 \in X_b$. If any of the tests fails, discard this quartet.
   b) For a quartet (P_1, P_2, P_3, P_4) which passes the above filtering we obtain from a precomputed table the possible values for the m_b subkey bits which enter E_b and affect the α difference. We also obtain from a precomputed table the possible values for the m_f subkey bits which enter E_f and affect the δ difference. The specific implementation aspects of this step are described later on. We increment counters which correspond to subkeys K_b, K_f for which $E_{b_{K_b}}(P_1) \oplus E_{b_{K_b}}(P_2) = E_{b_{K_b}}(P_3) \oplus E_{b_{K_b}}(P_4) = \alpha$ and $E_{f_{K_f}}^{-1}(C_1) \oplus E_{f_{K_f}}^{-1}(C_3) = E_{f_{K_f}}^{-1}(C_2) \oplus E_{f_{K_f}}^{-1}(C_4) = \delta$.
8. Repeat Steps 2–7 until a subkey is suggested 4 times.

We call Steps 2–5 an ϵ-shift as each plaintext is encrypted, then shifted by all possible ϵ's and the values of the shifted ciphertexts are decrypted.

The time complexity of Step 1 is equivalent to $2^{m_b+m_f}$ memory accesses. However, we can keep the counters in a more efficient data structures (like B-trees, or dynamic hash tables), for which Step 1 takes only one memory access.

Each F induces a set of 2^{r_b} ciphertexts in G, and each ciphertext is shifted by 2^{t_f} possible values, hence $|H| = |I| = 2^{r_b+t_f}$. Even though we expand the number of possible quartets (by multiplying the size of I by 2^{t_f}), the number of right quartets does not change. Hence, we still need about $\lceil 8(\hat{p}\hat{q})^{-2}/2^{r_b+1} \rceil$ structures.

The data complexity of the attack is $N = 2^{r_b + t_f} \cdot \lceil 8(\hat{p}\hat{q})^{-2} / 2^{r_b + 1} \rceil$. However, we might get that $N > 2^n$. We can implement these cases in one of two ways. The first way is to ask for the encryption/decryption N (not necessarily different) oracle queries. The second way to implement this is to store the already encrypted/decrypted values in a table, and test for each encryption/decryption if it is already in the table, in order to save most of the encryptions/decryptions. This way the attack requires 2^n known plaintexts and N memory accesses.

Like in the previous section we perform the time complexity analysis for each F independently of other F's, as each execution of Steps 6–7 for a given structure is independent of the execution of these steps for some other structure.

Like in the previous section, Step 6 inserts into a hash table the plaintexts. Thus, the time complexity of Step 6 is $2^{r_b + t_f}$ memory accesses per structure, and N memory accesses in total.

Since each collision in the hash table of Step 6 suggests a quartet, we have in this step about $2^{3r_b + 2t_f - n - 1}$ memory accesses for each structure F, and the expected number of remaining quartets is $2^{3r_b + 2t_f - n - 1} \cdot (2^{t_b - r_b})^2 = 2^{2t_b + r_b + 2t_f - n - 1}$. Since there are $N / 2^{r_b + t_f}$ structures we conclude that for the whole attack this step requires about $N \cdot 2^{2r_b + t_f - n - 1}$ memory accesses.

We recover subkey material both in E_b and E_f. By repeating the analysis from Section 3, each remaining quartet suggests $2^{m_b + m_f - 2t_b - 2t_f - 2}$ subkeys for E_b and E_f, and thus, we get $2^{r_b + m_b + m_f - n - 3}$ hits (on average) from each structure and $N \cdot 2^{m_b + m_f - n - t_f - 3}$ subkey hits in total. We conclude that Step 7(b) requires $N \cdot 2^{t_b + t_f - n - 1} \cdot (2^{m_b + t_f} + 2^{m_f + t_b})$ memory accesses for the whole attack.

As there are about $N \cdot 2^{m_b + m_f - t_f - n - 3}$ subkey hits in total, the expected number of hits per subkey is about $N \cdot 2^{-t_f - n - 3}$. Note that N might be bigger than 2^n but on the same time, $N \leq 2^{n + t_f}$ (as we take at most 2^n ciphertexts and shift them by 2^{t_f} values). We again find that the number of hits per wrong subkey is $\leq 1/8$.

In Step 8 we check whether one of the counters has the value of 4 (or more). Using the same methods as in the previous section, we can use more appropriate data structures which reduce the time complexity of this step to 1 memory accesses for each F, and for the entire attack $N / 2^{r_b + t_f}$ memory accesses.

The data complexity of the attack is $N = 2^{r_b + t_f} \cdot \lceil 8(\hat{p}\hat{q})^2 / 2^{r_b + 1} \rceil$ adaptive chosen plaintexts and ciphertexts (and as stated earlier if $N > 2^n$ we can replace it by 2^n known plaintexts using a table of size 2^n). The expected time complexity of the attack is $N(2 + 2^{2r_b + t_f - n - 1} + 2^{m_b + t_b + 2t_f - n - 1} + 2^{m_f + 2t_b + t_f - n - 1})$ memory accesses. The memory complexity of the attack is $2^{m_b + m_f} + 2^{r_b + t_f}$.

Table 3 summarizes the time complexity of each step and the number of plaintexts / pairs / quartets that remain after each step of the algorithm.

We use the same decomposition of 3.5-round SC2000 as in Section 3. For this decomposition the following properties were presented in [8]: $r_b = r_f = m_b = m_f = 40, t_b = 27, t_f = 27.9, n = 128, \hat{p} = 2^{-8.96}, \hat{q} = 2^{-9.16}$. Hence, the data complexity of the attack is $2^{67.9}$ adaptive chosen plaintexts and ciphertexts (this complexity can be reduced to 2^{67} by attacking E^{-1}). The attack requires $2^{67.9}$ memory cells (when we attack E^{-1} it requires only 2^{67} memory cells). The time

Table 3. The Steps of the Improved Boomerang Attack and Their Effect

Step No.	Short Description	Time Complexity	# Remaining Texts/ Pairs/ Quartets
1	Subkey counters' init.	1 MA	—
2+3	Data generation	$N/2^{t_f}$ encryptions	$N/2^{t_f}$ plaintexts
4	Data generation	N MA	No change
5	Data generation	N decryptions	N Plaintexts
6	Eliminating wrong pairs	N MA	$N2^{2r_b+2t_f-n-2}$ qts.
7(a)	Eliminating quartets	$N2^{2r_b+t_f-n-1}$ MA	$N2^{2t_b+2t_f-n-2}$ qts.
7(b)	Subkey detection	$N2^{t_b+t_f-n-2}(2^{m_b+t_f}+2^{m_f+t_b})$MA	No change
8	Printing subkey	$N/2^{r_b+t_f}$ MA	No change

MA - Memory Accesses

complexity of the attack is $2^{68.9}$ memory accesses (the attack on E^{-1} requires 2^{68} memory accesses).

We also use the same decomposition of 10-round Serpent like in Section 3. For this decomposition the following properties were presented in [7]: $r_b = m_b = 76, r_f = m_f = 20, t_b = 48.85, t_f = 13.6, n = 128, \hat{p} = 2^{-25.4}, \hat{q} = 2^{-34.9}$. The data complexity of the attack is $N = 2^{123.6+13.6} = 2^{137.2}$. As mentioned before, we can either treat this as $2^{137.2}$ queries to the encryption/decryption oracle (of course not distinct queries) or we can just ask the encryption/decryption of any plaintext/ciphertext we need, and store it in a table. The attack requires $2^{173.8}$ memory accesses.

Table 4. Comparison of the Boomerang and the Rectangle Generic Attacks

Attack	Rectangle (Section 3)	Boomerang (Section 4)	Enhanced Boomerang (Section 5)
Cipher's parts being attacked	$E_f \circ E_1 \circ E_0 \circ E_b$	$E_1 \circ E_0 \circ E_b$	$E_f \circ E_1 \circ E_0 \circ E_b$
Type of Attack	Chosen Plaintext	Adaptive Chosen Plaintext and Ciphert.	Adaptive Chosen Plaint. and Ciphertext
Data Complexity (N)	$\max\{2^{r_b}, 2^{n/2+2}/\hat{p}\hat{q}\}$	$\max\{8(\hat{p}\hat{q})^{-2}, 2^{r_b}\}$	$2^{t_f}\max\{2^{r_b}, 8(\hat{p}\hat{q})^{-2}\}$
Memory Accesses	$N^2(2^{r_f-n-1}+2^{t_f-n}$ $+2^{2t_f+2r_b-2n-2}$ $+2^{m_b+2t_f+t_b-2n-1}$ $+2^{m_f+2t_b+t_f-2n-1})$ $+N$	$N(1+2^{2r_b-n-2}+$ $2^{t_b+m_b-n-1})$	$N(2+2^{2r_b+t_f-n-1}+$ $2^{m_b+t_b+2t_f-n-1}+$ $2^{m_f+t_f+2t_b-n-1})$
Memory Cells	$2^{m_b+m_f}+N$	$2^{r_b}+2^{m_b}$	$2^{r_b+t_f}+2^{m_b+m_f}$
Subkey Bits	m_b+m_f	m_b	m_b+m_f
Hits per Wrong Subkey	$\leq 1/16$	$\leq 1/8$	$\leq 1/8$

Table 5. New Boomerang and Rectangle Results on SC2000 and Serpent

Cipher	Attack	Number of Rounds	Complexity		
			Data	Time	Memory
SC2000	Rectangle – this paper	3.5	$2^{84.6}$ CP	$2^{84.6}$ MA	$2^{84.6}$
	Boomerang – this paper	3	2^{41} ACPC	2^{41} MA	2^{40}
	Boomerang – this paper	3.5	2^{67} ACPC	2^{67} MA	2^{67}
best	Linear [19]	4.5	$2^{104.3}$ KP	$2^{83.3}$ MA	2^{80}
Serpent	Amp. Boomerang[10]	9	2^{110} CP	2^{252} MA	2^{208}
	Rectangle[7]	10	$2^{126.8}$ CP	2^{217} MA	2^{192}
	Rectangle[7]	10	$2^{126.8}$ CP	$2^{219.4}$ MA	$2^{126.8}$
	Boomerang – this paper	9	$2^{123.6}$ ACPC	$2^{123.6}$ MA	$2^{21.5}$
	Boomerang – this paper	10	2^{128} KP	$2^{173.8}$ MA	2^{96}
	Rectangle – this paper	10	$2^{126.3}$ CP	$2^{173.8}$ MA	$2^{126.3}$
best	Linear [6]	11	2^{118} KP	2^{214} MA	2^{85}

MA - Memory Accesses
CP - Chosen Plaintexts, KP - Known Plaintexts
ACPC - Adaptive Chosen Plaintexts and Ciphertexts

6   Summary

This paper presents several contributions. The first contribution is an improved generic rectangle attack. The improved attack algorithm can attack 10-round Serpent with data complexity of $2^{126.3}$ chosen plaintexts and time complexity of $2^{173.8}$ memory accesses. This new result enables attacking 10-round Serpent with 192-bit subkeys. We also have shown that the algorithm is very successful and almost always reduces the number of candidate subkeys.

The second contribution is a generic boomerang key recovery attack. The attack uses similar techniques as in the rectangle key recovery attack and the result is an efficient algorithm for retrieving subkey material. In the analysis of this attack we found out that this attack also almost always succeeds.

The third contribution is extending the generic boomerang key recovery attack to attack more rounds. This contribution allows for the boomerang attack to attack as many rounds as the rectangle attack despite its adaptive chosen plaintext and ciphertext nature. This allows to attack 10-round Serpent with the enhanced boomerang attack using $2^{137.2}$ adaptive chosen plaintexts and ciphertexts (or 2^{128} known plaintexts), and $2^{173.8}$ memory accesses.

In Table 4 we compare the requirements of the generic attacks, and in Table 5 we present our new results on Serpent and SC2000. For comparison, we also include the previous boomerang and rectangle results and the best known attacks against these ciphers.

References

1. Ross Anderson, Eli Biham, Lars R. Knudsen, *Serpent: A Proposal for the Advanced Encryption Standard*, NIST AES Proposal, 1998.

2. Eli Biham, *Higher Order Differential Cryptanalysis*, unpublished paper, 1994.
3. Eli Biham, Adi Shamir, *Differential Cryptanalysis of the Data Encryption Standard*, Springer-Verlag, 1993.
4. Eli Biham, Alex Biryukov, Adi Shamir, *Cryptanalysis of Skipjack reduced to 31 rounds*, Advances in Cryptology, proceedings of EUROCRYPT '99, LNCS 1592, pp. 12–23, Springer-Verlag, 1999.
5. Eli Biham, Alex Biryukov, Adi Shamir, *Miss in the Middle Attacks on IDEA and Khufu*, proceedings of Fast Software Encryption 6, LNCS 1636, pp. 124–138, Springer-Verlag, 1999.
6. Eli Biham, Orr Dunkelman, Nathan Keller, *Linear Cryptanalysis of Reduced Round Serpent*, proceedings of Fast Software Encryption 8, 2001, to appear.
7. Eli Biham, Orr Dunkelman, Nathan Keller, *The Rectangle Attack – Rectangling the Serpent*, Advances in Cryptology, proceedings of EUROCRYPT '01, LNCS 2045, pp. 340–357, Springer-Verlag, 2001.
8. Orr Dunkelman, Nathan Keller, *Boomerang and Rectangle Attacks on SC2000*, preproceedings of the NESSIE second workshop, 2001.
9. Louis Granboulan, *Flaws in Differential Cryptanalysis of Skipjack*, proceedings of Fast Software Encryption 8, 2001, to appear.
10. John Kelsey, Tadayoshi Kohno, Bruce Schneier, *Amplified Boomerang Attacks Against Reduced-Round MARS and Serpent*, proceedings of Fast Software Encryption 7, LNCS 1978, pp. 75–93, Springer-Verlag, 1999.
11. Lars Knudsen, *Truncated and Higher Order Differentials*, proceedings of Fast Software Encryption 2, LNCS 1008, pp. 196–211, Springer-Verlag, 1995.
12. Lars Knudsen, Håvard Raddum, *A Differential Attack on Reduced-Round SC2000*, preproceedings of the NESSIE second workshop, 2001.
13. Lars Knudsen, Matt J.B. Robshaw, David Wagner, *Truncated Differentials and Skipjack*, Advances in Cryptology, proceedings of CRYPTO '99, LNCS 1666, pp. 165–180, Springer-Verlag, 1999.
14. Suzan K. Langford, Martin E. Hellman, *Differential-Linear Cryptanalysis*, Advances in Cryptology, proceedings of CRYPTO '94, LNCS 839, pp. 17–25, Springer-Verlag, 1994.
15. Mitsuru Matsui, *Linear Cryptanalysis Method for DES Cipher*, Advances in Cryptology, proceedings of EUROCRYPT '93, LNCS 765, pp. 386–397, Springer-Verlag, 1994.
16. NESSIE - New European Schemes for Signatures, Integrity and Encryption. *http://www.nessie.eu.org/nessie/*.
17. Takeshi Shimoyama, Hitoshi Yanami, Kazuhiro Yokoyama, Masahiko Takenaka, Kouichi Itoh, Jun Yajima, Naoya Torii, Hidema Tanaka, *The Block Cipher SC2000*, proceedings of Fast Software Encryption 8, 2001, to appear.
18. David Wagner, *The Boomerang Attack*, proceedings of Fast Software Encryption 6, LNCS 1636, pp. 156–170, Springer-Verlag, 1999.
19. Hitoshi Yanami, Takeshi Shimoyama, Orr Dunkelman, *Differential and Linear Cryptanalysis of a Reduced-Round SC2000*, these proceedings.

Multiplicative Differentials

Nikita Borisov, Monica Chew, Rob Johnson, and David Wagner

University of California at Berkeley

Abstract. We present a new type of differential that is particularly suited to analyzing ciphers that use modular multiplication as a primitive operation. These differentials are partially inspired by the differential used to break Nimbus, and we generalize that result. We use these differentials to break the MultiSwap cipher that is part of the Microsoft Digital Rights Management subsystem, to derive a complementation property in the xmx cipher using the recommended modulus, and to mount a weak key attack on the xmx cipher for many other moduli. We also present weak key attacks on several variants of IDEA. We conclude that cipher designers may have placed too much faith in multiplication as a mixing operator, and that it should be combined with at least two other incompatible group operations.

1 Introduction

Modular multiplication is a popular primitive for ciphers targeted at software because many CPUs have built-in multiply instructions. In memory-constrained environments, multiplication is an attractive alternative to S-boxes, which are often implemented using large tables. Multiplication has also been quite successful at foiling traditional differential cryptanalysis, which considers pairs of messages of the form $(x, x \oplus \Delta)$ or $(x, x + \Delta)$. These differentials behave well in ciphers that use xors, additions, or bit permutations, but they fall apart in the face of modular multiplication. Thus, we consider differential pairs of the form $(x, \alpha x)$, which clearly commute with multiplication. The task of the cryptanalyst applying multiplicative differentials is to find values for α that allow the differential to pass through the other operations in a cipher.

It is well-known that differential cryptanalysis can be applied with respect to any Abelian group, with the group operation defining the notion of difference between texts. However, researchers have mostly ignored multiplicative differentials, i.e., differentials over the multiplicative group $(\mathbb{Z}/n\mathbb{Z})^*$, perhaps because it was not clear how to combine them with basic operations like xor. In this paper, we develop new techniques that make multiplicative differentials a more serious threat than previously recognized.

A key observation is that in certain cases, multiplicative differentials can be used to approximate bitwise operations, like xor, with high probability. As we will see in Section 4, for many choices of n there exists a Δ^n such that $-1 \cdot x \bmod n = x \oplus \Delta^n$ with non-negligible probability. Similarly, $2x \bmod 2^n$ is simply a left-shift operation. It is therefore possible to analyze how these differentials interact with other operations that are normally thought incompatible with multiplication, such as xor and bitwise permutations.

J. Daemen and V. Rijmen (Eds.): FSE 2002, LNCS 2365, pp. 17–33, 2002.
© Springer-Verlag Berlin Heidelberg 2002

Table 1. A summary of some cryptanalytic results using multiplicative differentials. The attacks on xmx are distinguishing attacks with advantages close to one; the remaining attacks are key-recovery attacks. All attacks are on the full ciphers; we do not need to consider reduced-round variants. "CP" denotes chosen plaintexts, and "KP" denotes known plaintexts.

Cipher	Complexity			Comments
	[Data]	[Time]	[Keys]	
Nimbus	2^8 CP	2^{10}	all	see [4] (previously known)
xmx (standard version)	2 CP	2	all	mult. complementation property (new)
xmx (challenge version)	2^{33} CP	2^{33}	2^{-8}	multiplicative differentials (new)
MultiSwap	2^{13} CP	2^{25}	all	multiplicative differentials (new)
MultiSwap	2^{22} KP	2^{27}	all	multiplicative differentials (new)
IDEA-X	2^{38} CP	2^{36}	2^{-16}	multiplicative differentials (new)

After reviewing previous work in Section 2, we give two examples using the ciphers xmx [11] and Nimbus [8] to convey the flavor of these attacks in Section 3. In Section 4, we generalize these ideas and catalogue several common cipher primitives that preserve multiplicative differentials. We then focus on specific ciphers. Section 5 presents many moduli, including the xmx challenge modulus, that admit large numbers of weak keys in xmx. In Section 6, we examine the MultiSwap cipher [12], which is used in Microsoft's Digital Rights Management system, and show that it is extremely vulnerable to multiplicative differential cryptanalysis. In Section 7, we study several IDEA [7] variants obtained by replacing additions with xors and show that these variants are vulnerable to weak key attacks using multiplicative differentials. As an example, we show that IDEA-X, a version of IDEA derived by replacing all the additions with xors, is insecure. This suggests that multiplicative differentials may yield new attacks on IDEA. Table 1 summarizes the attacks developed in this paper.

2   Related Work

In this paper, we analyze the xmx cipher, originally proposed by M'Raihi, Naccache, Stern and Vaudenay [11]. We also look at Nimbus, which was proposed by Machado [8] and broken by Furman [4]. IDEA was first proposed by Lai, Massey and Murphy [7]. Meier observed that part of the IDEA cipher often reduces to an affine transformation, and used this to break 2 rounds using differential cryptanalysis [10]. Daemen, Govaerts, and Vandewalle observed that $-x \bmod 2^{16}+1 = x \oplus 11 \cdots 101$ whenever x_1, the second least signicant bit of x, is 1[2]. They showed that if certain IDEA subkeys are ± 1, the algorithm can be broken with differential cryptanalysis. We use the same observation to find weak keys for a variant of IDEA in Section 7. The class of weak keys we find is much larger (2^{112} keys versus 2^{51} keys), but they are otherwise unrelated. The newest cipher we look at, MultiSwap, was designed by Microsoft and subsequently reverse-engineered and published on the Internet under the pseudonym Beale Screamer [12].

Differential cryptanalysis was invented by Biham and Shamir [1]. In the present paper, we apply the ideas of differential cryptanalysis using a non-standard group operation: multiplication modulo n. Daemen, van Linden, Govaerts, and Vandewalle have performed a very thorough analysis of multiplication mod $2^\ell - 1$, how it relates to elementary bit-operations, and its potential for foiling differential cryptanalysis [3].

In Section 6 we use the multiplicative homomorphism $(\mathbb{Z}/2^{32}\mathbb{Z})^* \to (\mathbb{Z}/2^{16}\mathbb{Z})^*$ to recover MultiSwap keys efficiently. This technique is the multiplicative equivalent of Matsui's linear cryptanalysis [9]. In a similar vein, Harpes, Kramer and Massey applied the quadratic residue multiplicative homomorphism QR: $(\mathbb{Z}/n\mathbb{Z})^* \to \mathbb{Z}/2\mathbb{Z}$, for $n = 2^{16} + 1$, to attack IDEA [5]. Kelsey, Schneier and Wagner used the reduction map $\mathbb{Z}/n\mathbb{Z} \to \mathbb{Z}/m\mathbb{Z}$ (a ring homomorphism), for $n = 2^\ell - 1$ and m dividing n, in cryptanalysis[6].

3   Two Examples

To illustrate some of the ideas behind our attacks, we give two examples of using multiplicative differentials to cryptanalyze simple ciphers. Throughout the paper, x_i will represent the ith bit of x, and x_0 will denote the least significant bit of x.[1]

Cryptanalysis of xmx. As a first example, we demonstrate a complementation property for the "standard" version of the xmx cipher [11], which operates on ℓ-bit blocks using two basic operations: multiplication modulo n and xor. The ith round of the cipher is

$$f(x, k_{2i-1}, k_{2i}) = (x \circ k_{2i-1}) \times k_{2i} \bmod n,$$

where the binary operator "$\circ$" is defined by

$$x \circ k_{2i-1} = \begin{cases} x \oplus k_{2i-1} & \text{if } x \oplus k_{2i-1} < n \\ x & \text{otherwise.} \end{cases}$$

The cipher has an output termination phase that may be viewed as an extra half-round, so the entire algorithm is

$$\mathsf{xmx}(x) = (f(f(\cdots f(x, k_1, k_2)\cdots), k_{2r-3}, k_{2r-2}), k_{2r-1}, k_{2r}) \circ k_{2r+1}.$$

where r counts the number of rounds.

In the paper introducing xmx [11], the designers recommend selecting $n = 2^\ell - 1$.[2] The curious thing about this choice of n is that for all x,

$$x \oplus n = -x \bmod n.$$

This is a consequence of the following simple observation: if $0 \le x, y < 2^\ell - 1$, then $x + y = 2^\ell - 1$ if and only if $x \oplus y = 2^\ell - 1$. As a result, this differential will be preserved with probability 1 through the entire cipher, giving a complementation property

$$\mathsf{xmx}(-x \bmod n) = -\mathsf{xmx}(x) \bmod n.$$

---

[1] However, for convenience, we will use k_i to denote a cipher's ith subkey, not the ith bit of k.

[2] Actually, at one point the authors suggest that n be secret, but later state, "Standard implementations should use ... $\ell = 512$, $n = 2^{512} - 1$." For this reason, we call this the "standard" version of xmx, as opposed to the "challenge" version discussed later in this paper.

After describing the basic cipher, the xmx designers suggest several possible extensions, including rotations and other bit permutations. None of these enhancements would destroy this complementation property.

We analyze other versions of xmx later; see Section 5.

Cryptanalysis of Nimbus. As a second example, we explain how the framework of multiplicative differentials can be used to better understand a previously known attack on Nimbus. Nimbus accepts 64-bit blocks, and its ith round is

$$f(x) = k_{2i+1} \times rev(x \oplus k_{2i}) \bmod 2^{64},$$

where $rev()$ reverses the bits in a 64-bit word. The subkeys k_{2i+1} must be odd for the cipher to be invertible.

At FSE2001, Furman used the xor differential $011 \cdots 10 \longrightarrow 011 \cdots 10$, which passes through one round of Nimbus whenever $t = rev(x \oplus k_{2i})$ is odd, to launch a devastating attack on this cipher [4].

Furman's xor differential may appear mysterious at first, but can be readily explained using the language of multiplicative differentials. Whenever t is odd,

$$t \oplus 11 \cdots 10 = -t \bmod 2^\ell.$$

(This is a standard fact from two's complement arithmetic, and follows from the earlier observation that $(t \oplus 11 \cdots 11) + t = 2^\ell - 1$.) So Furman's differential pairs $(x, x \oplus 011 \cdots 10)$ are in fact pairs (x, x^*) where $x^* = -x \bmod 2^{63}$ but $x^* \neq -x \bmod 2^{64}$, a property that obviously survives multiplication by k_{2i+1} whenever k_{2i+1} is odd. In other words, Furman's xor differential is equivalent to the multiplicative differential

$$-1 \longrightarrow -1 \qquad \text{(with probability } 1/2\text{)},$$

taken mod 2^{63}, with explicit analysis of the high bit to ease propagation through the $rev()$ operation.

Discussion. The complementation property of standard xmx has not been previously described, despite xmx's relative maturity. The attack on Nimbus was previously described using xor differentials, but is neatly summarized in our new framework for multiplicative differentials. We believe these two examples motivate further study of multiplicative differentials, and the remainder of this paper is dedicated to this task.

4   New Differentials

Most of the conclusions in this section are summarized in Table 2.

The xmx example in Section 3 used the multiplicative difference $\alpha = -1$, because $-x \bmod 2^\ell - 1 = x \oplus 11 \cdots 1$. Thus the multiplicative differential pair $(x, -x)$ is equivalent to the xor differential pair $(x, x \oplus 11 \cdots 1)$. In the Nimbus example, the modulus is of the form 2^ℓ instead of $2^\ell - 1$, so the identity between the multiplicative and xor differentials does not hold. However, there is an approximate identity $-x \bmod 2^\ell = x \oplus 11 \cdots 10$, which holds whenever x is odd, or equivalently, when $x_0 = 1$.

Table 2. A partial list of the operations we consider. Each entry in the table specifies the probability that the two operations commute. See Proposition 1 for an explanation of $c(n)$. See Proposition 3 for the definitions of $z(\sigma)$ and $\omega(\sigma)$. An entry of "0" indicates the probability is negligible, and a "–" means we do not investigate this combination.

Operation	Modulus	multiply by α	xor	rotate	bit perm σ
multiply by -1	$2^\ell - 1$	1	1	1	1
multiply by -1	2^ℓ	1	$\frac{1}{2}$	0	$z(\sigma)$
multiply by -1	n	1	$2^{-c(n)}$	–	–
multiply by 2	$2^\ell - 1$	1	0	1	–
multiply by 2	2^ℓ	1	0	$\frac{1}{4}$	$2^{-\omega(\sigma)-z(\sigma)}$
reduction mod 2^k	2^ℓ	1	1	–	–

n	1	1	1	1	0	0	0	0	0	1	1	1	0	0	1
x	x_{14}	x_{13}	x_{12}	0	x_{10}	x_9	x_8	x_7	1	x_5	x_4	0	x_2	1	x_0
Δ^n	1	1	1	0	1	1	1	1	0	1	1	0	1	0	1
$x + x \oplus \Delta^n$	1	1	1	1	0	0	0	0	0	1	1	1	0	0	1

Fig. 1. The modulus $n = 30777$, the bit-constraints on values of x for which $x + (x \oplus \Delta^n) = n$, and Δ^n. See Proposition 1 for a precise definition of Δ^n.

To generalize the multiplicative/xor correspondence exploited in these two examples, first observe that every ℓ-bit modulus, n, can be divided into strings of the form $11 \cdots 1$ and strings of the form $100 \cdots 0$. As an example, the 15-bit modulus $n = 30777$ is divided into such substrings in Figure 1.

For each segment of the modulus of the form $11 \cdots 1$, we use the xor differential $11 \cdots 1$. For the segments of the modulus of the form $100 \cdots 0$, we use the xor differential $011 \cdots 10$. Suppose $n_k \cdots n_j$ is one of the segments of n of the form $100 \cdots 0$. Then we also require that $x_j = 1$ and $x_k = 0$. The constraint that $x_j = 1$ serves the same purpose as the constraint that x be odd in the Nimbus differential: it ensures that when x and $x \oplus \Delta^n$ are added together, a chain of carries is started at bit j. The requirement that $x_k = 0$ assures that no carry bits propagate past bit k when x and $x \oplus \Delta^n$ are added together. In the example, bit i of x is constrained if and only if bit i of Δ^n is 0. This is always true because of the symmetry between x and $-x$.

The above scheme works by controlling the carry bits when x and $x \oplus \Delta^n$ are added together. It ensures that, for each substring of the modulus of the form $10 \cdots 0$, a carry chain is started at the low bit and terminated at the high bit. Starting and stopping carry chains necessitates imposing constraints on x, and if two substrings of the form $10 \cdots 0$ are adjacent, it is more efficient to simply ensure that the carry chain from the first substring propagates to the second. Analogously, if the modulus contains a substring of the form $11 \cdots 1011 \cdots 1$, then the above method will start a carry chain, only to terminate it at the next bit. A more efficient approach would ensure that no carry ever started. Algorithm 1, which computes an optimal value of Δ^n for a given n, incorporates these improvements. The algorithm also outputs ω^n and ν^n, which represent the bits of x constrained to 0 and 1, respectively.

Algorithm 1 *Compute the optimal $\Delta = \neg(\omega \vee \nu)$.*

```
best-differential(n)
   c ← 0, ω,ν ← 00···0
   for i = 0,... ,length(n)-2
      switch (n_{i+1},n_i,c)
         case (0,0,0) // Begin a carry chain by requiring x_i = 1.
            ν_i ← 1, c ← 1
         case (0,1,1) // Force carry propagation by requiring x_i = 1.
            ν_i ← 1
         case (1,0,0) // Force no carry by requiring x_i = 0.
            ω_i ← 1
         case (1,1,1) // End carry chain by requiring x_i = 0.
            ω_i ← 1, c ← 0
         default // No change to carry bit. No constraint on x.
   if c = 1 then ω_{ℓ-1} ← 1
   Δ = ¬(ω ∨ ν)
   output (Δ,ω,ν)
```

To determine the probability that a randomly selected $x \in \mathbb{Z}/n\mathbb{Z}$ satisfies the bit-constraints described above, let $c(n)$ be the number of 0 bits in Δ^n (i.e., the number of bits of x that are constrained). Then x will satisfy these constraints with probability at least $2^{-c(n)}$. To see why this is only a lower bound, consider the modulus $n = 1001$ (base 2). The constraints derived from this modulus are $x_3 = 0$ and $x_1 = 1$. However, only one value of $x \in \mathbb{Z}/n\mathbb{Z}$ fails to satisfy $x_3 = 0$, so this constraint is nearly vacuous. The following proposition formalizes this discussion:

Proposition 1. *Let n be an ℓ-bit modulus. Let the ℓ-bit words ω^n, ν^n be the result of Algorithm 1, and let $\Delta^n = \neg(\omega^n \vee \nu^n)$. Take any $x \in \mathbb{Z}/n\mathbb{Z}$. Define:*

$$C_n(x) = \begin{cases} -1 & \text{if } x \wedge \omega^n = 0 \text{ and } \neg x \wedge \nu^n = 0 \\ 1 & \text{otherwise.} \end{cases}$$

Then $C_n(x) = -1$ if and only if $-x \bmod n = x \oplus \Delta^n$. By symmetry, $C_n(-x) = C_n(x)$. Further, define $c(n)$ to be the number of 0 bits in Δ^n. Then, for a uniformly distributed $x \in \mathbb{Z}/n\mathbb{Z}$, $C_n(x) = -1$ with probability at least $2^{-c(n)}$. Finally, for any Δ', $\Pr[x \oplus (-x \bmod n) = \Delta'] \leq \Pr[x \oplus (-x \bmod n) = \Delta^n]$.

The Nimbus attack uses the slight tweak of considering pairs (x, x^*) such that $x^* = -x \bmod 2^{\ell-1}$ but not mod 2^ℓ. Generalizing this gives a truncated multiplicative differential.

Proposition 2. *Suppose*

$$x^* = x \oplus (a_{\ell-1}a_{\ell-2} \cdots a_m 11 \cdots 10),$$

where each a_i stands for any single bit, and suppose moreover that x is odd. If k is odd, then

$$k \times x^* = (k \times x) \oplus b_{\ell-1}b_{\ell-2} \cdots b_m 11 \cdots 10,$$

where the multiplication is modulo 2^ℓ. Additionally, $a_m = b_m$.

Until now, we have only discussed multiplicative differential pairs $(x, -x)$, but the cryptanalysis of MultiSwap uses pairs of the form $(x, 2x \bmod 2^{32})$. One of the basic operations in MultiSwap is to swap the two 16-bit halves of a 32-bit word. The multiplicative relation $(x, 2x)$ is preserved through this operation whenever $x_{15} = x_{31} = 0$.

An arbitrary bit permutation σ can cause two types of problems for the multiplicative differential 2. First, it can disturb the consecutive ordering of the bits. Because multiplication by 2 is just a left-shift, it's not surprising that the bit ordering comes into play. Second, σ may place some bit i in position 0. If σ is to commute with multiplication by 2, then the value of bit i must be 0. These notions are summarized in the following proposition:

Proposition 3. *Let σ be a permutation of the set $\{0, \dots, \ell - 1\}$, and let $\hat{\sigma}$ be the induced function on ℓ-bit words given by $\hat{\sigma}(x) = x_{\sigma(\ell-1)} x_{\sigma(\ell-2)} \cdots x_{\sigma(0)}$. Then*

$$\Pr\left[\hat{\sigma}(2x) = 2\hat{\sigma}(x)\right] = 2^{-\omega(\sigma) - z(\sigma)}$$

where

$$\omega(\sigma) = \#\{j \in 0, \dots, \ell - 2 | \sigma(j+1) \neq \sigma(j) + 1\}$$

and

$$z(\sigma) = \begin{cases} 0 & \text{if } \sigma(0) = 0 \\ 1 & \text{otherwise.} \end{cases}$$

Intuitively, $\omega(\sigma)$ counts the number of times that σ disturbs the consecutive ordering of the bits, and $z(\sigma)$ tests whether σ places bit $i \neq 0$ in position 0. So, for example, the $(x, 2x)$ differential survives rotations with probability $\frac{1}{4}$ independent of the amount of rotation. Also, dividing a word into k chunks, such as dividing a 32-bit word into 4 bytes, and permuting the chunks leaves the differential undisturbed with probability 2^{-k}.

Multiplicative differentials are compatible with many other operations. Reversing the bits in a word transforms the pair $(x, 2x \bmod 2^\ell)$ into $(x, x/2 \bmod 2^\ell)$ with probability 1. Multiplicative differentials may even survive addition in some cases, since $\alpha \times a + \alpha \times b = \alpha \times (a + b)$. Finally one may consider differentials in which part of the differential is defined using multiplication, and part is defined using some other operation. For example, if a cipher operates on 64-bit blocks (a, b, c, d), where a, b, c, and d are 16-bit subblocks, we may want to consider differential pairs (a, b, c, d) and (a^*, b^*, c^*, d^*) where $a^* = \alpha \times a, b^* = b \oplus \Delta, c^* = c \oplus \Delta$, and $d^* = \alpha \times d$. In other words, the differences $(\alpha, \Delta, \Delta, \alpha)$ are elements of the group $(\mathbb{Z}/2^{16}\mathbb{Z})^* \times (\mathbb{Z}/2\mathbb{Z})^{16} \times (\mathbb{Z}/2\mathbb{Z})^{16} \times (\mathbb{Z}/2^{16}\mathbb{Z})^*$. When there can be no confusion as to the groups in question, we simply refer to these as "hybrid" differentials.

5   xmx

We can now apply our new understanding to find differentials for a large class of moduli in the xmx cipher.

We describe the analysis using the parameters given in the "xmx challenge" [11]. This cipher has 8 rounds, 256-bit blocks, and modulus $n = (2^{80} - 1) \cdot 2^{176} + 157$, which is the smallest prime whose 80 most significant bits are all 1. Written in binary, this modulus is

$$n = \overbrace{11 \cdots 1}^{80} \overbrace{00 \cdots 0}^{168} 10011101,$$

which has $c(n) = 4$. From Proposition 1, whenever x is of the form

$$x = x_{255} x_{254} \cdots x_{177} 0 x_{175} \cdots x_8 1 x_6 1 x_4 x_3 x_2 0 x_0$$

(i.e. whenever $x_{176} = 0$, $x_7 = 1$, $x_5 = 1$ and $x_1 = 0$) then $C_n(x) = -1$ and therefore $x \oplus \Delta^n = -x \mod n$, where

$$\Delta^n = \overbrace{11 \cdots 1}^{79} 0 \overbrace{11 \cdots 1}^{167} 101011101.$$

Recall that Δ^n has a 0 bit in exactly those positions that are constrained in x. If $k \wedge \neg \Delta^n = 0$, then k has a 0 in each constrained bit position, and hence $C_n(x \oplus k) = C_n(x)$.

The key schedule for the xmx challenge cipher is

$$s, s, \ldots, s, s, s \oplus s^{-1}, s, s^{-1}, \ldots, s, s^{-1}$$

where s is a 256-bit number. Suppose $s \wedge \neg \Delta^n = 0$ and $s^{-1} \wedge \neg \Delta^n = 0$. The first equation is satisfied whenever bits 1,5,7 and 176 of s are 0, and hence will be satisfied with probability 2^{-4}. The second equation establishes similar requirements on the bits of s^{-1}, and will be satisfied with probability 2^{-4}. So about 2^{-8} of the keys s satisfy these constraints simultaneously. Obviously, if $s \wedge \neg \Delta^n = 0$ and $s^{-1} \wedge \neg \Delta^n = 0$, then $(s \oplus s^{-1}) \wedge \neg \Delta^n = 0$, as well.

Consider one round of xmx using such a weak key, and let a and b be the subkeys for the current round, so that $a = s$ or s^{-1} or $s \oplus s^{-1}$, but it doesn't matter which. Suppose we apply this round of the cipher to the differential pair (x, x^*), where $x \oplus x^* = \Delta^n$ and $C_n(x) = -1$. Then by Proposition 1, $x^* = x \oplus \Delta^n = -x \mod n$. Since $a \wedge \neg \Delta^n = 0$, $C_n(x \oplus a) = -1$, so $x^* \oplus a = x \oplus a \oplus \Delta^n = -(x \oplus a) \mod n$.

We would like to conclude that $-(x \circ a) = x^* \circ a \mod n$, but must consider the two different behaviors of the operator "$\circ$". From the definition,

$$x \circ a = \begin{cases} x \oplus a & \text{if } x \oplus a < n \\ x & \text{otherwise.} \end{cases}$$

But by assumption, $x_{176} = a_{176} = 0$. Thus bit 176 of $x \oplus a$ is also 0. Furthermore, bits 255 through 176 of n are all 1. This implies $x \oplus a < n$. Bit 176 of Δ^n is 0, so $x^*_{176} = 0$, and hence $x^* \oplus a < n$ for the same reasons. From this, $-(x \circ a) = -(x \oplus a) = x \oplus a \oplus \Delta^n = x^* \oplus a = x^* \circ a$.

So $x^* \circ a = -(x \circ a) \mod n$. The next step in a round of xmx is multiplication by the second subkey, b, which will preserve this multiplicative relationship. So at the end of one round of xmx, with probability 1, the outputs $y = (x \circ a) \times b \mod n$ and $y^* = (x^* \circ a) \times b \mod n$ will satisfy $y^* = -y \mod n$. However, it's not clear that $C_n(y) = -1$. Since multiplication by b affects each bit of the output in a complicated way, we can assume that y is randomly distributed, and therefore $C_n(y) = -1$ with probability $2^{-c(n)} = 2^{-4}$, by Proposition 1. When $C_n(y) = -1$, $y^* = y \oplus \Delta^n$. Thus an input pair $(x, x \oplus \Delta^n)$ becomes an output pair of the form $(y, y \oplus \Delta^n)$ after one round of encryption with probability $\frac{1}{16}$. This yields the following 1-round iterative xor-differential:

$$\Delta^n \longrightarrow \Delta^n \qquad \text{(with probability } 1/16\text{)},$$

or equivalently, the 1-round iterative multiplicative differential

$$-1 \longrightarrow -1 \qquad \text{(with probability } 1/16\text{)}.$$

The probability of the differential may be much higher for many keys, because there are many $-1 \leftrightarrow \Delta$ correspondences that hold with high probability. For example,

$$x \oplus (-x \mod n) = \overbrace{11 \cdots 1}^{78} \, 00 \, \overbrace{11 \cdots 1}^{167} \, 101011101.$$

whenever $x_1 = 0$, $x_5 = 1$, $x_7 = 1$, $x_{176} = 1$ and $x_{177} = 0$. Thus, if, in addition to the weak key constraints described above, $s_{177} = s_{177}^{-1} = 0$, then the multiplicative differential -1 survives one round of the cipher with probability $2^{-4} + 2^{-5}$. There are many other very similar differentials, and if s satisfies even more weak key constraints, then the -1 differential will survive with even greater probability.

This differential survives 8 rounds of the cipher with probability 2^{-32}. The last half round of the cipher consists of only the "$\circ$" operator, and we've already seen that this differential passes through that operation with probability 1, so the differential survives the whole cipher with probability 2^{-32}. Each right pair, (x, x^*), yields 4 constraints on the bits of $(x \oplus s) \times s \mod n$, the output of the first round of the cipher. Although a careful analysis of multiplication mod n may reveal an efficient key recovery attack, we leave this as a distinguishing attack.

This analysis easily generalizes to other instances of xmx with different parameters. For any ℓ-bit modulus n that is not a power of 2, we can compute Δ^n and $c(n)$ as described in the previous section. Consider a single round of xmx that uses modulus n, subkeys k_{2i-1} and k_{2i} in the $\circ$ and multiply steps respectively, and suppose $k_{2i-1} \wedge \neg\Delta^n = 0$. Given an input pair $(x, x \oplus \Delta^n)$ where $C_n(x) = -1$, with probability $2^{-c(n)}$ the output of the round for the pair is of the form $(y, y \oplus \Delta^n)$, with $C_n(y) = -1$, by an analysis similar to the one above. Therefore, the differential survives r rounds of the cipher with probability $2^{-c(n)r}$, as long as each subkey used in the "$\circ$" operation satisfies $k_{2i-1} \wedge \neg\Delta^n = 0$. If independent subkeys are used, $2^{-c(n)r}$ of all keys satisfy this weak key condition. If the xmx key schedule is used, $2^{-2c(n)}$ of all keys are weak, since only s and s^{-1} must satisfy the condition.

Whenever the modulus n used in xmx has a highly regular bit pattern—in particular, long sequences of 1's and 0's—$c(n)$ will be small and therefore such a weak key analysis may be of significantly lower complexity than an exhaustive search.

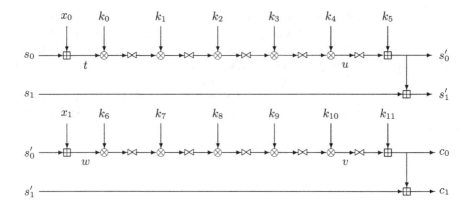

Fig. 2. The MultiSwap cipher. Processing begins with $s_0 = s_1 = 0$ on plaintext x_0, x_1 and proceeds from left to right. The output of the first half, s_0' and s_1', is fed into the second half to produce ciphertext c_0, c_1. The "⋈" operator swaps the 16-bit halves of a 32-bit word, "⊞" represents addition modulo 2^{32}, and "⊗" represents multiplication modulo 2^{32}.

6 MultiSwap

The MultiSwap cipher is used in Microsoft's Digital Rights Management subsystem and was first described in a report published on the Internet under the pseudonym Beale Screamer [12]. The cipher, shown in Figure 2, operates entirely on 32-bit words, maintains two words of internal state, s_0 and s_1, and uses 12 32-bit subkeys $k_0, \dots, k_{11}$. The subkeys $k_0, \dots, k_4, k_6, \dots, k_{10}$ must be odd if the cipher is to be invertible. Unless the cipher is being used in some sort of feedback mode, $s_0 = s_1 = 0$; we will assume this in the analysis. This analysis is also applicable when s_0 and s_1 are non-zero but their values are known. No key schedule is described, so we assume the subkeys are all independent. The cipher operates on 64-bit blocks (x_0, x_1) to produce ciphertext (c_0, c_1).

We first present a chosen-plaintext attack, and then describe how to convert this to a known-plaintext attack. Consider the algorithm operating on input $(0, x_1)$. Since $s_0 = s_1 = 0$, $t = s_0 + x_0 = 0$. Since $u = 0$ if and only if $t = 0$, u is also 0. Thus $s_0' = s_1' = k_5$. After the second half, regardless of the input x_1 the output satisfies $c_1 = c_0 + k_5$. Thus one can derive $k_5 = c_1 - c_0$ with one chosen-plaintext message of the form $(0, x_1)$. Given k_5, one additional message suffices to recover k_{11}. With input $(0, -k_5)$, it is still the case that $s_0' = s_1' = k_5$. In the second half, though, since $x_1 = -k_5$, $w = s_0' + x_1 = k_5 + (-k_5) = 0$, which propagates through the multiplications and swaps as before. Thus the output is $c_0 = k_{11}$ and $c_1 = k_5 + k_{11}$. So a 2-message adaptive chosen-plaintext attack exposes k_5 and k_{11}.

Given k_5, we can control the input to the second half of the cipher. To make $w = a$, query the encryption oracle with the plaintext $(0, a - k_5)$. With k_{11}, we can partially decrypt a ciphertext to obtain the intermediate value v. Therefore, we only have to analyze the sequence of multiplications and swaps in the second half of the cipher between w and v. Similarly, we can analyze the sequence between t and u using knowledge of k_5 and the fact that $s_0' = c_1 - c_0 - s_1$. Because this is a chosen-plaintext attack, we have

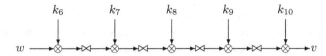

Fig. 3. The second half of the MultiSwap cipher.

reduced the problem to the system in Figure 3 for which the input, w, can be controlled and the output, v, can be observed. The goal is to recover $k_6, \ldots, k_{10}$.

So we focus only on this fragment of MultiSwap. If this fragment operates on inputs w and $w^* = 2w$, then $k_6 \cdot w^* = k_6 \cdot (2w) = 2(k_6 \cdot w)$. From Proposition 3, $\bowtie(k_6 \cdot w^*) = 2 \cdot \bowtie(k_6 \cdot w)$ whenever bits 15 and 31 of $k_6 \cdot w$ are 0, or $\frac{1}{4}$ of the time. Analyzing the rest of Figure 3 in the same way shows that $v^* = 2v$ with probability $\frac{1}{256}$.

If this condition holds, call $(w, 2w)$ a right pair. Then with high probability bits 15 and 31 of $k_6 \cdot w$ are 0. This is a two-bit condition on $k_6 \cdot w$ that we can use to filter the set of potential values of k_6; $\frac{1}{4}$ of all k_6 values will pass this test. We can repeat this test for 16 right input pairs $(w_1, 2w_1), \ldots, (w_{16}, 2w_{16})$ chosen uniformly at random, and the probability of a given k_6 value surviving all 16 tests is roughly $(\frac{1}{4})^{16} = 2^{-32}$, so with high probability only one value of k_6 survives.

If $(w, 2w)$ is a right pair, then the multiplicative differential of $w^* = 2w$ must survive each one of the $\bowtie$ operations. Therefore, $k_7 \cdot \bowtie(k_6 \cdot w)$ must have bits 15 and 31 set to 0. Thus the same right pairs can determine k_7, and then k_8 and k_9. At this point we can determine k_{10} from any known-plaintext. Thus 16 right pairs are enough to recover $k_6, \ldots, k_{10}$, and we can obtain the pairs with about 2^{12} chosen plaintexts. Repeating the analysis for $k_0, \ldots, k_4$ breaks the whole cipher with 2^{13} chosen plaintexts. This is surprisingly small considering the large key size.

The work factor of breaking the cipher is quite low. Let $(w_1, 2w_1), \ldots, (w_{16}, 2w_{16})$ be right pairs that determine k_6. By definition of being right, bits 15 and 31 of $k_6 \cdot w_i$ are 0 for all i. Observe that bit 15 of $k_6 \cdot w_i$ is independent of bits 16 through 31 of k_6. Thus we can determine the value of the low 16 bits of k_6 independently of the high bits. After discovering the low 16 bits, we can then do the same thing for the upper 16 bits. Since we have to test each half of a key against each right pair, the total number of tests performed is $2 \cdot 2^{16} \cdot 16 = 2^{21}$. Repeating for $k_7, \ldots, k_9$, and then again for $k_0, \ldots, k_3$ yields a break on the whole cipher requiring $8 \cdot 2^{21} = 2^{24}$ tests. Each test is quite cheap, involving only a multiply, bit-mask, and test for equality.

To convert this to a known-plaintext attack, observe that even without knowledge of k_5 and k_{11}, we can derive the input to the second half of the cipher via $w = c_1 - c_0 + x_1$. Consider a pair of inputs such that the differential $(w, 2w) \longrightarrow (v, 2v)$ holds. Suppose further that $\bowtie(2v) = 2 \cdot \bowtie(v)$. In this case, $c_0 = \bowtie(v) + k_{11}$ and $c_0^* = 2 \cdot \bowtie(v) + k_{11}$; hence, $k_{11} = 2c_0 - c_0^*$. If, on the other hand, $\bowtie(2v) \neq 2 \cdot \bowtie(v)$, there are three possible values for c_0^*: $2c_0 + k_{11} - 1, 2c_0 + k_{11} + 65536, 2c_0 + k_{11} + 65535$. Each of these possibilities suggests an equation for k_{11}; we can try all four equations and see which makes $v^* = 2v$ hold under partial decryption of c_0, c_0^*. Therefore, each right pair suggests the correct value for k_{11}.

So collect 2^{22} known plaintexts which by the birthday paradox will contain 2^{12} pairs whose input to the second half of the cipher is of the form $(w, 2w)$. Each pair is a right pair with probability 2^{-8}, so the correct k_{11} value will be suggested 16 times. Most wrong pairs will suggest a random value for k_{11}, but, because the sequence of multiplies and swaps maintains sufficient structure, some incorrect values of k_{11} will be suggested with a lower, but still significant probability. In practice, the correct k_{11} will be among the top few, say 8, suggested; since the rest of the analysis is fast, we can repeat it for each of the top 8 suggested values of k_{11} and use trial encryptions to detect the correct one.

With k_{11}, we can now use the same set of pairs to recover $k_6, \ldots, k_{10}$. A similar attack reveals k_5, and then $k_0, \ldots, k_4$. Except for having to repeat the attack for several possible values of k_{11} and k_5, the work factor is about the same as for the previous attacks. Hence the total work for the known plaintext attack is 2^{27}. The storage is also quite small, since we don't have to keep a counter for every possible value of k_{11}, only the ones suggested by a pair. Since the attack uses only about 2^{12} pairs, the storage requirement is about 2^{15} bytes.

To summarize, MultiSwap can be broken with a 2^{13} chosen-plaintext attack requiring 2^{25} work or a 2^{22} known-plaintext attack requiring a work factor of about 2^{27}.

7   IDEA Variants

The IDEA cipher designers deliberately used incompatible group operations to destroy any algebraic relations among the inputs, and this strategy has proven very successful. The basic operations used in IDEA are addition modulo 2^{16}, xor of 16-bit words, and multiplication modulo $2^{16} + 1$.

IDEA uses the addition operation in two places: key mixing and the MA-structure. As has been noted before [10], if $a+b < 2^{16}$, then $a+b \mod 2^{16} = a+b \mod 2^{16}+1$, and thus the MA-structure is linear about $\frac{1}{4}$ of the time. We consider a variant of IDEA, which we call IDEA-X, in which all the additions have been replaced by xors. Because of the observations above, it may appear that IDEA-X is an improvement over IDEA, but we show below that IDEA-X has a large class of weak keys for which it is susceptible to multiplicative differential cryptanalysis.

Because of the heavy use of the xor operation in IDEA-X, we use the multiplicative differential $(-1, -1, -1, -1)$. Let $n = 2^{16} + 1$, and $\Delta = 11 \cdots 101$. By Proposition 1, $-x \mod n = x \oplus \Delta$ if and only if $C_n(x) = -1$. For $n = 2^{16} + 1$, $C_n(x) = -1$ if and only if $x_1 = 1$.[3] The analysis maintains, with non-negligible probability, an invariant on all the intermediate values z and z^* in the cipher. The invariant is that all the intermediate values will satisfy the relation $z^* = (-1)^{z_1} z$. This condition may look mysterious, but it simply means that either $z^* = z$, or $z^* = -z = z \oplus \Delta$. The rest of the analysis is essentially repeated application of the following two rules.

Rule 1 *If* $x^* = (-1)^{x_1} \cdot x$, *then* $kx^* = (-1)^{(kx)_1} \cdot kx$ *with probability* $\frac{1}{2}$. *This is the multiplication rule.*

Rule 2 *If* $a^* = (-1)^{a_1} \cdot a$ *and* $b^* = (-1)^{b_1} \cdot b$, *then* $a^* \oplus b^* = (-1)^{(a \oplus b)_1} \cdot (a \oplus b)$ *with probability 1. This is the xor rule.*

---

[3] Technically, $C_n(x) = -1$ if and only if $x_1 = 1$ and $x_{16} = 0$, but since x is a 16-bit number, the latter condition is vacuous.

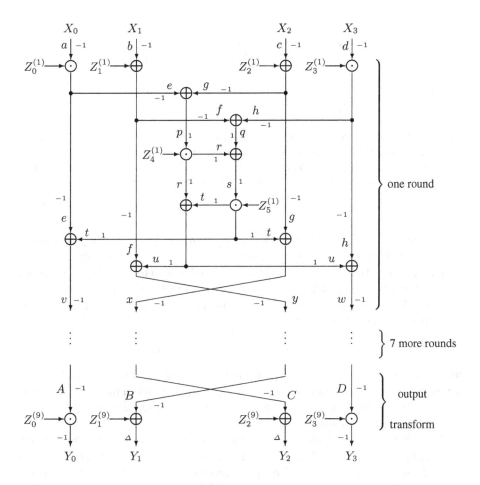

Fig. 4. The IDEA-X cipher. All the adds in IDEA have been changed to xors. The diagram is annotated with the path of the multiplicative differential $(-1, -1, -1, -1)$.

Both rules are easy to prove. Figure 5 explains the xor rule in more detail.

To demonstrate the use of these rules, consider one round of IDEA-X in which bit 1 of $Z_2^{(1)}$ is 0. This is a weak key condition. We'll look only at the inputs X_0 and X_2. Referring to Figure 4, consider two different executions of the round, one with inputs

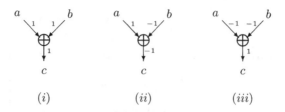

$$(i) \qquad\qquad (ii) \qquad\qquad (iii)$$

Fig. 5. The three cases of the xor rule. The edges are labeled with the multiplicative differentials, e.g. x^*/x. The probability in each case is 1. Recall that differential pairs always satisfy $x^* = (-1)^{x_1} \cdot x$. So in case (iii), $a_1 = b_1 = 1$, and hence $a^* = -a = a \oplus \Delta$ and $b^* = -b = b \oplus \Delta$. Thus $c^* = -a \oplus -b = a \oplus \Delta \oplus b \oplus \Delta = a \oplus b = c$. Furthermore, $a_1 = b_1 = 1$, so $c_1 = 0$. Hence $c^* = (-1)^{c_1} \cdot c$. The other cases are similar.

$X_0 = a$ and $X_2 = c$, the other with inputs $X_0 = -a$ and $X_2 = -c$. Suppose also that $a_1 = c_1 = 1$. By the xor rule, $g_1 = 1$ and $g^* = -g = (-1)^{g_1} g$ with probability 1. By the multiplication rule, $e_1 = 1$ and $e^* = -e = (-1)^{e_1} e$ with probability $\frac{1}{2}$. Combining these two results and applying the xor rule to $e \oplus g$ shows that $p^* = p$ whenever $e_1 = 1$.

If we also assume that bit 1 of $Z_1^{(1)} = 0$ then more of the same sort of reasoning shows that the multiplicative differential

$$(-1, -1, -1, -1) \longrightarrow (-1, -1, -1, -1)$$

survives one round of IDEA-X with probability $\frac{1}{16}$. In order for this differential to work, the input (a, b, c, d) must satisfy $a_1 = b_1 = c_1 = d_1 = 1$. When the differential does successfully pass through the round, the output (v, w, x, y) satisfies $v_1 = w_1 = x_1 = y_1 = 1$. Thus the differential can be iterated.

So we have found an iterative 1-round multiplicative differential that works for keys in which bit 1 of $Z_1^{(i)}$ is 0 and bit 1 of $Z_2^{(i)}$ is 0 in every round. This differential survives 8 rounds of IDEA-X with probability 2^{-32}, and works against 2^{-16} of the keys. The only thing left to consider is the output phase. This phase uses multiplications, which will not disturb the -1 differential, and xors. The differential will survive the xors even without weak key constraints on the subkeys used in the output phase. To see this, consider a differential that has passed 8 rounds; then we have a pair of intermediate texts (A, B, C, D) and (A^*, B^*, C^*, D^*) where $(A^*, B^*, C^*, D^*) = (-A, -B, -C, -D)$. Recall that the -1 multiplicative differential is equivalent to the Δ xor differential. Therefore, $B^* = B \oplus \Delta$ and $C^* = C \oplus \Delta$. This differential survives the final xor of the output phase, giving a hybrid differential $(Y_0^*, Y_1^*, Y_2^*, Y_3^*) = (-Y_0, Y_1 \oplus \Delta, Y_2 \oplus \Delta, -Y_3)$ that survives the whole cipher with probability 2^{-32} for 2^{-16} of the keys.

Using this differential to recover keys is relatively straightforward. An attack using 2^{38} chosen plaintexts yields 32 right pairs with high probability. Each right pair (a, b, c, d) and (a^*, b^*, c^*, d^*) establishes the condition that bit 1 of $Z_0^{(1)} \cdot a$ is 1. Just as in the MultiSwap attack, we can use this condition to filter the possible values of $Z_0^{(1)}$, and given 32 right pairs only two values of $Z_0^{(1)}$ will survive. Unfortunately, whenever $Z_0^{(1)} \cdot a$ satisfies this constraint, so will $-Z_0^{(1)} \cdot a$, so this filter will leave us with two choices for $Z_0^{(1)}$ which differ by a factor of -1. We can recover $\pm Z_3^{(1)}$ in a similar manner. Each

Table 3. A characterization of many IDEA variants which are susceptible to multiplicative differential cryptanalysis.

Round Modifications	
A	The additions in the MA-structure are changed to xors or multiplications
B	The subkey Z_2 is mixed with input X_2 using xor or multiplication
C	The subkey Z_1 is mixed with input X_1 using xor or multiplication
	The additions in the MA-structure are changed to xors or multiplications

right pair also yields a constraint $r_1 = 0$. Observe that $r = Z_4^{(1)} \cdot (e \oplus Z_2^{(1)} \oplus c)$. After recovering $\pm Z_0^{(1)}$, we can compute $\pm e$. Thus we can compute the correct value of e or $e \oplus \Delta$. Hence we can use this condition to filter possible values for $(Z_2^{(1)}, Z_4^{(1)})$ and given 32 right pairs only two values will survive. If we guess the wrong value for $Z_0^{(1)}$, we will perform this filtering with intermediate value $e \oplus \Delta$, and hence will compute $Z_2^{(1)} \oplus \Delta$ instead of $Z_2^{(1)}$. As before, the filter can only determine $Z_4^{(1)}$ up to a factor of ± 1. We now guess the correct value of $Z_4^{(1)}$ and recover $Z_1^{(1)}$ and $\pm Z_5^{(1)}$ in a manner similar to the recovery of $Z_2^{(1)}$ and $\pm Z_4^{(1)}$. We then guess the correct value of $Z_5^{(1)}$, and recover $\pm Z_0^{(2)}$ the same way we recovered $Z_0^{(1)}$. We finish by making guesses for $\pm Z_0^{(1)}$, $\pm Z_3^{(1)}$, and $\pm Z_0^{(2)}$ and using trial encryptions to recover $Z_1^{(2)}$ and verify our guesses.

Recovering $Z_1^{(1)}$ and $Z_5^{(1)}$, dominates the analysis time, requiring $2 \cdot 2^{32} \cdot 32 \approx 2^{38}$ trials. Each trial involves one round of IDEA-X, so the work required is equivalent to about 2^{35} IDEA-X encryptions. A generous estimate of the rest of the work easily gives a work factor of 2^{36} IDEA-X encryptions.

Many other variants of IDEA are also vulnerable to multiplicative differential attacks. Table 3 characterizes a large class of weak IDEA variants by showing the minimum changes necessary to IDEA to render it vulnerable. These IDEA variants have three different round functions, A, B, and C. The output function, D, is exactly as in the original IDEA cipher. The cipher can be any number of rounds, and can begin with A, B, or C, but must cycle through the rounds in the order A, B, C. The round functions are almost identical to the IDEA round functions, except that some of the additions have been changed to xors or multiplications. Each of the specified additions may be replaced with either an xor or a multiply, independent of the other additions. The other additions in the cipher may also be replaced, but this isn't necessary.

This class of weak IDEA variants generalizes our results on IDEA-X: it is only necessary to remove half the additions from the cipher to render it vulnerable to multiplicative differential attacks. From this we conclude that multiplicative differentials can be applicable even to some ciphers with three incompatible group operations.

8   Experimental Verification

We performed several experiments to verify our claims. We first tested the differential probabilities derived in this paper; Table 4 summarizes the results. With the exception of the xmx challenge cipher, all the measurements agree with the theory.

Table 4. Experimental verification of differential probabilities. We use reduced-rounds variants of the xmx challenge cipher and IDEA-X to make the measurement feasible.

Cipher	Rounds	Probability	Pairs	Right pairs [Expected]	[Actual]
Nimbus	5	2^{-5}	10^6	31250	31245
xmx (standard version)	8	1	10^5	10^5	10^5
xmx (challenge version)	4	2^{-16}	10^6	15.3	562
IDEA-X	4	2^{-16}	10^8	1525.9	1537
MultiSwap	all	2^{-8}	10^8	390625	390532

The experiments show that the differential -1 survives 4 rounds of xmx with much higher probability than expected. Part of this discrepancy can be explained by observing that a randomly chosen weak key may allow many $-1 \leftrightarrow \Delta$ correspondences, for different choices of Δ, increasing the probability of the -1 differential. We did further experiments to verify that this was indeed the source of the discrepancy, and were able to predict the experimentally observed probability for a given key to within a factor of 4 (for 4 rounds). We leave it as an open question to explain the remaining error. Since the differential actually survives four rounds with probability about 2^{-11}, we estimate that the xmx challenge cipher can be distinguished using only 2^{23} chosen plaintexts.

We next verified the claim, made in Section 5, that approximately 2^{-8} keys for the xmx challenge cipher are weak. Recall that a key s is weak if $s \wedge \neg \Delta^n = s^{-1} \wedge \neg \Delta^n = 0$. These comprise 4 bit constraints on s and 4 bit constraints on s^{-1}. It is not clear that this will be satisfied by 2^{-8} keys, so we tested 10^6 randomly generated keys, from which we expected to find 3906 weak keys. The actual number of weak keys was 3886, confirming our analysis.

Next we implemented the chosen-plaintext key recovery attack on IDEA-X. In order to make the data requirements feasible, we only attacked 4 rounds of IDEA-X, and only ran 10 trials. The IDEA-X experiments required an average of 124 right pairs to recover the key, about 4 times as many right pairs as we predicted in Section 7. Observing the attack in action reveals that frequently a small number of right pairs—around 30 or 40—are sufficient to eliminate all but 2 or 3 candidates for a particular subkey. A more efficient attack would simply try each candidate, an approach we did not implement.

We implemented the known-plaintext attack on MultiSwap, but since this attack involves repeating the same attack on the two halves of the cipher, we only attacked the latter half. The attack worked as described in Section 6; however, on some trials there were too few right pairs to perform the analysis. Nonetheless, 70 out of 100 runs using 2^{22} known plaintexts were able to successfully recover the key. Increasing the number of plaintexts to 5000000 increased the success rate to 99%. Recall that the attack required guessing the correct value of k_{11} from a list sorted by likelihood. The average position of the correct k_{11} in this list was 2.

9 Conclusion

In this paper we have defined the concept of a multiplicative differential. We described several particular differentials and analyzed how they interact with standard operations used in cryptography such as xor and bit permutations. We then used these differentials to cryptanalyze two existing ciphers and variants of IDEA.

Our results demonstrate that the modular multiplication operation by itself is insufficient to prevent differential attacks. Further, multiplicative differentials can be surprisingly resilient in the presence of incompatible group operations. Therefore, multiplication needs to be carefully combined with other group operations to destroy these differential properties. We are hopeful that this paper will help further the understanding of how to use the multiply operator to build secure cryptographic algorithms.

Acknowledgements. We thank Joan Daemen for providing LaTeX source for a diagram of IDEA from which Figure 4 is derived. We also thank John Kelsey for suggesting that we implement our attacks, an exercise which led to many improvements to our results.

References

1. Eli Biham and Adi Shamir. Differential cryptanalysis of DES-like cryptosystems. *Journal of Cryptology*, 4(1):3–72, 1991.
2. Joan Daemen, Rene Govaerts, and Joos Vandewalle. Weak keys for IDEA. In *CRYPTO*, pages 224–231, 1993.
3. Joan Daemen, Luc van Linden, Rene Govaerts, and Joos Vandewalle. Propagation properties of multiplication modulo $2^n - 1$. In G. H. L. M. Heideman et.al., editor, *Thirteenth Symp. on Information Theory in the Benelux*, pages 111–118, Enschede (NL), 1-2 1992. Werkgemeenschap Informatie- en Communicatietheorie, Enschede (NL).
4. Vladimir Furman. Differential cryptanalysis of Nimbus. In *Fast Software Encryption*. Springer-Verlag, 2001.
5. Carlo Harpes, Gerhard G. Kramer, and James L. Massey. A Generalization of Linear Cryptanalysis and the Applicability of Matsui's Piling-up Lemma. In *EUROCRYPT '95*. Springer-Verlag, May 1995.
6. John Kelsey, Bruce Schneier, and David Wagner. Mod n cryptanalysis, with applications against RC5P and M6. In *Fast Software Encryption*, pages 139–155, 1999.
7. Xuejia Lai, James L. Massey, and Sean Murphy. Markov ciphers and differential cryptanalysis. In *EUROCRYPT '91*. Springer-Verlag, 1991.
8. Alexis Warner Machado. The Nimbus cipher: A proposal for NESSIE. NESSIE Proposal, September 2000.
9. Mitsuru Matsui. Linear cryptanalysis method for DES cipher. In T. Helleseth, editor, *EUROCRYPT '93*, volume 765, pages 386–397, Berlin, 1994. Springer-Verlag.
10. Willi Meier. On the security of the IDEA block cipher. In *EUROCRYPT '93*, pages 371–385. Springer-Verlag, 1994.
11. David M'Raihi, David Naccache, Jacques Stern, and Serge Vaudenay. XMX: a firmware-oriented block cipher based on modular multiplications. In *Fast Software Encryption*. Springer-Verlag, 1997.
12. Beale Screamer. Microsoft's digital rights management scheme—technical details. http://cryptome.org/ms-drm.htm, October 2001.

Differential and Linear Cryptanalysis of a Reduced-Round SC2000

Hitoshi Yanami[1], Takeshi Shimoyama[1], and Orr Dunkelman[2]

[1] FUJITSU LABORATORIES LTD.
1-1, Kamikodanaka 4-Chome, Nakahara-ku, Kawasaki, 211-8588, Japan
{yanami,shimo}@flab.fujitsu.co.jp
[2] COMPUTER SCIENCE DEPARTMENT, TECHNION.
Haifa 32000, Israel
orrd@cs.technion.ac.il

Abstract. We analyze the security of the SC2000 block cipher against both differential and linear attacks. SC2000 is a six-and-a-half-round block cipher, which has a unique structure that includes both the Feistel and Substitution-Permutation Network (SPN) structures. Taking the structure of SC2000 into account, we investigate one- and two-round iterative differential and linear characteristics. We present two-round iterative differential characteristics with probability 2^{-58} and two-round iterative linear characteristics with probability 2^{-56}. These characteristics, which we obtained through a search, allowed us to attack four-and-a-half-round SC2000 in the 128-bit user-key case. Our differential attack needs 2^{103} pairs of chosen plaintexts and 2^{20} memory accesses and our linear attack needs $2^{115.17}$ known plaintexts and $2^{42.32}$ memory accesses, or $2^{104.32}$ known plaintexts and $2^{83.32}$ memory accesses.

Keywords: Symmetric block cipher, SC2000, differential attack, linear attack, characteristic, probability

1 Introduction

Differential cryptanalysis was initially introduced by Murphy [10] in an attack on FEAL-4 and was later improved by Biham and Shamir [1,2] to attack DES. Linear cryptanalysis was first proposed by Matsui and Yamagishi [6] in an attack on FEAL and was extended by Matsui [7] to attack DES. Both methods are well known and often provide very effective means for attacking block ciphers. One of the many steps in establishing a cipher's security is to evaluate its strength against these attacks. The respective degrees of security of a cipher against differential attacks and linear attacks can be estimated from the maximum differential probability and the maximum linear probability. A cipher is considered to be secure against attacks of both types if both probabilities are low enough to make the respective forms of attack impractical.

SC2000 is a block cipher which was submitted to the NESSIE [11] and CRYPTREC [3] projects by Shimoyama et al. [13]. In the *Self-Evaluation Report*, one

J. Daemen and V. Rijmen (Eds.): FSE 2002, LNCS 2365, pp. 34–48, 2002.

of the submitted documents, the security against differential and linear cryptanalysis was evaluated by estimating the number of active S-boxes in differential
and linear characteristics. The strength of the SC2000 cipher has been evaluated
in other published work on attacking SC2000 [3,4,5,12,17].

This paper is based on the work of Yanami and Shimoyama [17] at the 2nd
NESSIE workshop, which is a report by the authors on investigation they carried
out with both differential and linear attacks on a reduced-round SC2000. We use
the same differential characteristics as was used in the above work in the work
we describe here. This has a slightly higher probability than the characteristics
that have been found by Raddum and Knudsen [12]. The linear cryptanalysis
by Yanami and Shimoyama [17] contained some incorrect calculations. We have
corrected these and re-examined the linear characteristics. Moreover, in this
paper we use a better method of deducing the subkey bits. This new technique
reduces the time complexity of the attacks by several orders of magnitude.

In this paper we investigate the one- and two-round iterative differential/lin-
ear characteristics of SC2000. By iterating the differential/linear characteristic
obtained by our search, we construct a longer characteristic and utilize it to
attack four-and-a-half-round SC2000.

The paper is organized as follows. We briefly describe the encryption algorithm for SC2000 in Section 2. In Section 3, we illustrate our search method
and show our search results. We present our differential and linear attacks on
four-and-a-half-round SC2000 in Sections 4 and 5, respectively. We summarize
our paper in Section 6.

2 Description of SC2000

SC2000 is a block cipher which was submitted to the NESSIE [11] and CRYPT-
REC [3] projects by Shimoyama et al. [13]. SC2000 has a 128-bit block size and
supports 128-/192-/256-bit user keys, and in these ways is the same as the AES.

Before proceeding further, we need to make two remarks: Firstly, we mainly
take up the case of 128-bit user keys. Secondly, we omit the description of the
SC2000 key schedule as it has no relevance to the attacks presented in this paper.
The key schedule generates sixty-four 32-bit subkeys from a 128-bit user key.

2.1 The Encryption Algorithm

Three functions are applied in the SC2000 encryption algorithm: the I, R and
B functions. Each function has 128-bit input and output. The R functions used
are of two types, which only differ in terms of a single constant (0x55555555 or
0x33333333). When we need to distinguish between the two types, we use R_5
and R_3 to indicate the respective constants.

The encryption function may be written as:

$$I\text{-}B\text{-}I\text{-}R_5{\times}R_5\text{-}I\text{-}B\text{-}I\text{-}R_3{\times}R_3\text{-}I\text{-}B\text{-}I\text{-}R_5{\times}R_5\text{-}$$
$$I\text{-}B\text{-}I\text{-}R_3{\times}R_3\text{-}I\text{-}B\text{-}I\text{-}R_5{\times}R_5\text{-}I\text{-}B\text{-}I\text{-}R_3{\times}R_3\text{-}I\text{-}B\text{-}I,$$

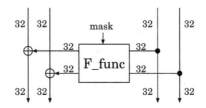

Fig. 1. The R function

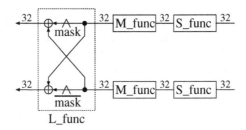

Fig. 2. The F function

where $\times$ stands for the exchange of the left and right 64 bits. We define -I-B-I-$R\times R$- as the round in SC2000. The round is iterated six times by the cipher and the final set of operations, -I-B-I, is then applied to obtain symmetry for encryption and decryption. For the sake of simplicity, we refer to the last part -I-B-I as half a round. SC2000 involves six and a half rounds with a 128-bit user key, and seven and a half rounds with a 192-/256-bit user key.

2.2 The I Function

The I function XORs a 128-bit input with four 32-bit subkeys. The I function divides the input into four 32-bit words, and then applies an XOR to each of these words and a corresponding 32-bit subkey. These subkeys are only used in the I function.

2.3 The R Function

The R function has a conventional Feistel structure, except for the swapping of the left and right 64 bits in its last part (Fig. 1). The F function is applied to the right 64 bits of the input, and the output of the function is XORed with the left 64 bits. The result of the XOR becomes the left half of the output of the R function. The right half of the output of the R function is the same as the right half of the input.

The input and output of the F function in the R function are both 64 bits; the function consists of three subfunctions, the S, M and L functions (Fig. 2).

The F function divides its input into two 32-bit variables and each variable is successively dealt with by the S and M functions. The two outputs become the input value for the L function, the output of which becomes the output of the F function. Note that the F function is bijective. We describe the S, M and L functions below.

The S function is a 32-bit input/output nonlinear function. The 32-bit input is divided into groups, in order, 6, 5, 5, 5, 5 and 6 bits. These groups of bits enter corresponding S-boxes; each of the two 6-bit groups enters S_6, while each of the four 5-bit groups enters S_5. The output of the S function is the concatenation of the outputs of the S-boxes. We refer to Shimoyama et al. [13] for the values in the S_5 and S_6 tables.

The M function is a 32-bit input/output linear function. The output b for an input a is the product of a and a matrix M,

$$b = a \cdot M,$$

where M is a square matrix of order 32 with entries that are elements of $GF(2)$, the Galois field of order two, and a and b are row vectors with entries from $GF(2)$. We refer to Shimoyama et al. [13] for the entries of the matrix M.

The L function has two 32-bit variables as its input and output. We use (a, b) to denote the input and (c, d) to denote the corresponding output. The variables c and d are obtained by applying the following formulae:

$$c = (a \wedge mask) \oplus b; \quad d = (b \wedge \overline{mask}) \oplus a,$$

where $mask$ is the constant we earlier mentioned, 0x55555555 or 0x33333333, $\overline{mask}$ is the bitwise-NOT of the $mask$, and the symbol $\wedge$ represents the bitwise-AND operation.

2.4   The B Function

The B function has an SPN structure with 128-bit input/output which contains the thirty-two 4-bit input/output S-boxes. This structure is similar to the one that is used in the Serpent block cipher. We can use a bitslice approach to implement the B function. We represent the input to the B function as (a, b, c, d), where each variable has 32 bits; the i-th bit of a is a_i, and equivalent notation applies to b, c and d. For $i = 0, 1, \ldots, 31$, the i-th bit is taken from each of the variables a, b, c and d, and the resulting four bits (a_i, b_i, c_i, d_i) are replaced by (e_i, f_i, g_i, h_i) according to the S_4 table, where the notation is analogous to that for (a, b, c, d) above, and (e, f, g, h) denotes the four 32-bit words which compose the output of the B function. We refer to Shimoyama et al. [13] for the values in the S_4 table. Note that if the B functions in SC2000 are all replaced by the swapping of the left and right 64 bits, the resulting structure becomes the classical Feistel structure.

3   The Differential and Linear Characteristics of SC2000

In investigating the differential/linear characteristics of a block cipher, it is very hard to compute the probability of every characteristic by applying an exhaustive search to the cipher itself. Roughly speaking, the following strategy is often used to construct a long characteristic that has a high probability. 1) Examine few-round iterative characteristics, i.e., characteristics of the same input and output differences/masks appearing at intervals of a few rounds. 2) Iterate the one with the highest differential/linear probability which was found by the search to make a characteristic for larger numbers of rounds. We will follow this strategy in examining the differential/linear characteristics.

In the SC2000 encryption algorithm, the I functions are merely used for XORing data with subkeys, so we can eliminate them from our examination of differential/linear relationships. Removing these functions leaves the following sequence:

$$B\text{-}R_5{\times}R_5\text{-}B\text{-}R_3{\times}R_3\text{-}B\text{-}R_5{\times}R_5\text{-}B\text{-}R_3{\times}R_3\text{-}B\text{-}R_5{\times}R_5\text{-}B\text{-}R_3{\times}R_3\text{-}B.$$

It can be seen that $\text{-}B\text{-}R{\times}R\text{-}$ is a period. It is repeated six times until the final B function is added to preserve symmetry of the encryption and decryption procedures. Taking the period into consideration, we investigate the one- and two-round differential/linear characteristics with certain patterns of differences and masks.

3.1   Differential Characteristics

We explain our efficient way of searching for an iterative differential characteristic that has a high probability. We start by reviewing the nonlinear functions of SC2000. They are all realized by the S-boxes, S_4, S_5 and S_6. There are thirty-two S_4 boxes in the B function and eight S_5 boxes and four S_6 boxes in the R function. The S function is made up of four S_5 boxes and two S_6 boxes; there are two S functions in the R function. In the differential distribution tables for the S-boxes, we see that given a pair of nonzero input and output differences, the differential probability for S_4 is either 2^{-2} or 2^{-3}, while for S_5 it is 2^{-4} and for S_6 it is either 2^{-4} or 2^{-5}. These facts suggest that the number of nonzero differences for the S_5 and S_6 boxes in the S function will have a stronger effect on the overall differential probability of characteristics than the number for the S_4 boxes in the B function.

Taking this into consideration, we decide to investigate those differential characteristics that have a nonzero difference in a single one of the four S functions in a $\text{-}B\text{-}R{\times}R\text{-}$ cycle, which enables us to efficiently find those differential characteristics that have high probabilities.

One-Round Characteristics. We investigate those one-round iterative characteristics that have a nonzero difference in a single one of the four S functions

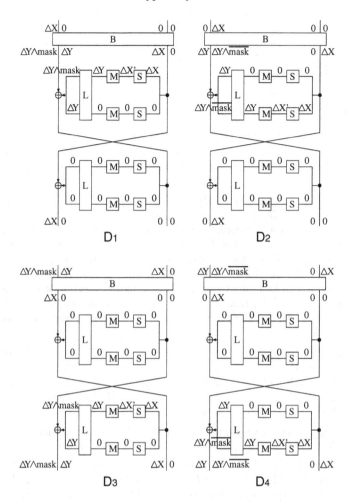

Fig. 3. Patterns of differences

in a $-B\text{-}R{\times}R\text{-}$ cycle. We illustrate the differential patterns of the characteristics we need to investigate in Fig. 3.

We call the respective types of differential pattern D_1, D_2, D_3 and D_4, according to the position of the S function that has a nonzero difference.

We have investigated differential characteristics that have these patterns for both $-B\text{-}R_5{\times}R_5\text{-}$ and $-B\text{-}R_3{\times}R_3\text{-}$. We have found differential characteristics that have a probability of 2^{-33} in both cycles. This is the highest probability for differential characteristics of the four types mentioned above. These characteristics are of type D_3. We give an example of one such differential characteristic:

$$-B\text{-}R{\times}R\text{-}$$

$$
B \left\{
\begin{array}{llll}
(& 0 \;\; \text{0x00080008} \;\; \text{0x08090088} & & 0) \\
& \qquad\qquad \downarrow B & & \\
(\text{0x08090088} & 0 & 0 & 0)
\end{array}
\right\} 2^{-15}
$$

$$R \; (\qquad\quad 0 \qquad\qquad 0) \xleftarrow{F} (\qquad\quad 0 \qquad\qquad 0) \; 1$$

$$R \; (\qquad 0 \;\; \text{0x00080008}) \xleftarrow{F} (\text{0x08090088} \qquad\qquad 0) \; 2^{-18}.$$

Note that this characteristic has a probability of 2^{-33} regardless of the constant in the R function, and that we can construct an n-round differential characteristic with probability 2^{-33n} by concatenating this characteristic n times.

Two-Round Characteristics. By distinguishing R_5 from R_3, we are also able to treat $-B\text{-}R_5{\times}R_5\text{-}B\text{-}R_3{\times}R_3\text{-}$ as a cycle. Turning our attention to this cycle, we have investigated those two-round iterative characteristics which have a nonzero difference in a single one of the four S functions in each $-R{\times}R\text{-}$ part. It would appear that we are able to independently choose differential patterns from among D_i $(i = 1, 2, 3, 4)$ for the former $-B\text{-}R_5{\times}R_5\text{-}$ and latter $-B\text{-}R_3{\times}R_3\text{-}$ sequence, but some patterns cannot be concatenated. We can judge whether or not it is possible to concatenate D_i and D_j from the differential distribution table of S_4 (see the Appendix). Below we list the pairs that may be concatenated and thus need to be investigated:

$-B\text{-}R_5{\times}R_5\text{-}$	$-B\text{-}R_3{\times}R_3\text{-}$		$-B\text{-}R_5{\times}R_5\text{-}$	$-B\text{-}R_3{\times}R_3\text{-}$
$\Delta A \to (D_1) \to \Delta B$	$\to (D_1) \to \Delta A$		$\Delta A \to (D_1) \to \Delta B$	$\to (D_2) \to \Delta A$
$\Delta A \to (D_2) \to \Delta B$	$\to (D_2) \to \Delta A$		$\Delta A \to (D_2) \to \Delta B$	$\to (D_1) \to \Delta A$
$\Delta A \to (D_3) \to \Delta B$	$\to (D_3) \to \Delta A$		$\Delta A \to (D_3) \to \Delta B$	$\to (D_4) \to \Delta A$
$\Delta A \to (D_4) \to \Delta B$	$\to (D_4) \to \Delta A$		$\Delta A \to (D_4) \to \Delta B$	$\to (D_3) \to \Delta A$.

Note that differences ΔA's and ΔB's in the above list should be adjusted as required: When, for example, the former pattern is D_1 and the latter is D_2, we think of ΔA as $(0, \Delta X, 0, 0)$ and ΔB as $(\Delta X, 0, 0, 0)$, adopting the respective output differences of the preceding R functions. We have investigated characteristics of the above types and found that the differential characteristics with the highest probability have the pattern

$$\Delta A \to (D_4) \to \Delta B \to (D_3) \to \Delta A$$

with probability 2^{-58}. An example of such a characteristic is given below:

$$-B\text{-}R_5{\times}R_5\text{-}B\text{-}R_3{\times}R_3\text{-}$$

$$
B \left\{
\begin{array}{llll}
(\text{0x01120000} \;\; \text{0x01124400} \;\; \text{0x01124400} & & 0) \\
\qquad\qquad\qquad \downarrow B & & \\
(\qquad 0 \;\; \text{0x01124400} & 0 & 0)
\end{array}
\right\} 2^{-15}
$$

$$R_5 \; (\qquad\quad 0 \qquad\qquad 0) \xleftarrow{F} (\qquad\quad 0 \qquad\qquad 0) \; 1$$

$$R_5 \; (\text{0x01124400} \;\; \text{0x00020000}) \xleftarrow{F} (\qquad 0 \;\; \text{0x01124400}) \; 2^{-16}$$

$$
B \left\{
\begin{array}{llll}
(\text{0x01124400} \;\; \text{0x00020000} & 0 \;\; \text{0x01124400}) \\
\qquad\qquad \downarrow B & & \\
(\text{0x01124400} & 0 & 0 & 0)
\end{array}
\right\} 2^{-11}
$$

$$R_3 \; (\qquad\quad 0 \qquad\qquad 0) \xleftarrow{F} (\qquad\quad 0 \qquad\qquad 0) \; 1$$

$$R_3 \; (\text{0x01120000} \;\; \text{0x01124400}) \xleftarrow{F} (\text{0x01124400} \qquad\qquad 0) \; 2^{-16}.$$

The probability 2^{-58} of this characteristic is higher than 2^{-66}, the probability obtained for the two-round differential characteristic from the one-round iterative characteristic with the highest probability that we found. We will later use the former characteristic in our differential attack on a reduced-round SC2000.

3.2 Linear Characteristics

As with the differential probability, the linear distribution tables tell us that the number of nonzero masks for the S_5 and S_6 boxes in the S function have a stronger effect on the overall linear probability than the number for the S_4 boxes in the B function; given a pair of nonzero input/output masks, the linear probability for S_4[†] is either 2^{-2} or 2^{-4}, while for S_5 it is 2^{-4} and for S_6 it is between 2^{-4} and 2^{-8}.

We investigate the linear characteristics in the same way as the iterative differential characteristics, i.e., we examine the linear characteristics with a nonzero mask in a single one of the four S functions in a $-B-R\times R-$ cycle.

One-Round Characteristics. We investigate those one-round iterative characteristics whose masks have a nonzero value in a single one of the four S functions in a $-B-R\times R-$ cycle. We illustrate the mask patterns of the characteristics we investigate in Fig. 4.

We call the respective types of mask pattern L_1, L_2, L_3 and L_4, according to the position of the S function that has a nonzero mask.

We investigated characteristics of these types and found that the linear characteristics with the highest probability have probabilities of $2^{-28.83}$ for $-B-R_5\times R_5-$ and of 2^{-28} for $-B-R_3\times R_3-$. All of them are of type L_2. Below, we give examples of such linear characteristics for both $-B-R_5\times R_5-$ and $-B-R_3\times R_3-$:

$$-B-R_5\times R_5-$$

$$
B \left\{
\begin{array}{l}
(\quad\quad 0 \quad\quad\quad 0\ \text{0x84380080 0x04100000}) \\
\quad\quad\quad\quad\quad\quad \downarrow B \\
(\text{0x84380080 0x04100000} \quad\quad 0\ \text{0x84180000})
\end{array}
\right\} 2^{-20}
$$

$$R_5\ (\text{0x84380080 0x04100000}) \xleftarrow{F} (\quad\quad 0\ \text{0x84180000})\ 2^{-8.83}$$

$$R_5\ (\quad\quad 0 \quad\quad\quad 0)\xleftarrow{F} (\quad\quad 0 \quad\quad\quad 0)\ 1,$$

$$-B-R_3\times R_3-$$

$$
B \left\{
\begin{array}{l}
(\quad\quad 0 \quad\quad\quad 0\ \text{0x12020040 0x12020000}) \\
\quad\quad\quad\quad\quad\quad \downarrow B \\
(\text{0x12020040 0x12020000} \quad\quad 0\ \text{0x12020040})
\end{array}
\right\} 2^{-10}
$$

$$R_3\ (\text{0x12020040 0x12020000}) \xleftarrow{F} (\quad\quad 0\ \text{0x12020040})\ 2^{-18}$$

$$R_3\ (\quad\quad 0 \quad\quad\quad 0)\xleftarrow{F} (\quad\quad 0 \quad\quad\quad 0)\ 1.$$

[†] In Yanami and Shimoyama [17], the authors used 2^{-2} or 2^{-3} for this probability, which turned out to be wrong. We have re-examined linear characteristics with the correct values.

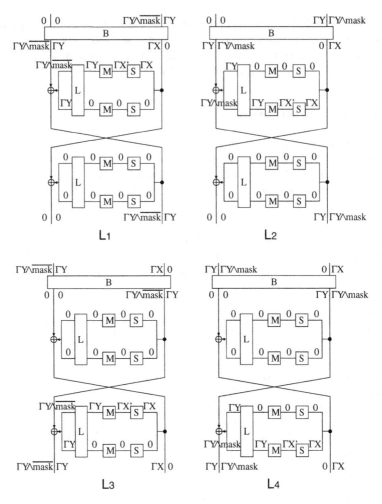

Fig. 4. Patterns of masks

It is not possible for these characteristics to pass through the sequence when the constant is changed into the other one. The highest probability for any linear characteristics which do pass through both constants is $2^{-36.83}$. An example of such a linear characteristic is:

$$\textit{-B-R×R-}$$

$$B \left\{ \begin{array}{l} (\qquad 0 \qquad\qquad 0\ \text{0x11108008 0x11100000}) \\ \qquad\qquad\qquad \downarrow B \\ (\text{0x11108008 0x11100000} \qquad\qquad 0\ \text{0x11108008}) \end{array} \right\} 2^{-14}$$

$$R\ (\text{0x11108008 0x11100000}) \xleftarrow{F} (\qquad\qquad 0\ \text{0x11108008})\quad 2^{-22.83}$$

$$R\ (\qquad 0 \qquad\qquad 0) \xleftarrow{F} (\qquad 0 \qquad\qquad 0)\quad 1.$$

We are able to construct an n-round linear characteristic with probability $2^{-36.83n}$ by concatenating the above characteristic n times.

Two-Round Characteristics. We can apply the same method as we used for the differential case. We can use the linear distribution table of S_4 to judge whether or not L_i and L_j can be concatenated (see the Appendix). Below, we list those pairs that need to be investigated:

$$-B\text{-}R_5 \times R_5\text{-} \qquad -B\text{-}R_3 \times R_3\text{-} \qquad\qquad -B\text{-}R_5 \times R_5\text{-} \qquad -B\text{-}R_3 \times R_3\text{-}$$

$$\Gamma A \to (L_1) \to \Gamma B \to (L_1) \to \Gamma A \qquad \Gamma A \to (L_1) \to \Gamma B \to (L_2) \to \Gamma A$$
$$\Gamma A \to (L_2) \to \Gamma B \to (L_2) \to \Gamma A \qquad \Gamma A \to (L_2) \to \Gamma B \to (L_1) \to \Gamma A$$
$$\Gamma A \to (L_3) \to \Gamma B \to (L_3) \to \Gamma A \qquad \Gamma A \to (L_3) \to \Gamma B \to (L_4) \to \Gamma A$$
$$\Gamma A \to (L_4) \to \Gamma B \to (L_4) \to \Gamma A \qquad \Gamma A \to (L_4) \to \Gamma B \to (L_3) \to \Gamma A \,.$$

The masks ΓA and ΓB should be considered as the masks output by the immediately preceding R functions, respectively. We have investigated characteristics of the above types and found that the linear characteristics with the highest probability have the pattern

$$\Gamma A \to (L_4) \to \Gamma B \to (L_4) \to \Gamma A.$$

The probability of these linear characteristics is 2^{-56}. We list a example of such a characteristic below:

$$-B\text{-}R_5 \times R_5\text{-}B\text{-}R_3 \times R_3\text{-}$$

$$B \left\{ \begin{array}{l} (\text{0x204000a2 0x20000022} \qquad\quad 0\ \text{0x20400022}) \\ \qquad\qquad\qquad\quad \downarrow B \\ (\qquad\quad 0 \qquad\qquad 0\ \text{0x204000a2 0x00400000}) \end{array} \right\} 2^{-12}$$

$$R_5\ (\qquad\quad 0 \qquad\qquad 0) \xleftarrow{F} (\qquad\qquad 0 \qquad\qquad 0)\ 1$$

$$R_5\ (\text{0x204000a2 0x00400000}) \xleftarrow{F} (\qquad 0\ \text{0x20400022})\ 2^{-16}$$

$$B \left\{ \begin{array}{l} (\text{0x204000a2 0x00400000} \qquad\quad 0\ \text{0x20400022}) \\ \qquad\qquad\qquad\quad \downarrow B \\ (\qquad\quad 0 \qquad\qquad 0\ \text{0x204000a2 0x20000022}) \end{array} \right\} 2^{-12}$$

$$R_3\ (\qquad\quad 0 \qquad\qquad 0) \xleftarrow{F} (\qquad\qquad 0 \qquad\qquad 0)\ 1$$

$$R_3\ (\text{0x204000a2 0x20000022}) \xleftarrow{F} (\qquad 0\ \text{0x20400022})\ 2^{-16}.$$

The probability 2^{-56} of this characteristic is much higher than $2^{-73.66}$, the probability of the two-round linear characteristic obtained from any one of the one-round iterative characteristics with the highest probability which we had previously found. We use this characteristic in our linear attack on a reduced-round SC2000.

4   A Differential Attack on 4.5-Round SC2000

We now present our attack on a reduced-round SC2000 with 128-bit user key. By using a differential or linear characteristic with the highest probability which we had obtained in our search, we were able to attack the following four-and-a-half-round SC2000:

$$I\text{-}B\text{-}I\text{-}R_5 \times R_5\text{-}I\text{-}B\text{-}I\text{-}R_3 \times R_3\text{-}I\text{-}B\text{-}I\text{-}R_5 \times R_5\text{-}I\text{-}B\text{-}I\text{-}R_3 \times R_3\text{-}I\text{-}B\text{-}I.$$

In this section, we illustrate how we use our differential attack to guess some bits in subkeys. Our linear attack will be described in the next section.

Our differential attack utilizes the two-round iterative differential characteristic with probability 2^{-58}, which we mentioned in Section 3.1. By concatenating the two-round differential characteristic we mentioned in Section 3.1 twice (and removing one B function), we obtain a three-and-a-half-round differential characteristic with probability 2^{-101}. We apply the characteristic in the following way:

$$\overbrace{\text{Input-} \oplus_{K_1} - \underbrace{B}_{(1)} \overset{0}{-}R\overset{16}{-}R\overset{11}{-}B\overset{0}{-}R\overset{16}{-}R\overset{15}{-}B\overset{0}{-}R\overset{16}{-}R\overset{11}{-}B\overset{0}{-}R\overset{16}{-}R\underbrace{-B}_{(2)} - \oplus_{K_2} - \text{Output,}}^{101}$$

where the numeral 15 above the B is read as "the differential probability for the B function is 2^{-15}." We present a table of total differential probabilities according to the number of functions in the Appendix.

By using this differential characteristic, we are able to deduce 40 bits in the subkeys K_1 and K_2. These subkey bits correspond to the five active S-boxes of the first B function and the five active S-boxes of the last B function. We used the following algorithm to retrieve these 40 bits in the subkeys K_1 and K_2:

- We start by encrypting 2^{84} structures, where each structure contains 2^{20} plaintexts which have the same value in the 108 inactive bits in the first B function, and with the 20 active bits varying across all possible values.
- In each structure we look for collisions in the 108 inactive output bits of the last B function.
- When a collision is detected, we analyze the pair of plaintexts and ciphertexts, and check for the subkey values where the pair satisfied the differential characteristic. For each subkey which satisfies the characteristic, we increment the counter by 1.
- In the end, we go over all of the subkey counters and output the subkey that corresponds to counter with the highest number.

As each structure induces 2^{19} pairs, we have in total 2^{103} pairs which have the same input difference as the input of the characteristic above after the B function. We would expect the right subkey to be suggested about four times. Since in each structure the chance that two of the 2^{20} ciphertexts will agree is about $(2^{20})^2/2 \cdot 2^{-108} = 2^{-69}$, we would also expect $2^{-69} \cdot 2^{84} = 2^{15}$ false hits varying over all the 2^{40} possible subkeys, These false hits have few effects on subkey-counting. Thus, we are assured with a very high probability that the suggested subkey is the correct one.

The time complexity of this attack (aside from the 2^{104} encryptions) is the time taken to hash the ciphertexts in each structure according to the 108 inactive bits, plus the time taken to analyze the 2^{15} suggested pairs. The first term may be neglected as representing part of the encryption, and the analysis can be done in about 2^{20} memory accesses.

5   Linear Attacks on 4.5-Round SC2000

We now present our linear attack on a reduced-round SC2000 with 128-bit user key. By using the linear characteristic with the highest probability which we had obtained in our search, we are able to attack the following four-and-a-half-round SC2000:

$$I\text{-}B\text{-}I\text{-}R_5{\times}R_5\text{-}I\text{-}B\text{-}I\text{-}R_3{\times}R_3\text{-}I\text{-}B\text{-}I\text{-}R_5{\times}R_5\text{-}I\text{-}B\text{-}I\text{-}R_3{\times}R_3\text{-}I\text{-}B\text{-}I.$$

We now illustrate how we use the linear attack to guess subkeys. We use the two-round iterative linear characteristic with probability 2^{-56} from Section 3.2. By concatenating this characteristic twice, we obtain a four-round linear characteristic with probability 2^{-112}. We illustrate two types of attacks; in one, the four-round linear characteristic is used; in the other, the three-and-a-half-round characteristic obtained by eliminating the first B function from the four-round one is used. We illustrate both attacks in due order.

5.1   Attack Using a Four-Round Characteristic

We use the following four-round linear characteristic with probability 2^{-112}:

$$\text{Input-}\underbrace{\overset{12}{B}\text{-}\overset{16}{R}\text{-}\overset{0}{R}\text{-}\overset{12}{B}\text{-}\overset{16}{R}\text{-}\overset{0}{R}\text{-}\overset{12}{B}\text{-}\overset{16}{R}\text{-}\overset{0}{R}\text{-}\overset{12}{B}\text{-}\overset{16}{R}\text{-}\overset{0}{R}\text{-}}_{112}\underbrace{B}_{K_1}\text{-}\underset{K_1}{\oplus}\text{-Output,}$$

$$(1)$$

where the numeral 12 above the B is read as "the linear probability of passage through the B function is 2^{-12}." We present a table of total linear probabilities according to the number of functions in the Appendix.

We are able to deduce 20 bits in the subkey K_1, which consists of four 32-bit subkeys. In the last B function (1), output mask is only related to the five S_4 S-boxes. As there are only 20 ciphertext bits which interest us, we are able to count the number of occurrences of each case, and do this analysis once for each 20-bit ciphertext value. The following algorithm is capable of extracting the 20 subkey bits:

- Initialize a 2^{20} array of counters (corresponding to the 20 ciphertext bits which are related to the characteristic).
- Encrypt $2^{112} \cdot 9 = 2^{115.17}$ plaintexts.
- For each plaintext and its corresponding ciphertext add or subtract 1 (according to the parity of the input subset) to/from the counter related to the given 20 ciphertext bits.
- After counting all occurrences of the given 20 ciphertext bits, for each subkey and for each 20 bit value, calculate the parity of the output subset.
- Rank the subkey candidates according to their respective biases from 1/2.

We expect the right subkey to be highly ranked, and on this basis guess that the top-ranked candidate is the right subkey. The time complexity of the above algorithm is at most $2^{40} \cdot 5 = 2^{42.32}$ S_4 calls. The success rate for the above algorithm is at least 62.3%. We can use key ranking to improve the success rate without affecting the complexity of the data. We conclude that our linear attack requires $2^{115.17}$ known plaintexts and $2^{42.32}$ S_4 calls.

5.2   Attack Based on a 3.5-Round Characteristic

We use the following linear characteristic with probability 2^{-100}:

$$\text{Input-} \oplus \underset{K_1}{-} \underbrace{B}_{(1)} \overset{16}{-} R \overset{0}{-} R \overset{12}{-} B \overset{16}{-} R \overset{0}{-} R \overset{12}{-} B \overset{16}{-} R \overset{0}{-} R \overset{12}{-} B \overset{16}{-} R \overset{0}{-} R \underbrace{-B}_{(2)} \underset{K_2}{-} \oplus \text{- Output.}$$

We need to infer 20 bits in each of K_1 and K_2.

By making a small change to the above algorithm (taking the 40 plaintext and ciphertext bits into consideration, and trying 40 subkey bits) we obtain the result that, given $2^{104.32}$ known plaintexts, the attack requires $2^{83.32}$ S_4 calls.

6   Conclusions

We have studied the security of SC2000 against differential and linear cryptanalysis. Taking the periodic structure of SC2000 into consideration, we have investigated two-round iterative characteristics in which the differences or masks have a nonzero value in only one of the four S functions in each -B-$R \times R$- cycle, and found iterative differential characteristics with probability 2^{-58} and iterative linear characteristics with probability 2^{-56}.

We respectively utilized the best differential and best linear characteristic we found. We have presented both differential and linear attacks on the four-and-a-half-round SC2000. Our differential attack needs 2^{104} pairs of chosen plaintexts and 2^{20} memory accesses, and our linear attack requires $2^{115.17}$ known plaintexts and $2^{42.32}$ S_4 calls, or $2^{104.32}$ known plaintexts and $2^{83.32}$ S_4 calls. Either attack is capable of deducing 40 bits in the subkeys used in the first and last I functions.

We stress that neither our differential nor our linear attack would work on the full-round SC2000, which has six and a half rounds. The equivalent differential and linear characteristics needed to attack 6.5-round SC2000 has respective probabilities of 2^{-159} and 2^{-156}. We conclude that these figures show that these attacks are not applicable to the full-round SC2000.

References

1. E. Biham and A. Shamir, *Differential Cryptanalysis of DES-like Cryptosystems,* CRYPTO '90, LNCS 537, pp.2-21, 1991.
2. E. Biham and A. Shamir, *Differential Cryptanalysis of the Full 16-round DES,* CRYPTO '92, LNCS 740, pp.487-496, 1993.

3. CRYPTREC project – Evaluation of Cryptographic Techniques.
   (http://www.ipa.go.jp/security/enc/CRYPTREC/index-e.html)
4. O. Dunkelman and N. Keller, *Boomerang and Rectangle Attack on SC2000*, Proceedings of Second Open NESSIE Workshop, September 12-13, 2001.
5. L. R. Knudsen and H. Raddum, *A first report on Whirlpool, NUSH, SC2000, Noekeon, Two-Track-Mac and RC6*, 2001.
   (http://www.cosic.esat.kuleuven.ac.be/nessie/reports/)
6. M. Matsui and A. Yamagishi, *A new method for known plaintext attack of FEAL cipher*, EUROCRYPT '92, LNCS 658, pp.81-91, 1993.
7. M. Matsui, *Linear Cryptanalysis Method for DES Cipher*, EUROCRYPT '93, LNCS 765, pp.386-397, 1994.
8. M. Matsui, *The First Experimental Cryptanalysis of the Data Encryption Standard*, CRYPTO '94, LNCS 839, pp.1-11, 1994.
9. M. Matsui, *On Correlation Between the Order of S-boxes and the Strength of DES*, EUROCRYPT '94, LNCS 950, pp.366-375, 1995.
10. S. Murphy, *The cryptanalysis of FEAL-4 with 20 chosen plaintexts*, Journal of Cryptology 2(3), pp.145-154, 1990.
11. NESSIE - New European Schemes for Signatures, Integrity and Encryption.
    (http://www.nessie.eu.org/nessie)
12. H. Raddum, L. Knudsen, *A Differential Attack on Reduced-Round SC2000* Proceedings of Second Open NESSIE Workshop, September 12-13, 2001.
13. T. Shimoyama, H. Yanami, K. Yokoyama, M. Takenaka, K. Itoh, J. Yajima, N. Torii, H. Tanaka, *The SC2000 Block Cipher*, Proceedings of First Open NESSIE Workshop, November 13-14, 2000.
14. T. Shimoyama, H. Yanami, K. Yokoyama, M. Takenaka, K. Itoh, J. Yajima, N. Torii, H. Tanaka, *The Block Cipher SC2000*, Preproceedings of 8th Fast Software Encryption Workshop, April 2-4, 2001.
15. H. Yanami, T. Shimoyama, *Differential/Linear Characteristics of the SC2000 Block Cipher*, Proceedings of the 2001 Symposium on Cryptography and Information Security, SCIS2001-12A-2, pp.653-658, 2001, in Japanese.
16. H. Yanami, T. Shimoyama, *Differential/Linear Characteristics of the SC2000 Block Cipher (II)*, IEICE Technical Report, ISEC2001-10, pp.63-70, 2001, in Japanese.
17. H. Yanami, T. Shimoyama, *Differential and Linear Cryptanalysis of Reduced-Round SC2000*, Proceedings of Second Open NESSIE Workshop, September 12-13, 2001.

Appendix

Differential distribution table of S_4

ΔIn							ΔOut									
	0x00	0x10	0x20	0x30	0x40	0x50	0x60	0x70	0x80	0x90	0xa0	0xb0	0xc0	0xd0	0xe0	0xf
0x0	16	0	0	0	0	0	0	0	0	0	0	0	0	0	0	0
0x1	0	0	0	0	0	2	2	2	2	2	2	2	2	0	0	0
0x2	0	0	0	0	2	0	4	2	2	2	0	0	0	2	0	2
0x3	0	0	0	0	2	0	2	0	0	0	2	2	2	0	4	2
0x4	0	0	0	2	0	2	0	4	0	2	2	2	0	0	2	0
0x5	0	2	4	0	0	2	0	0	0	0	2	0	0	4	2	0
0x6	0	2	0	4	2	0	0	0	2	0	0	0	0	2	4	0
0x7	0	0	0	2	2	4	0	0	2	2	0	2	0	2	0	0
0x8	0	0	2	4	0	4	0	2	0	0	2	0	0	0	0	2
0x9	0	0	0	2	2	0	0	0	4	0	0	2	2	0	0	4
0xa	0	2	2	2	2	2	0	2	0	0	0	2	0	2	0	0
0xb	0	2	0	0	0	2	0	0	0	4	0	2	4	0	0	2
0xc	0	2	4	0	0	0	2	0	0	4	2	0	0	2	0	0
0xd	0	4	2	0	2	0	0	0	2	0	2	2	0	0	0	2
0xe	0	2	0	0	2	0	2	2	0	0	0	2	2	2	2	0
0xf	0	0	2	0	0	0	4	2	2	0	0	0	2	0	2	2

(Prob $= \{\Delta In \to \Delta Out\} = x/16$)

Linear distribution table of S_4

ΓIn							ΓOut									
	0x00	0x10	0x20	0x30	0x40	0x50	0x60	0x70	0x80	0x90	0xa0	0xb0	0xc0	0xd0	0xe0	0xf
0x0	8	0	0	0	0	0	0	0	0	0	0	0	0	0	0	0
0x1	0	0	0	0	2	2	-2	-2	4	0	0	4	2	-2	2	-2
0x2	0	0	0	0	-4	4	0	0	2	2	-2	-2	-2	-2	-2	-2
0x3	0	0	0	0	-2	-2	-2	-2	-2	2	-2	2	0	4	0	-4
0x4	0	2	2	-4	0	2	-2	0	0	2	-2	0	0	2	2	4
0x5	0	2	-2	0	2	4	0	2	-4	2	2	0	2	0	0	-2
0x6	0	-2	-2	-4	0	-2	-2	4	2	0	0	-2	2	0	0	0
0x7	0	-2	2	0	2	0	0	-2	-2	0	-4	-2	4	-2	-2	0
0x8	0	-2	-2	0	0	2	-2	-4	0	-2	2	-4	0	2	2	0
0x9	0	2	-2	-4	2	0	4	-2	0	-2	-2	0	-2	0	0	-2
0xa	0	-2	-2	0	0	2	2	0	2	0	0	2	2	4	-4	2
0xb	0	2	-2	4	2	0	0	2	2	0	-4	-2	0	2	2	0
0xc	0	4	0	0	0	0	-4	0	0	-4	0	0	0	0	-4	0
0xd	0	0	-4	0	2	-2	-2	-2	0	4	0	0	-2	-2	-2	2
0xe	0	0	4	0	4	0	0	0	2	2	2	-2	-2	2	-2	-2
0xf	0	4	0	0	-2	-2	2	-2	2	2	2	-2	4	0	0	0

(Prob$\{In \cdot \Gamma In + Out \cdot \Gamma Out = 0\} - 1/2 = x/16$)

Probability list obtained from our best differential characteristic

Number of functions	1	2	3	4	5	6	7	8	9	10	11	12	13	14	15	16	17
Function name	B	R_5	R_5	B	R_3	R_3	B	R_5	R_5	B	R_3	R_3	B	R_5	R_5	B	R_3
Probability	15	0	16	11	0	16	15	0	16	11	0	16	15	0	16	11	0
Total probability	15	15	31	42	42	58	73	73	89	100	100	116	131	131	147	158	158

Probability list obtained from our best linear characteristic

Number of functions	1	2	3	4	5	6	7	8	9	10	11	12	13	14	15	16	17
Function name	B	R_5	R_5	B	R_3	R_3	B	R_5	R_5	B	R_3	R_3	B	R_5	R_5	B	R_3
Probability	12	0	16	12	0	16	12	0	16	12	0	16	12	0	16	12	0
Total probability	12	12	28	40	40	56	68	68	84	96	96	112	124	124	140	152	152

Impossible Differential Cryptanalysis of Reduced Round XTEA and TEA

Dukjae Moon, Kyungdeok Hwang, Wonil Lee, Sangjin Lee, and Jongin Lim

Center for Information and Security Technologies(CIST),
Korea University, Anam Dong, Sungbuk Gu,
Seoul, KOREA
{djmoon, kdhwang, wonil, sangjin, jilim}@cist.korea.ac.kr

Abstract. We present the impossible differential cryptanalysis of the block cipher XTEA[7] and TEA[6]. The core of the design principle of these block ciphers is an easy implementation and a simplicity. But this simplicity dose not offer a large diffusion property. Our impossible differential cryptanalysis of reduced-round versions of XTEA and TEA is based on this fact. We will show how to construct a 12-round impossible characteristic of XTEA. We can then derive 128-bit user key of the 14-round XTEA with $2^{62.5}$ chosen plaintexts and 2^{85} encryption times using the 12-round impossible characteristic. In addition, we will show how to construct a 10-round impossible characteristic of TEA. Then we can derive 128-bit user key of the 11-round TEA with $2^{52.5}$ chosen plaintexts and 2^{84} encryption times using the 10-round impossible characteristic.

1 Introduction

In 1990, E. Biham and A. Shamir proposed the differential cryptanalysis[1]. Later, it was regarded as a very useful method in attacking the known block ciphers. For these reasons, block ciphers have been designed to be secure against the differential cryptanalysis since the middle of 1990's. The differential cryptanalysis has also been advanced variously - truncated differential cryptanalysis[4], higher order differential cryptanalysis[5], impossible differential cryptanalysis[3], and so on.

In 1998, the impossible differential cryptanalysis[3] was proposed by E. Biham *et al.* This cryptanalysis is a chosen plaintext attack and applied to the reduced 31 rounds of Skipjack. The traditional differential cryptanalysis finds a key using the differential characteristic with high probability. But the impossible differential cryptanalysis uses the differential characteristic with probability zero. The general impossible differential cryptanalysis can be briefly described as follows: First of all, we must find an impossible differential characteristic. We then choose any plaintext pairs with the input difference of the impossible differential characteristic, and obtain the corresponding ciphertext pairs. We eliminate the ciphertext pairs which are not satisfied with a special property derived from the impossible differential characteristic. For each key value in the key space of the last one or two rounds, we decrypt the ciphertext pairs with

J. Daemen and V. Rijmen (Eds.): FSE 2002, LNCS 2365, pp. 49–60, 2002.
© Springer-Verlag Berlin Heidelberg 2002

that key value and if the differences of the decrypted ciphertext pairs satisfy the output difference of the impossible differential characteristic, then we eliminate the key value from the key space. We repeat the above process until the only one key value remains with very high probability.

In this paper, we describe an impossible differential cryptanalysis of 14-round XTEA. It is based on a 12-round impossible differential characteristic. We will be able to find the 128-bit user key of 14-round XTEA with $2^{62.5}$ chosen plaintexts and 2^{85} encryption times. In addition, we present an impossible differential cryptanalysis of 11-round TEA. To find the 128-bit user key of 11-round TEA, this cryptanalysis uses a 10-round impossible differential characteristic and requires $2^{52.5}$ chosen plaintexts and 2^{84} encryption times.

The paper is organized as follows: In Section 2, we introduce notation. The description of algorithms of TEA and XTEA is briefly given in Section 3. In Section 4, we explain how to construct a 12-round impossible differential characteristic of XTEA and a 10-round impossible differential characteristic of TEA. In Section 5, we describe our attack on 14-round XTEA and 11-round TEA. Finally, we summarize our results and conclude this paper.

2   Notation

We introduce notation as follows for some binary operations.

Exclusive-OR :
The operation of addition of n-tuples over the field $\mathbb{F}_2$ (also known as exlusive-or) is denoted by $x \oplus y$.

Integer Addition :
The addition operation of integers under modulo 2^n is denoted by $x \boxplus y$ (where x , $y \in \mathbb{Z}_{2^n}$). The value of n will be clear from the context.

Bitwise Shifts :
The logical left shift of x by y bits is denoted by $x \ll y$. The logical right shift of x by y bits is denoted by $x \gg y$.

3   Description of TEA and XTEA

TEA(Tiny Encryption Algorithm)[6] was presented by David J. Wheeler *et al.* TEA was designed for a short program which would run on most machines and encrypt safely. This cipher used a simple key schedule. But because of the simple key schedule, the related key attack[8] was possible. To repair this weakness, designers of TEA proposed XTEA(TEA Extensions)[7] which evolved from TEA.

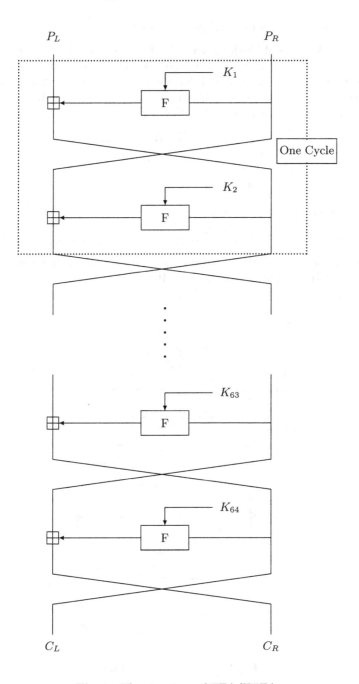

Fig. 1. The structure of TEA/XTEA

3.1   The Tiny Encryption Algorithm (TEA)

TEA is a 64-bit iterated block cipher with 64 rounds, as shown in Fig. 1. TEA can be represented as 32 cycles in which one cycle is two rounds. The round function F consists of the key addition, bitwise XOR and left and right shift operation. We can describe the output (Y_{i+1}, Z_{i+1}) of the i-th cycle of TEA with the input (Y_i, Z_i) as follows (See Fig. 2):

$$Y_{i+1} = Y_i \boxplus F(Z_i, K[0, 1], \delta_i),$$
$$Z_{i+1} = Z_i \boxplus F(Y_{i+1}, K[2, 3], \delta_i),$$
$$\delta_i = \lfloor \frac{i+1}{2} \rfloor \cdot \delta,$$

where the round function F is defined by

$$F(X, K[j, k], \delta_i) = ((X \ll 4) \boxplus K[j]) \oplus (X \boxplus \delta_i) \oplus ((X \gg 5) \boxplus K[k]).$$

The constant $\delta = 9E3779B9_h$ is derived from the golden number, where 'h' represents the hexadecimal number, e.g., $10_h = 16$.

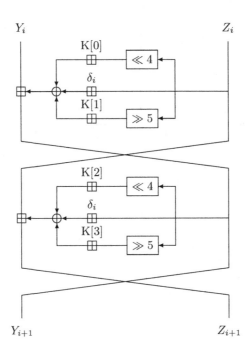

Fig. 2. The i-th cycle of TEA

The key schedule algorithm is very simple. The 128-bit user key K is split into four 32-bit blocks $K = (K[0], K[1], K[2], K[3])$. Then the round keys K_r are as follows:

$$K_r = \begin{cases} (K[0], K[1]) & \text{if } r \text{ is odd} \\ (K[2], K[3]) & \text{if } r \text{ is even,} \end{cases}$$

where $r = 1, \cdots, 64.$

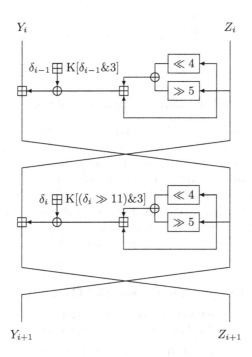

Fig. 3. The i-th cycle of XTEA

3.2   The TEA Extensions (XTEA)

XTEA[7] is an evolutionary improvement of TEA. XTEA makes essentially use of arithmetic and logic operations like TEA. New features of XTEA are to use two bits of δ_i and the shift of 11 . This adjustments cause the indexes of round keys to be irregular. We can describe the output (Y_{i+1}, Z_{i+1}) of the i-th cycle of XTEA with the input (Y_i, Z_i) as follows (See Fig.3):

$$Y_{i+1} = Y_i \boxplus F(Z_i, K_{2i-1}, \delta_{i-1}),$$
$$Z_{i+1} = Z_i \boxplus F(Y_{i+1}, K_{2i}, \delta_i),$$
$$\delta_i = \lfloor \frac{i+1}{2} \rfloor \cdot \delta,$$

where the round function F is defined by

$$F(X, K_*, \delta_{**}) = ((X \ll 4) \oplus (X \gg 5)) \boxplus X \oplus \delta_{**} \boxplus K_*.$$

The constant $\delta = 9E3779B9_h$ is derived from the golden number.

To generate the round keys, first the 128-bit key K is split into four 32-bit blocks $K = (K[0], K[1], K[2], K[3])$, and then the round keys K_r are determined by

$$K_r = \begin{cases} K[\delta_{\frac{r-1}{2}} \ \& \ 3] & \text{if } r \text{ is odd} \\ K[(\delta_{\frac{r}{2}} \gg 11) \ \& \ 3] & \text{if } r \text{ is even,} \end{cases}$$

where $r = 1, \cdots, 64$.

4   Impossible Differential Characteristics

We found a 12-round impossible differential characteristic of XTEA, and a 10-round impossible differential characteristic of TEA. The round functions of XTEA and TEA use the addition as a nonlinear part that is due to the carry which only propagates upwards(diffusion in only one direction). Therefore, we will construct impossible differential characteristics in consideration of this fact and the structure of XTEA and TEA. For this work, we use notation as follows: an α_i is used to denote an arbitrary bit, where i is a positive integer and α is any small letter. And we denote by Λ_x an arbitrary 4-bit , where Λ is any capital letter.

4.1   A 12-Round Impossible Differential Characteristic of XTEA

In this subsection, we show how to construct a 12-round impossible differential characteristic of XTEA. As shown in Fig. 4, if the form of an input difference is

$$(A_x \ a_1a_210 \ 0_x0_x0_x0_x0_x0_x \ \| \ b_1000 \ 0_x0_x0_x0_x0_x0_x0_x), \tag{1}$$

then the difference after round 6 must be of the form

$$(N_xO_xP_xQ_xR_xS_x \ f_1f_210 \ 0_x \ \| \ T_xU_xV_xW_xX_xY_xZ_x \ g_1g_2g_31). \tag{2}$$

On the other hand, we can predict the difference after round 6 from the output difference of round 12, i.e., to consider the differentials in the backward direction. Similarly to the 6-round differential characteristic with peobability 1, there is a backward 6-round differential characteristic with probability 1. It has the difference

$$(a_1000 \ 0_x0_x0_x0_x0_x0_x0_x \ \| \ A_x \ b_1b_210 \ 0_x0_x0_x0_x0_x0_x), \tag{3}$$

after round 12, and then it is clear that the difference after round 6 must be of the form

$$(T_xU_xV_xW_xX_xY_xZ_x \ g_1g_2g_31 \ \| \ N_xO_xP_xQ_xR_xS_x \ f_1f_210 \ 0_x). \tag{4}$$

Combining these two differential characteristics, we conclude that any pair with input difference (1) before round 1 and output difference (3) after round 12 must have differences of the form (2) = (4) after round 6. But this event never occurs. Therefore, this characteristic is a 12-round impossible characteristic of XTEA.

4.2   A 10-Round Impossible Differential Characteristic of TEA

Similarly, if the form of an input difference is $(A_x B_x\ a_1 a_2 a_3 1\ 0_x 0_x 0_x 0_x 0_x\ \|$ $C_x\ b_1 b_2 0 0\ 0_x 0_x 0_x 0_x 0_x 0_x)$, then the difference after round 10 cannot be of the form $(A_x\ a_1 a_2 0 0\ 0_x 0_x 0_x 0_x 0_x 0_x\ \|\ B_x C_x\ b_1 b_2 b_3 1\ 0_x 0_x 0_x 0_x 0_x)$. See Fig.5 for a detailed depiction of the 10-round impossible differential characteristic of TEA.

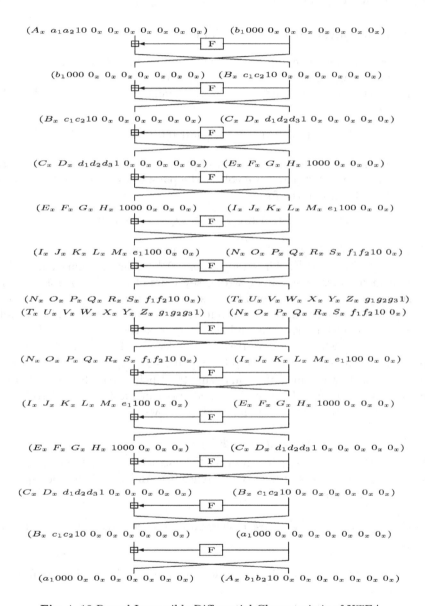

Fig. 4. 12-Round Impossible Differential Characteristic of XTEA

5   Impossible Differential Cryptanalysis

In this section, we will analyze 14-round XTEA and 11-round TEA using the impossible differential characteristics above. Our cryptanalysis of the 14-round XTEA uses a 2R-attack with the 12-round impossible differential characteristic (Fig. 4). And our crypanalysis of the 11-round TEA uses a 1R-attack with the 10-round impossible differential characteristic (Fig. 5).

5.1   Cryptanalysis of 14-Round XTEA

We present an impossible differential cryptanalysis of the 14-round XTEA using the 12-round impossible differential characteristic (Fig. 4). We use structures of 2^7 plaintexts, where every differences of this plaintext pairs matches the difference of the form

$$(A_x \ a_1 a_2 10 \ 0_x 0_x 0_x 0_x 0_x 0_H \ \| \ b_1 000 \ 0_x 0_x 0_x 0_x 0_x 0_x 0_x). \tag{5}$$

Such structures propose about 2^{13} pairs of plaintexts. Given such $2^{55.5}$ structures ($2^{62.5}$ plaintexts , $2^{68.5}$ pairs), we collect the pairs whose ciphertext differences match the difference of the form

$$(A_x B_x \ a_1 a_2 a_3 1 \ 0_x 0_x 0_x 0_x 0_x \ \| \ C_x D_x E_x F_x \ 1000 \ 0_x 0_x 0_x). \tag{6}$$

Then, the probability that a plaintext pair satisfies this condition is $(2^{-4})^9 \times (2^{-1}) \simeq 2^{-37}$. Thus only about $2^{31.5}$ pairs remain. For each of the remained pairs, we eliminate wrong key pairs of the 2^{64} possible values of the key space of the last two rounds by examining whether the decrypted values of the last two rounds have the output difference of the 12-round impossible differential characteristic. The probability that a key pair in the key space survives the test when each ciphertext pair is decrypted is $(1 - 2^{-26})(1 - 2^{-31})$. So, there remain only about $2^{64} \times \{(1 - 2^{-26})(1 - 2^{-31})\}^{2^{31.5}} < 2^{-3.4}$ wrong key pair of the last two rounds after analyzing $2^{31.5}$ pairs. It is thus expected that only one key pair (K_{13}, K_{14}) remains, and this pair must be the correct round key pair. This attack can be summarized as follows;

Goal: Finding round key pair (K_{13}, K_{14}).

1. Choose $2^{68.5}$ plaintext pairs from the $2^{55.5}$ structures, which the difference of each plaintext pair satisfies the difference (5). And ask for the corresponding ciphertext pairs.

2. Collect plaintext pairs whose ciphertext difference agrees with the difference (6).

3. Select a plaintext pair in the collected pairs which are derived from the process **2**.

4. For each key pair (K_{13}, K_{14}) in the key space,

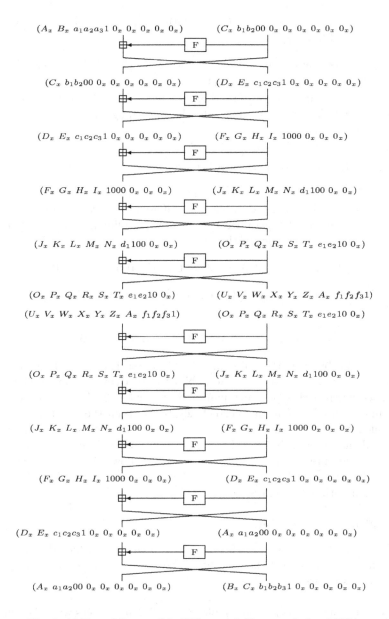

Fig. 5. 10-Round Impossible Differential Characteristic of TEA

(a) Decrypt the corresponding ciphertext pair up to the position of 12-round output.
(b) Compute the difference of the decrypted pair in (a).
(c) Compare the result of (b) to the output difference of the 12-round impossible differential characteristic. If the two results are the same, then remove the selected key pair from the key space.

5. If $|K_{13}| \leq \varepsilon$ and $|K_{14}| \leq \varepsilon'$, stop. Otherwise, go to the process **3**. (ε and ε' are small integers.)

The required work of this attack is about $2^{85} \simeq 2^{64} \cdot 2^{-5} + 2^{64}\{(1 - 2^{-26})(1 - 2^{-31})\} \cdot 2^{-5} + \cdots + 2^{64}\{(1 - 2^{-26})(1 - 2^{-31})\}^{(2^{31.5}-1)} \cdot 2^{-5}$ encryptions where 2^{-5} means two round operations of XTEA encryption. Hence, we can find 64 bits of the user key with our method. The remaining 64 bits will be found by exhaustive search.

5.2   Cryptanalysis of 11-Round TEA

We can also find 128-bit user key of 11-round TEA similarly with the above method. We use the 10-round impossible differential characteristic (Fig. 5). We obtain structures of 2^{17} plaintexts, where every differences of this plaintexts matches the input difference of the form

$$(A_x B_x \ a_1 a_2 a_3 1 \ 0_x 0_x 0_x 0_x 0_x \ \| \ C_x \ b_1 b_2 00 \ 0_x 0_x 0_x 0_x 0_x 0_x) \tag{7}$$

and collect the pairs whose ciphertext differences match the difference of the form

$$(A_x B_x \ a_1 a_2 a_3 1 \ 0_x 0_x 0_x 0_x 0_x \ \| \ C_x D_x E_x F_x \ 1000 \ 0_x 0_x 0_x). \tag{8}$$

Then, the probability that any plaintext pair satisfies the difference (8) is $(2^{-4})^9 \times (2^{-1}) \simeq 2^{-37}$, and the probability that a key is eliminated from the key space when each ciphertext pair is decrypted is 2^{-26}. Let the number of the pairs required to find one correct key be N, then the number N is about $2^{31.5}$ such that $2^{64} \times (1 - 2^{-26})^N < 1$. Therefore, the required number of chosen plaintext pairs is $2^{37} \times 2^{31.5} = 2^{68.5}$. Since we can collect 2^{33} pairs from one structure, we need $2^{35.5}$ structures. It follows that the attack requires $2^{35.5} \times 2^{17} = 2^{52.5}$ chosen plaintexts. We can get 64 bits of K_{11}. This attack can be summarized as follows;

Goal: Finding 11 round key.

1. Choose $2^{68.5}$ plaintext pairs from the $2^{35.5}$ structures, which the difference of each plaintext pair satisfies the difference (7). And ask for the corresponding ciphertext pairs.

2. Collect plaintext pairs whose ciphertext difference agrees with the difference (8).

3. Select a plaintext pair in the collected pairs which are derived from the process **2**.

4. For each key $K_{11}(= K[0], \ K[1])$ in the key space,
   (a) Decrypt the corresponding ciphertext pair up to the position of 10-round output.

(b) Compute the difference of the decrypted pair in (a).

(c) Compare the result of (b) to the output difference of the 10-round impossible differential characteristic. If the two results are the same, then remove the selected key from the key space.

5. If $\mid K_{11} \mid \leq \varepsilon$, stop. Otherwise, go to the process **3**. (ε is small integer.)

The required work of this attack is about $2^{84} \simeq 2^{64} \cdot 2^{-6} + 2^{64}(1 - 2^{-26}) \cdot 2^{-6} + \cdots + 2^{64}(1 - 2^{-26})^{2^{31.5}-1} \cdot 2^{-6}$ encryptions where 2^{-6} means one round operation of TEA encryption. Hence, we can find 64 bits of the user key with our method. The remaining 64 bits will be found by exhaustive search.

6   Conclusion

We described the algorithms of XTEA and TEA, and found the 12-round impossible differential characteristic of XTEA and the 10-round impossible differential characteristic of TEA. Using the 12-round impossible differential characteristic we attacked the 14-round XTEA with $2^{62.5}$ chosen plaintexts and 2^{85} encryptions. In TEA, using the 10-round impossible differential we attacked the 11-round with $2^{52.5}$ chosen plaintexts and 2^{84} encryptions.

TEA is designed for software implementation and XTEA is evolved from TEA to remove the simplicity of key schedule. However, XTEA is weaker than TEA in the impossible differential cryptanalysis although the former is stronger than the latter in the related attack[2].

Acknowledgment. We would like to thank Seokhie Hong and Jaechul Sung for many helpful discussions.

References

1. E. Biham and A. Shamir, *Differential Cryptanalysis of DES-like cryptosystems*, Advances in Cryptology – CRYPTO'90, LNCS 537, Springer-Verlag,1991, pp.2–21.
2. E. Biham,*New Types of Cryptanalytic Attacks Using Related Keys*, Advances in Cryptology – EUROCRYT'93, Springer-Verlag,1994, pp.398–409.
3. E. Biham, A. Biryukov and A. Shamir, *Cryptanalysis of skipjack reduced to 31 round using impossible differentiala*, Advances in Cryptology – EUROCRYT'99, LNCS 1592, Springer-Verlag,1999, pp.12–23. Available at http://www.cs.technion.ac.il/ biham/Reports/Skipjack/.
4. L. R. Knudsen, *Truncated and Higher Order Differential*, Fast Software Encrytion Workshop 94, LNCS 1008, Springer-Verlag, 1995, pp.229–236.
5. S. Moriai, T. Shimoyama and T. Kaneko, *Higher Order Differential Attack of a CAST cipher*,Fast Software Encrytion Workshop 98, LNCS 1372, Springer-Verlag, 1998, pp.17–31.
6. D. Wheeler and R. Needham, *TEA, a Tiny Encryption Algorithm*, Fast Software Encryption, Second International Workshop Proceedings, Springer-Verlag, 1995, pp. 97-110.

7. D. Wheeler and R. Needham, *TEA Extensions*, October 1997.
8. J. Kelsey, B. Schneier and D. Wagner, *Related-Key Cryptanalysis of 3-WAY, Biham-DES, CAST, DES-X, NewDES, RC2, and TEA*, In Information and Communications Security-Proceedings of ICICS 1997, Lecture Notes in Computer Science 1334, Springer-Verlag, 1997.

Improved Cryptanalysis of MISTY1

Ulrich Kühn

Dresdner Bank AG
IS-STA Software Technology and Architecture
Research & Innovations
Jürgen-Ponto-Platz 1
D-60301 Frankfurt, Germany
Ulrich.Kuehn@dresdner-bank.com
ukuehn@acm.org

Abstract. The block cipher MISTY1 [9] proposed for the NESSIE project [11] is a Feistel network augmented with key-dependent linear FL functions. The proposal allows a variable number of rounds provided that it is a multiple of four.
Here we present a new attack – the Slicing Attack – on the 4-round version, which makes use of the special structure and position of these key-dependent linear FL functions. While the FL functions were introduced to make attacks harder, they also present a subtle weakness in the 4-round version of the cipher.

Keywords: Block cipher, Cryptanalysis, Impossible Differential, Slicing Attack.

1 Introduction

The MISTY1 block cipher [9] is a proposal for the NESSIE project [11] in the class *Normal-Legacy* with a block size of 64 bits and a key length of 128 bits. It is designed to be resistant against differential [2] and linear [8] cryptanalysis. Another feature of the design is the use of key-dependent linear functions FL to avoid possible attacks other than differential and linear cryptanalysis.

The best previous attacks on versions of MISTY1 without the linear functions attack were on 5 rounds by higher-order differentials [12] and 6 rounds with impossible differentials [7]. Additionally, the 4-round version including most of the linear functions, leaving out the layer of final applications of the FL functions, has been attacked by impossible differentials as well as collision searching [7]. Very recent results [6] using integral cryptanalysis yield attacks on 5 rounds of MISTY1 without the final FL layer as well as on 4 rounds, also without the final FL layer, having a very low data complexity.

In this paper we present attacks on the 4-round version of MISTY1 with all FL functions by impossible differentials and by a new method called the Slicing Attack. The slicing attack makes use of the position and the structure of the key-dependent linear functions to derive knowledge about the key; further

J. Daemen and V. Rijmen (Eds.): FSE 2002, LNCS 2365, pp. 61–75, 2002.
© Springer-Verlag Berlin Heidelberg 2002

key bits can then be found with impossible differentials, or, in the chosen plaintext/ciphertext model, by the meet-in-the-middle technique. Table 1 shows a summary of the attacks.

While the computational effort for the attack using only impossible differentials is very high, the slicing attack is surprisingly efficient; the existence of this attack shows that augmenting the Feistel network with the linear FL functions, which makes some attacks much harder, also introduces a new line of attack that has to be considered a subtle weakness not being present in the underlying Feistel network.

Table 1. Summary of the new and the best previously known attacks on MISTY1. A memory unit is one block of 64 bits. Versions of MISTY1 with "most" FL functions do not have the final FL layer.

Rounds	FL	Complexity			Comments
		Time	Data	Memory	
5	none	2^{17}	11×2^7		[12]
6	none	2^{61}	2^{54}		[7], Section 4.1
4	most	$2^{90.4}$	2^{23}		[7], Section 4.2
4	most	2^{62}	2^{38}		[7], Section 4.2
4	most	2^{89}	2^{20}		[7], Section 4.2
4	most	2^{76}	2^{28}		[7], Section 4.2
4	most	2^{27}	25		[6]
5	most	2^{48}	2^{34}		[6]
4	all	2^{116}	$2^{27.5}$	$2^{29.5}$	Impossible diff. (this paper)
4	all	2^{45}	$2^{22.25}$	$2^{31.2}$	Slicing Attack, preprocessing (this paper)
4	all	$2^{81.6}$	$2^{27.2}$	$2^{31.2}$	Slicing + impossible diff. (this paper)
4	all	2^{48}	$2^{23.25}$	2^{33}	Slicing Attack in chosen plaintext / ciphertext model (this paper)

This paper is outlined as follows. In Section 2 the MISTY1 design is described, Section 3 presents the attack on 4-round MISTY1 using impossible differentials alone, Section 4 introduces the Slicing Attack, and finally Section 5 draws some conclusions.

2   The Structure of MISTY1

The MISTY1 [9] proposal for the NESSIE project [11] is a block cipher with a 64-bit block and a 128-bit key. It consists of a Feistel network augmented by applying key-dependent linear functions FL to the left resp. right half of the data in every second round, starting with the first, and additionally after all the rounds (see left half of Figure 1). While the cipher is proposed with 8 rounds, the proposal allows a variable number of rounds provided that it is a multiple of four. In this paper we will only consider the 4-round version.

The bijective round function FO is a 3-round network with a structure shown in the right half of Figure 1. This network uses a bijective inner round function FI, which itself is a 3-round network with the same structure, employing two bijective S-boxes $S9$ and $S7$, which are 9 bits resp. 7 bits wide; the key to FI is 16 bits wide. The details of the internal structure of FI will be of no further concern in this paper.

The FL function is a linear or affine function for any fixed key; its internal structure is a 2-round Feistel network (see Figure 2) with the round functions being bitwise boolean AND resp. bitwise OR with key material.

The key scheduling takes a 128-bit key consisting of 16-bit values $K_1, \ldots, K_8$ and, as a first step, computes additional 16-bit values $K'_t = \mathrm{FI}_{K_{t+1}}(K_t)$, $1 \leq t \leq 8$, $K_9 := K_1$. It produces three streams of sub-keys $\mathrm{KO}_i = (\mathrm{KO}_{i1}, \ldots, \mathrm{KO}_{i4})$, $\mathrm{KI}_i = (\mathrm{KI}_{i1}, \ldots, \mathrm{KI}_{i3})$, and $\mathrm{KL}_i = (\mathrm{KL}_{i1}, \mathrm{KL}_{i2})$ as follows (i is identified with $i - 8$ for $i > 8$):

KO_{i1}	KO_{i2}	KO_{i3}	KO_{i4}	KI_{i1}	KI_{i2}	KI_{i3}	KL_{i1}	KL_{i2}
K_i	K_{i+2}	K_{i+7}	K_{i+4}	K'_{i+5}	K'_{i+1}	K'_{i+3}	$K_{\frac{i+1}{2}}$ (odd i) $K'_{\frac{i}{2}+2}$ (even i)	$K'_{\frac{i+1}{2}+6}$ (odd i) $K_{\frac{i}{2}+4}$ (even i)

3 Differential Attack on 4-Round MISTY1

The attack given in this section works against the 4-round version of MISTY1 with all FL functions, improving the result of [7] as there the final applications of the FL functions were left out. The attack applies impossible differentials [1, 5] and uses particular properties of the key scheduling, i.e. the fact that the keys for the final FL functions and the fourth round have some key bits in common.

To be concrete, these sub-keys are $\mathrm{KO}_4 = (K_4, K_6, K_3, K_8)$ and $\mathrm{KI}_4 = (K'_1, K'_5, K'_7)$ for the fourth round's FO resp. $\mathrm{KL}_5 = (K_3, K'_1)$ and $\mathrm{KL}_6 = (K'_5, K_7)$ for the final FLs. The values K'_1, K'_5, and K_3 are used twice.

For the attack we use Property 1 of FO from [7]:

Property 1. If the output difference of FO in round i is of the form (β, β) with nonzero β from input with difference (α_l, α_r), then the input and output differences of FI in the third round are zero; thus the sub-keys KO_{i3}, KO_{i4}, and KI_{i3} cannot influence the output difference and are consequently of no concern. The inputs to the first FI with difference α_l yield an output difference α_r under the keys KO_{i1} and KI_{i1}, while the second FI yields output difference β from the inputs with difference α_r under the keys KO_{i2} and KI_{i2}.

The attack makes use of the 3-round impossible differential $(0, \alpha) \overset{3R}{\nrightarrow} (0, \beta)$ with nonzero α, β; this impossible differential works for any Feistel network with bijective round functions, even when FL functions are used [7, Section 4.2]. The attack proceeds as follows.

MISTY1

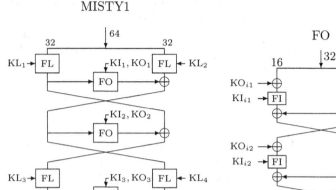

Fig. 1. Global structure of MISTY1 with four rounds (left) and structure of outer round functions FO (right).

1. *Data Collection.* Build a structure of $2^{27.5}$ chosen plaintexts $P_i = (x, y, a_i, b_i)$ where all the (a_i, b_i) are different and obtain the corresponding ciphertexts $C_i = (c_i, d_i, e_i, f_i)$ by encryption under the unknown key.

2. *Processing.* After guessing $KL_6 = (K'_5, K_7)$ and $KL_5 = (K_3, K'_1)$ obtain $\tilde{C}_i = (\tilde{c}_i, \tilde{d}_i, \tilde{e}_i, \tilde{f}_i)$ by $(\tilde{c}_i, \tilde{d}_i) = FL_{KL_6}^{-1}(c_i, d_i)$, $(\tilde{e}_i, \tilde{f}_i) = FL_{KL_5}^{-1}(e_i, f_i)$, and compute $w_i = \tilde{e}_i \oplus \tilde{f}_i$. Every matching pair (i, j) with $w_i = w_j$ results in a difference $\tilde{C}_i \oplus \tilde{C}_j = (\alpha_l, \alpha_r, \delta, \delta)$. For each such matching pair (i, j) do the following steps.

    - (Round 1 of FO) Guess the value of $K_4 = KO_{41}$ ($K'_1 = KI_{41}$ is already known) and check if

$$FI_{K'_1}(\tilde{c}_i \oplus K_4) \oplus FI_{K'_1}(\tilde{c}_j \oplus K_4) = \tilde{d}_i \oplus \tilde{d}_j, \tag{1}$$

    where K'_1 is known from the guess of KL_5. Expect a single guess for K_4 to fulfill this condition.

    - (Round 2 of FO) Independently, guess the value of $K_6 = KO_{42}$ ($K'_5 = KI_{42}$ is already known) and check the condition

$$FI_{K'_5}(\tilde{d}_i \oplus K_6) \oplus FI_{K'_5}(\tilde{d}_j \oplus K_6) = \tilde{e}_i \oplus \tilde{e}_j = \tilde{f}_i \oplus \tilde{f}_j, \tag{2}$$

where K_5' is known from the guess of KL_6. Again, expect a single guess for K_6 fulfilling this condition.

Any values of K_4 and K_6 that satisfy (1) and (2) must be wrong as they would cause the impossible differential to hold. Use a map of 2^{32} bits – which can be reused for each guess of KL_5, KL_6 – to mark these wrong guesses of (K_4, K_6).

Analysis. First, we determine the work needed for a structure of size 2^m with m to be determined later; note that m is necessarily bounded by $m \leq 32$ due to the block-size of 64 bits. For each C_i and all guesses of KL_5 and KL_6 the decryption through FL^{-1} takes $2^{32}(1+2^{32}) \approx 2^{64}$ computations of FL^{-1} so for the structure about 2^{64+m} computations of FL^{-1} are needed. Checking the conditions on K_4 and K_6 for each matching pair needs work of $2 \cdot 2 \cdot 2^{16}$ computations of FI.

For the structure we expect about $\binom{2^m}{2}/2^{16} \approx 2^{2m-17}$ matching pairs for each guess of the 32 bits in KL_5. Each matching pair is expected to discard a single wrong guess of 32 bits (K_4, K_6) for each guess of the 32 bits of KL_6.

Thus, for the whole structure we expect in total about $2^{2m-17} \cdot 2^{32} \cdot 2^{32} = 2^{2m+47}$ wrong keys of 96 bit to be discarded. Assuming that the wrong keys appear at random with equal probability, finding all wrong keys is the coupon collector's problem [3,10]. Therefore, with about $2^{96} \ln(2^{96}) \approx 67 \cdot 2^{96}$ keys of 96 bits being discarded we expect only the right key to remain. Thus, $m = 27.5$ yielding a structure of size $2^{27.5}$ is sufficient. As only the right key is expected to remain, the bitmap – which is reused for each guess of KL_5 and KL_6 – is expected to contain only a single unmarked position for the correct guess of KL_5 and KL_6.

The attack needs a single structure of $2^{27.5}$ chosen plaintexts, $2^{64+27.5} = 2^{91.5}$ computations of FL^{-1}, and $2^{32} \cdot 2^{32} \cdot 2^{18} \cdot 2^{2 \cdot 27.5 - 17} = 2^{120}$ computations of FI, roughly equivalent to 2^{116} encryptions.

As we need to store only the ciphertexts C_i for the structure, a working copy of all $\tilde{C}_i$, and the w_i, the memory consumption can be bounded by $2^{28.5}$ blocks of 64 bits each. In the processing step the map of 2^{32} bits to mark wrong guesses of (K_4, K_6) needs much less memory than the working copy of the ciphertexts.

Remark 1. The reduced number of chosen plaintexts required for this attack in comparison to the attack given in [7, Section 4.2] (which did not include the final FL functions) is due to the fact that here the use of one single structure allows to make efficient use of the plaintexts; this technique can also be applied to [7, Section 4.2] with a significant reduction in the plaintext requirements.

Remark 2. While the chosen plaintext requirements as well as the memory consumption are well in reach of today's attackers, the work factor makes the attack only an academic possibility. But nevertheless, it is much faster than guessing the 128-bit key by brute force.

Experimental results. This attack has been in part verified experimentally. All key words except K_6 (used as KO_{42}) and K_7 (second half of KL_6) were

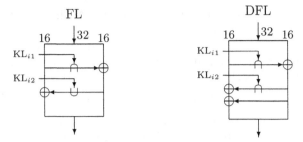

Fig. 2. Structure of FL (left), ∩ denotes the bitwise AND operation, ∪ the bitwise OR operation. When looking only at differential behavior, the structure on the right results (note the changed operation in the second round).

assumed to be known, thus reducing the work factor involved and also reducing the map to 2^{16} bits. Due to memory constraints only $N = 5 \cdot 2^{23}$ chosen plaintexts were used; due to time constraints only 511 passes with wrong values of K_7 in addition to the correct value of K_7 have been tested.

Assume now a pass with a fixed guess for K_7. As K_4 is known and thus fixed, we expect only a fraction of 2^{-16} of the pairs to fulfill (1), thus we expect about $M = \frac{N(N-1)}{2} \cdot 2^{-16} \cdot 2^{-16} = 204800$ pairs to have output XOR (β, β) and to fulfill (1); each of these pairs is expected to remove one guess of K_6 from the map. With $r = M/2^{16}$ we expect $2^{16} \exp(-r) \approx 2880$ candidates for K_6 to remain unmarked in the map (see [10, Theorem 4.18] for this instance of the occupancy problem).

The observed mean of removed guesses for K_6 (including collisions) in the experiments was 204759, the mean of remaining candidates for K_6 was 2883, thus matching the theory very accurately. The correct K_6 was still in the map when using the correct value of K_7. For each guess of K_7 about 80 minutes of CPU-time were needed on a PC with Pentium III (800MHz), 256 MBytes of RAM plus 512 MBytes of swap space; about 640 MBytes of memory were used.

4   The Slicing Attack on 4-Round MISTY1

In this section a new kind of attack is presented that makes essential use of presence, position and structure of the key-dependent FL functions. This attack bypasses the components of the cipher that provide the provable security against differential and linear cryptanalysis.

4.1   Differential Properties of the FL Function

The FL function is a linear (or affine) function for any fixed key. It consists of a 2-round Feistel network with the round function being a bitwise AND resp. OR operation with the key bits (see Figure 2).

Notation. Denote bit i of a value a by $a[i]$ counting the bits from LSB to MSB starting with 0.

As only bitwise operations without any shifts or other means of diffusion are involved, FL basically consists of 16 parallel versions of a cipher with a 2-bit block. Let (a, b) be the input and (c, d) be the output of FL with 16-bit values a, b, c, d. Then block i consists of the bits $(a[i], b[i])$ and $(c[i], d[i])$.

In the following the round functions are analysed algebraically in order to obtain a closed description for the differential behavior of FL. Let k denote a key bit and x an input. Then the round functions are as follows:

$$x \cap k := xk$$
$$x \cup k := x \oplus k \oplus xk.$$

Now let x^* denote a second input and let $x' = x \oplus x^*$; then the differential behavior of these operations is as follows:

$$(x \cap k) \oplus (x^* \cap k) = xk \oplus x^*k = x'k$$
$$(x \cup k) \oplus (x^* \cup k) = (x \oplus k \oplus xk) \oplus (x^* \oplus k \oplus x^*k) = x' \oplus x'k.$$

Therefore, for differences the FL function has the effective description given on the right hand side of Figure 2. Call this function DFL.

4.2 Slicing 4-Round MISTY1

The attack in the previous section employed the 3-round impossible differential $(0, \alpha) \nrightarrow (0, \beta)$. Any sub-key for the last round (including the FL functions that follow) that yields the output difference of the impossible differential must be wrong. This is used to discard all the wrong keys.

Another view on this situation is as follows. It focuses on the changes to the nonzero difference in the half of the data that is the right half of the input. This is shown in Figure 3. The input difference in the right half is $\alpha \neq 0$, causing an nonzero output difference α' of the FL function in round 1. The first round's FO has a zero output difference, so no further change occurs here. The difference is modified again in round 3 by an FL function ($\alpha'' \neq 0$) and by XOR with the output difference γ of FO, yielding δ'. Finally, it is modified through the output transformation by FL yielding a difference δ in the left half of the ciphertext. This is shown in the right part of Figure 3.

The output difference γ of FO in round 3 must be nonzero, as can be seen as follows. The input difference of the FO in round 2 is $\alpha' \neq 0$, so $\beta' \neq 0$, as FO is bijective. Therefore, in round 3, the input to FO is also nonzero, thus causing $\gamma \neq 0$.

It should be noted that the difference of concern here – right half of plaintext difference, left half of ciphertext difference – is changed only by the keys to FL with the single exception of the XOR with the difference γ in round 3.

Therefore, any set of keys (KL_2, KL_4, KL_6) to a stack of three instances of DFL that yields δ from α implies that $\gamma = 0$, thus it must be wrong. The

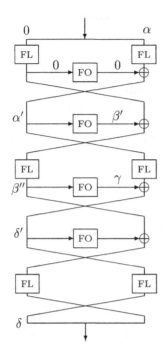
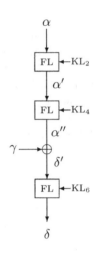

Fig. 3. Slicing MISTY1. The differential path of the data from the right half of the input to the left half of the output is shown on the right side. The difference γ is known to be nonzero.

result is that we are dealing only with a slice of 4-round MISTY1. Note that this property does not hold for the underlying Feistel network without the key-dependent linear functions and that an extension to more rounds seems not to work.

Definition 1. *The slice of three instances of* DFL *consists of 16 parallel instances of the same key-dependent function. Denote it by F, indexed by the 6-bit key k, i.e. $F_k : \{0,1\}^2 \to \{0,1\}^2$. The blocks are located in the same places as for* FL.

In the following some properties of F are shown that will subsequently be used for the attack.

Lemma 1. *Depending on the key, F realises one of six different bijective functions. Thus F has six classes of equivalent keys. There are four classes with 11 keys and two with 10 keys.*

Proof. From the structure of F it is clear that F is a bijective function, therefore $F_k(a) \neq 0$ for nonzero a. As the input and output of F are differences, it follows that $F_k(0) = 0$ for any k. On the remaining three inputs F realises a permutation, of which there are $3! = 6$ different. Checking all 64 possible keys gives the number of keys per class showing that all of these functions are indeed realised.   □

Notation. Let $\mathcal{K}_1, \ldots, \mathcal{K}_6$ denote the six classes of equivalent keys for F and $\mathcal{F}_i$ denote the function realised by any of the keys in $\mathcal{K}_i$ for $i \in \{1, \ldots, 6\}$.

Proposition 1. *For any nonzero a, b there are exactly two $i \in \{1, \ldots, 6\}$ such that $\mathcal{F}_i(a) = b$ holds.*

4.3   Attacking the Slice

As a consequence of these classes of equivalent keys it should be clear that the best one can hope for is to find a vector of 16 functions that never implies the output difference of the third round's FO being zero. Further conditions to distinguish right vs. wrong keys must come from the key scheduling or other means besides the slicing attack (see Section 4.4).

Definition 2. *Let $\alpha = (\alpha_l, \alpha_r)$ be the input to the slice of three DFL functions and $\delta = (\delta_l, \delta_r)$ its output. As both α and δ come from plaintext resp. ciphertext differences, still call this a* pair *and denote it as $\alpha \rightarrow \delta$. A vector $(f_{15}, \ldots, f_0) \in \{\mathcal{F}_1, \ldots, \mathcal{F}_6\}^{16}$ is called* valid *for $\alpha \rightarrow \delta$ if for each $i \in \{0, \ldots, 15\}$ the 2-bit block $(\alpha_l[i], \alpha_r[i])$ is mapped to $(\delta_l[i], \delta_r[i])$ by f_i.*

As one cannot distinguish between functions with a zero input and output, any pair that causes a zero input / output to any of the 16 parallel instances of F cannot be used. Such a pair is called a *bad pair* whereas a pair with only nonzero input and output blocks for each of the 16 instances of F is called a *good pair*.

From Proposition 1 it follows that each good pair has 2^{16} valid vectors of functions while there are $6^{16} \approx 2^{41.4}$ vectors in total. With $2^{41.4}/2^{16} = 2^{25.4}$ good pairs there are $2^{41.4}$ valid vectors. Assuming that the valid vectors appear at random with equal probability, this is the coupon collector's problem [3,10]. Therefore, with about $2^{41.4} \ln(2^{41.4}) \approx 2^{46.2}$ valid vectors from about $2^{30.2}$ good pairs all valid vectors are expected to be found. As it is known that an invalid vector exists, this one is expected be be singled out.

The chance that a random pair is a good pair is about $(9/16)^{16} \approx 2^{-13.3}$. Therefore about $2^{43.5}$ pairs are needed which can be gained from about $2^{22.25}$ chosen plaintexts.

Now we are ready to state the actual Slicing Attack:

1. *Data Collection.* Build a structure of $2^{22.25}$ plaintexts $P_i = (r, s, t_i, u_i)$ with constant (r, s) and (t_i, u_i) being arbitrary but all different; obtain the ciphertexts $C_i = (v_i, w_i, x_i, y_i)$ encrypted under the unknown key.
2. *Filtering.* For each pair (i, j), $i < j$, check if $\alpha = (\alpha_l, \alpha_r) = (t_i \oplus t_j, u_i \oplus u_j)$, $\delta = (\delta_l, \delta_r) = (v_i \oplus v_j, w_i \oplus w_j)$ form a good pair, i.e. $(\alpha_l[m], \alpha_r[m]) \neq 0$, $(\delta_l[m], \delta_r[m]) \neq 0$ for all $0 \leq m \leq 15$. For all good pairs store (α, δ) in a table T.
3. *Processing, Outer Loop.* For each of the 6^6 assignments of $(f_5, \ldots, f_0)$ do the following. First, select all those good pairs $\alpha \rightarrow \delta$ such that $(f_5, \ldots, f_0)$ is valid for the corresponding six blocks; store the selected good pairs in a table T'.
   Initialise a bit map B of $6^{10} < 2^{26}$ bits, then execute the inner loop:

– *Processing, Inner Loop.* For all good pairs in T' set the bits in B that correspond to the valid vectors $(f_{15}, \ldots, f_6)$ for the rightmost 10 blocks. Finding these can be done by using a preprocessed table to get the possibilities for each of the 10 blocks.

After all pairs in T' are processed check which bits in B are still cleared. These correspond to possibly invalid vectors.

Analysis. The filtering is expected to keep $\binom{2^{22.25}}{2}/2^{13.3} \approx 2^{30.2}$ good pairs, thus from the discussion above it is clear that the algorithm is expected to single out one invalid vector $(f_{15}, \ldots, f_0)$.

As by Proposition 1 exactly 2 out of 6 functions are valid for each given input and output of a 2 bit block the chance that a good pair in T is included in T' is about $\left(\frac{1}{3}\right)^6 \approx 2^{-9.5}$. Therefore T' is expected to have a size of $2^{30.2}/2^{9.5} < 2^{21}$.

The total running time consists of three components: first, the time for filtering, second, the time for constructing table T' in step 3, and, third, the time spent in the inner loop.

The filtering takes $2^{43.5}$ checks to find the good pairs. The building of T' is done 6^6 times where each time about $2^{30.2}$ checks have to be done (a check can be done with look-up tables, taking only constant time). This step thus takes a total of roughly 2^{46} checks.

Each execution of the inner loop sets about $2^{10} \cdot 2^{21} = 2^{31}$ bits, so the total time spent here in all iterations of step 3 is roughly $6^6 \cdot 2^{31} \approx 2^{47}$ elementary operations like computing indices plus setting bits in the bitmap etc.

This sums up to running time roughly equivalent to 2^{45} encryptions. The memory consumption can be bounded by the size of the plaintexts and ciphertexts, the tables T and T', and the bitmap in the inner loop, totaling to about $2^{31.2}$ blocks.

Remark 3. While the slicing attack does not directly reveal key bits, it gains knowledge about the class of equivalent keys of the real key $(\mathrm{KL}_2, \mathrm{KL}_4, \mathrm{KL}_6)$. This class contains at most $11^{16} \approx 2^{55.4}$ keys. Comparing this to the initial set of 2^{96} keys shows a gain in knowledge of about 40 bits.

Remark 4. When also considering the key scheduling, the real key is (K_3', K_5), (K_4', K_6), (K_5', K_7) with $K_5' = \mathrm{FI}_{K_6}(K_5)$, so that the real entropy is only 80 bits. But for the keys in the equivalence class the same 16-bit condition holds, so that about 2^{40} keys are expected to remain. This is also a gain in knowledge about the key of about 40 bits.

4.4   Finding the Real Key Bits

When using the knowledge gained in the slicing attack in a subsequent step of analysis the work factor of the slicing attack is only involved as additive work to what follows.

A simple way is brute force, namely enumeration of the about 2^{40} keys in the equivalence class and guessing the remaining 48 key-bits, requiring expected $\frac{1}{2}2^{88}$ encryptions, about 2 known plaintexts / ciphertexts and de facto no memory.

Better methods are given below. One uses impossible differentials in the usual chosen plaintext model of attack, the other uses the chosen plaintext/ciphertext model to efficiently find the complete key.

Improving the impossible differential attack. The differential attack of Section 3 can be improved significantly by using the information from the slicing attack; this is faster than the brute force method at the cost of more chosen plaintexts. It makes use of the fact that only 16 key-bits (K_1' in KL_5) have to be guessed in addition to enumerating the about 2^{40} keys in the equivalence class.

The attack proceeds as follows after having knowledge about the correct equivalence class, again using Property 1 of FO and the 3-round impossible differential $(0, \alpha) \not\to (0, \beta)$ (see Section 3).

1. *Data Collection.* Build a structure of 2^m plaintexts $P_i = (x, y, a_i, b_i)$ with constant x, y and random but different (a_i, b_i) and obtain the corresponding ciphertexts $C_i = (c_i, d_i, e_i, f_i)$. The number m is determined later in the analysis to be $m = 27.2$.
2. *Enumerate the Keys.* The keys in the equivalence class can be enumerated by stepping through all 2^{32} assignments of K_5' and K_6; then set $K_5 = \mathrm{FI}_{K_6}^{-1}(K_5')$ and enumerate all possible assignments of K_3', K_4', and K_7 by considering separately each 2-bit block using a precomputed table. For each assignment compute $K_4 = \mathrm{FI}_{K_5}^{-1}(K_4')$ and $K_3 = \mathrm{FI}_{K_4}^{-1}(K_3')$.
   For each 16-bit value for K_1' set $KL_5 = (K_3, K_1')$, $KL_6 = (K_5', K_7)$ and do the following step:
   a) *Find wrong keys.* For all ciphertexts C_i compute $(\tilde{e}_i, \tilde{f}_i) = \mathrm{FL}_{KL_5}^{-1}(e_i, f_i)$, $(\tilde{c}_i, \tilde{d}_i) = \mathrm{FL}_{KL_6}^{-1}(c_i, d_i)$, and build two lists

$$u_i = \tilde{e}_i \oplus \tilde{f}_i$$
$$w_i = (\mathrm{FI}_{K_1'}(K_4 \oplus \tilde{c}_i) \oplus \tilde{d}_i, \mathrm{FI}_{K_5'}(K_6 \oplus \tilde{d}_i) \oplus \tilde{e}_i).$$

If for any i, j there is a match $u_i = u_j$ and $w_i = w_j$, go to the next guess of the keys, otherwise keep the guessed keys. The rest of the key bits, i.e. K_2 and K_8, can be found by brute force and inverting the key schedule (to find K_1).

Analysis. The outer loop for the enumeration of the keys in the equivalence class has 2^{32} iterations with a single application of FI^{-1} taking place. For each assignment of K_5' and K_6 about 2^8 values are expected to be found for (K_3', K_4', K_7). While for some functions and fixed key-bits no suitable keys exist, this is not a problem because these events can be found efficiently with the precomputed table.

The costs here are 2^{32} computations of FI^{-1} and some table operations which is much less than the enumeration of all about 2^{40} keys. For each of these keys two FI^{-1} computations take place; in total this is about $2^{32} + 2 \cdot 2^{40} \approx 2^{41}$ computations of FI^{-1}, independent of the size 2^m of the structure.

It is easy to see that a pair (i, j) with $u_i = u_j$ yields a symmetric difference (β, β). Each pair (i, j) with $u_i = u_j$ and additionally $w_i = w_j$ fulfills the two conditions

$$FI_{K_1'}(K_4 \oplus \tilde{c}_i) \oplus FI_{K_1'}(K_4 \oplus \tilde{c}_j) = \tilde{d}_i \oplus \tilde{d}_j \quad \text{(first round of FO)}$$

$$FI_{K_5'}(K_6 \oplus \tilde{d}_i) \oplus FI_{K_5'}(K_6 \oplus \tilde{d}_j) = \tilde{e}_i \oplus \tilde{e}_j \quad \text{(second round of FO)},$$

thus fulfilling Property 1 of FO. Therefore this guess of the key must be wrong and is discarded; a correct key never fulfills these conditions. This ensures the correctness of the algorithm.

Per guessed key about $2 \cdot 2^m$ applications of FL^{-1} and the same number of applications of FI are done, for all key guesses (about 2^{40} from the equivalence class times 2^{16} from K_1') in total about 2^{57+m} applications of FL^{-1} and 2^{57+m} applications of FI.

For each pair (i, j) there is a chance of about 2^{-16} to fulfill $u_i = u_j$ and a chance of about 2^{-32} to fulfill $w_i = w_j$, thus a chance of 2^{-48} to discard a key of 56 bits. Modeling the keys discarded by each pair as random and assuming an equal probability, we expect about $2^{56} 2^{-48} = 2^8$ keys being discarded by any pair. The task of discarding all wrong keys is the coupon collector's problem [3, 10]. Therefore, with about $2^{56} \ln(2^{56}) \approx 2^{61.3}$ keys discarded by about $2^{61.3}/2^8 = 2^{53.3}$ pairs only the right key is expected to remain. This implies a choice of $m = 27.2$ and a structure of $2^{27.2}$ chosen plaintexts.

The work needed sums up to 2^{41} applications of FI^{-1}, $2 \cdot 2^{83.2}$ applications of FL^{-1}, and $2^{84.2}$ applications of FI. This is roughly equivalent to about $2^{81.6}$ encryptions.

The memory consumption can be bounded by the number of ciphertexts, a working copy for decryptions by FL^{-1}, and the tables for the u_i and w_i. This sums to roughly $2^{29.2}$ blocks which is less than needed for the slicing attack.

Attack in the chosen plaintext / ciphertext model. In this model the slicing attack can also be used to find an equivalence class for the sub-keys $KL_1 = (K_1, K_7')$, $KL_3 = (K_2, K_8')$, and $KL_5 = (K_3, K_1')$ with chosen ciphertext queries; call this the *backward slice* and denote its equivalence class by $\mathcal{K}_b$ in contrast to the *forward slice* with class $\mathcal{K}_f$ of the chosen plaintext attack.

This preprocessing steps together take $2 \cdot 2^{45}$ work, $2^{22.25}$ chosen plaintexts queries, $2^{22.25}$ chosen ciphertext queries and $2^{31.2}$ blocks of memory. Note that adaptiveness of the queries is not necessary here.

Now we use the fact that K_3 in KL_5 can be computed from K_3', K_4', and K_5 from the forward slice; a similar property holds for K_7. This can be used with the meet-in-the-middle technique parametrised by $0 \le N \le 16$ to allow a time/memory tradeoff:

1. *Global Loop.* Step through all values for the highest N bits of K_6.

a) *Enumerate forward slice.* Step through all values of the lower $16 - N$ bits of K_6 and the 16 bits of K_5'; compute $K_5 = \mathrm{FI}_{K_6}^{-1}(K_5')$. Enumerate all values for K_3', K_4', K_7 that are in $\mathcal{K}_f$ for the fixed sub-key values using a precomputed table. Compute $K_4 = \mathrm{FI}_{K_5}^{-1}(K_4')$, and $K_3 = \mathrm{FI}_{K_4}^{-1}(K_3')$. Store the 128 bits K_3, K_3', K_4, K_4', K_5, K_5', K_6, K_7 in a hash table T indexed by (K_3, K_7) allowing later to retrieve all entries with the same index.

b) *Enumerate backward slice.* For all values of K_1' and K_2 compute $K_1 = \mathrm{FI}_{K_2}^{-1}(K_1')$ and enumerate the values for K_8', K_7', and K_3 in $\mathcal{K}_b$, again using a precomputed table. For each of these also compute $K_8 = \mathrm{FI}_{K_1}^{-1}(K_8')$, $K_7 = \mathrm{FI}_{K_8}^{-1}(K_7')$.

   i. *Check the keys.* Retrieve all entries with the same (K_3, K_7) from the hash table T. Complete the key scheduling and do one or if necessary two trial encryptions to check whether it is the correct key.

Analysis. First look at the steps (a) and (b) that are executed inside the global loop. Step (a) is expected to enumerate about $2^{40}/2^N = 2^{40-N}$ sub-key values while performing about $2^{16-N} \cdot 2^{16} + 2 \cdot 2^{40-N}$ computations of FI^{-1} (like in the analysis above there might be values for K_5' such that no valid sub-keys are found, but the total work to find these is much less than the enumeration of all the 2^{40-N} sub-keys). The expected size of T is 2^{40-N} values à 128 bits which is 2^{41-N} blocks.

Step (b) is enumerating the about 2^{40} keys in $\mathcal{K}_b$ with about $2^{32} + 2 \cdot 2^{40}$ computations of FI^{-1}. In T we expect to find $2^{40} \cdot 2^{40-N} \cdot 2^{-32} = 2^{48-N}$ matches, therefore the completion of the key schedule and the trial encryption is expected to be done about 2^{48-N} times.

In total, taking the global loop into account, the time needed is about $2^{41-N+N} + 2^{41+N} = 2^{41} + 2^{41+N}$ computations of FI^{-1} roughly equivalent to about $2^{39} + 2^{39+N}$ encryptions. The full cipher is expected to be run about $2^N \cdot 2^{48-N} = 2^{48}$ times.

With $N = 10$ we can efficiently reuse the memory used in the slicing attack, needing about $1.5 \cdot 2^{49}$ work and 2^{31} memory, With $N = 8$ the time is dominated by the 2^{48} trial encryptions with a memory requirement of 2^{33} blocks.

4.5 MISTY Variants and the Slicing Attack

As the slicing attack allows to attack the 4-round version of MISTY1 very efficiently, one might ask whether this attack applies also to the MISTY1 variant KASUMI [4] which is used in 3rd generation cellular phones. In comparison to MISTY1 the FO and FI functions are modified and, more important here, the FL functions – with bit-rotations added in the round function – are moved to be part of KASUMI's round function. As KASUMI is a plain 8-round Feistel network with no key-dependent operations being performed outside the round function, the slicing attack does not apply.

In MISTY1 the slicing attack is possible because of the position of the FL functions; avoiding this requires to move the FL functions. The attack is also efficient because it is easy to determine all keys resp. vectors of parallel functions in the slice that map an input XOR to an output XOR. To prevent the slicing attack it would thus be necessary to add a new design criterion besides those given in [9]. A possible fix might be adding bit-rotations to FL's round functions (like in KASUMI's FL) to avoid the parallelism, but whether this prevents the attack is left to future research. On the other hand the slicing attack seems to work only for the 4-round version of MISTY1, thus using more than 4 rounds should prevent this attack.

5 Conclusion

While for the impossible differential attack on the 4-round version of MISTY1 presented in this paper the chosen plaintext requirements and the memory consumption are certainly in range of today's attackers, the high work factor involved does not threaten the cipher.

On the other hand the slicing attack is made possible by the position and structure of the FL functions. It shows that augmenting the Feistel network with key-dependent functions can introduce subtle weaknesses that are not present in the Feistel network itself; one special feature is that the slicing attack completely bypasses the components that provide the provable security of the cipher. Furthermore, this is surprisingly efficient, it is clearly in range of today's possibilities.

While the MISTY1 proposal allows any multiple of four as the number of rounds, the results in this paper show that the 4-round version should be avoided, thus leaving the recommended number of 8 rounds as a minimum.

The author would like to thank David Wagner for helpful discussions and for suggesting the use of the chosen plaintext / ciphertext model. Thanks are also due to the anonymous referees of the 2nd NESSIE workshop and FSE 2002 whose comments helped to improve the paper.

References

[1] E. Biham, A. Biryukov, and A. Shamir. Miss in the middle attacks on IDEA and Khufu. In L. Knudsen, editor, *Fast Software Encryption, 6th international Workshop*, volume 1636 of *Lecture Notes in Computer Science*, pages 124–138, Rome, Italy, 1999. Springer-Verlag.

[2] E. Biham and A. Shamir. *Differential Cryptanalysis of the Data Encryption Standard*. Springer Verlag, Berlin, 1993.

[3] K. L. Chung. *Elementary Probability Theory with Stochastic Processes*. Springer Verlag, 1979.

[4] ETSI/SAGE. Specification of the 3GPP Confidentiality and Integrity Algorithms – Document 2: KASUMI Specification, Version 1.0. 3G TS 35.202, December 23, 1999. http://www.etsi.org/dvbandca/3GPP/3GPPconditions.html.

[5] L. R. Knudsen. DEAL — A 128-bit block cipher. Technical Report 151, Department of Informatics, University of Bergen, Bergen, Norway, Feb. 1998.

[6] L. R. Knudsen and D. Wagner. Integral cryptanalysis. These Proceedings, pages 114–129.

[7] U. Kühn. Cryptanalysis of Reduced-Round MISTY. In B. Pfitzmann, editor, *Advances in Cryptology - EUROCRYPT 2001*, volume 2045 of *Lecture Notes in Computer Science*, pages 325–339. Springer Verlag, 2001.

[8] M. Matsui. Linear cryptanalysis method for DES cipher. In T. Helleseth, editor, *Advances in Cryptology - EUROCRYPT '93*, pages 386–397, Berlin, 1993. Springer-Verlag. Lecture Notes in Computer Science Volume 765.

[9] M. Matsui. New block encryption algorithm MISTY. In E. Biham, editor, *Fast Software Encryption: 4th International Workshop*, volume 1267 of *Lecture Notes in Computer Science*, pages 54–68, Haifa, Israel, 20–22 Jan. 1997. Springer-Verlag.

[10] R. Motwani and P. Raghavan. *Randomized Algorithms*. Cambridge University Press, New York, NY, 1995.

[11] NESSIE. New European Schemes for Signature, Integrity, and Encryption. http://www.cryptonessie.org.

[12] H. Tanaka, K. Hisamatsu, and T. Kaneko. Strength of MISTY1 without FL function for higher order differential attack. In M. Fossorier, H. Imai, S. Lin, and A. Poli, editors, *Proc. Applied algebra, algebraic algorithms, and error-correcting codes: 13th international symposium, AAECC-13*, volume 1719 of *Lecture Notes in Computer Science*, pages 221–230, Hawaii, USA, 1999. Springer Verlag.

Multiple Linear Cryptanalysis of a Reduced Round RC6

Takeshi Shimoyama, Masahiko Takenaka, and Takeshi Koshiba

Secure Computing Lab., Fujitsu Laboratories Ltd.,
4-1-1, Kamikodanaka Nakahara-ku Kawasaki 211-8588, Japan
{shimo,takenaka,koshiba}@flab.fujitsu.co.jp

Abstract. In this paper, we apply multiple linear cryptanalysis to a reduced round RC6 block cipher. We show that 18-round RC6 with weak key is breakable by using the multiple linear attack.

1   Introduction

The block cipher RC6 was proposed by Rivest et al. in [17] to meet the requirements of the Advanced Encryption Standard (AES) and is one of the finalists of the AES candidates. It has been admired for its high-level security and high-speed software implementation especially on Intel CPU. RC6 enters also the NESSIE Project selection and it has been nominated to the Phase II evaluation.

RC6 is designed based on the block cipher RC5 [16] which makes essential use of arithmetic key additions and data-dependent rotations. Kaliski and Robshaw [7] evaluated the resistance of RC5, which introduced data-dependent rotations as primitive operations, against Linear Attack [14]. Borst, Preneel, and Vandewalle [2] refined the linear attack of RC5. As additional primitive operations to RC6, the inclusion of arithmetic multiplications and fixed rotations is believed to contribute the strength of the security of RC6. There are some cryptanalyses of RC6: resistance against Differential Attack, Linear Attack, and Related Key Attack by Contini *et al.* [3], Mod n Attack [11], Linear Attack [2], and Statistical Attack [5]. One of most effective attacks is χ^2 attack by Knudsen and Meier [13] which can break up to 15-round RC6 with general keys and 17-round RC6 with weak keys. We note that their estimation is inferred from experimental results for at most 6-round RC6 and is not relied on any theoretical evidence. The cryptanalysis by Contini *et al.* [4] is actually is not only of RC6 itself but also of reduced variants of RC6. We enumerate attacks on RC6 in Table 1.

In [3], Contini *et al.* showed some upper bound of complexity to break RC6 against the linear attack on the assumption that the attacker uses the bias of the linear equations with respect to 1-bit masks both on input and output to arithmetic additive operations and the number of equations among multiple linear approximation, which are derived based on the notion of "linear hulls", to advantage.

In this paper, we evaluate the resistance of RC6 with 256-bit key against multiple linear attack. In order to do this, we use the technique of the linear

J. Daemen and V. Rijmen (Eds.): FSE 2002, LNCS 2365, pp. 76–88, 2002.

Table 1. Attacks on RC6

Attack	Rounds	Data size	Comments
Linear Attack [2]	16	2^{119}	Upper bound of complexity
Differential Attack [3]	12	2^{117}	Upper bound of complexity
Mod n Cryptanalysis [11]	—	—	—
χ^2 Cryptanalysis [13]	15	$2^{119.0}$	Lower bound of complexity (estimation)
	17	$\leq 2^{118}$	Lower bound (estimation, $1/2^{80}$ weak keys)
Multiple Linear Attack [This paper]	18	$2^{127.423}$	Lower bound ($1/2^{90}$ weak keys)

probability that we obtain by taking multiple paths into account and the theory of multiple linear approximation and evaluate rigorously the complexity to break RC6. To do that, we introduce a novel technique to use a "Matrix Representation" that is a generalization of the piling up lemma to obtain the linear probability. This technique ease us to count the multiple path and to estimate more exactly the linear probability that might depend on the extended-keys. As a result, we show that the target key of 14-round RC6 can be recovered and also that the target key of 18-round RC6 with weak keys, which exists with probability $1/2^{90}$ at least, can be recovered.

2   Preliminary

For any function $Y = F(X)$, input mask ΓX and output mask ΓY, we define the bias of linear equations $Bias_F()$ and the linear probability $LP_F()$ as follows.

$$Bias_F(\Gamma X \to \Gamma Y) = 2 \cdot \frac{\#\{X|(\Gamma X \cdot X) \oplus (\Gamma Y \cdot F(X)) = 0\}}{\#\{X\}} - 1$$
$$LP_F(\Gamma X \to \Gamma Y) = (Bias_F(\Gamma X \to \Gamma Y))^2$$

It is well known that for any r functions $X_{i+1} = F_i(X_i)$ ($i = 1, ..., r$) the composite function $H(X) = F_r \circ \cdots \circ F_1(X)$ has the expected (w.r.t. keys) linear probability satisfying[1]

$$LP_H(\Gamma X_1 \to \Gamma X_{r+1}) = \sum_{\Gamma X_2, ..., \Gamma X_r} \{\prod_{i=1}^{r} LP_{F_i}(\Gamma X_i \to \Gamma X_{i+1})\}.$$

Let $e_0, ..., e_{31}$ denote unit vectors over $GF(2)^{32}$ such that the ith element of e_i is 1 and the other elements of e_i are 0. Here we adopt the description of the descending order for vectors (e.g., $e_0 = (0, ..., 0, 1)$).

In this paper, we identify 32-bit values, which are used in RC6 encryption, with elements of $GF(2)^{32}$, unless otherwise specified.

---

[1] Strictly speaking, we need more structural information of functions F_i for holding the equation.

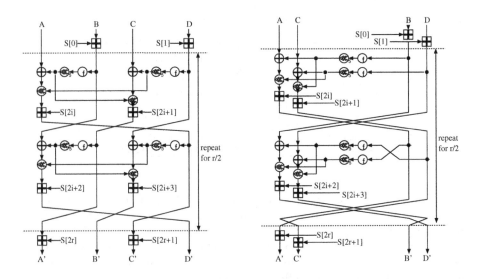

Fig. 1. Original RC6 and its equivalent transformation

3   RC6 and Its Equivalent Transformation

RC6 is a block cipher proposed by Rivest *et al.* [17]. It has a variable number of rounds denoted r and key size of $8b$ bits. The design is word-oriented for word sizes w and the block size is $4w$. Currently RC6 with $r = 20$, $4w = 128$ and $8b = 128, 192, 256$ is recommended to give sufficient resistance against possible attacks. (See Figure 1). In this paper, we refer n-round RC6 to $\mathrm{RC6}_{(n)}$. We leave the key scheduling of RC6, which generates extended keys from private keys, out of consideration.

In this paper, we use a Feistel-like description of RC6 which is obtained by exchanging input-output words B and C equivalently. It is easy to see that the new description help us to capture structural properties of RC6. (As long as the authors know, the new description is not shown.)

We consider a block cipher RC6⊕ that is obtained by replacing arithmetic additions of extended-keys of RC6 by exclusive-oring of extended-keys.

Moreover, we consider weak-keys of $2r$-round RC6. We define two types of weak keys. "Type I weak keys" are ones such that $lsb_5(S[4i-3]) = lsb_5(S[4i-4]) = 0$, and "Type II weak keys" are ones such that $lsb_4(S[4i-3]) = lsb_4(S[4i-4]) = 0$. It is easy to see that Type I weak-keys is of the fraction 2^{-10r}, and Type II weak-keys is of the fraction 2^{-8r}. Later, we will show that those of weak-keys are actually "weak".

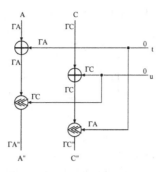

Fig. 2. Linear mask of R

4   Linear Probabilities of Data-Dependent Rotation

We consider a partial function R of $RC6$ as follows. (See also Figure 2).

$$R : (A, C, t, u) \in (GF(2)^{32})^4 \to (A'', C'') \in (GF(2)^{32})^2.$$
$$A'' = (A \oplus t) \lll u$$
$$C'' = (C \oplus u) \lll t$$

Let $\Gamma A, \Gamma C, \Gamma t, \Gamma u, \Gamma A'', \Gamma A''$ be masks for the variables A, C, t, u, A'', C'' of R, respectively. We consider linear approximation with significant linear probability such that $\Gamma t = \Gamma u = 0$.

Let us consider, for example, the case where $\Gamma A = \Gamma C = \Gamma A'' = \Gamma C'' = e_0$. Then, if $lsb_5(t) = lsb_5(u) = 0$ then $Ae_0 \oplus A'' e_0 = 0$ and $Ce_0 \oplus C'' e_0 = 0$. Only if the case where $lsb_5(t) = lsb_5(u) = 0$ occurs (its probability is 2^{-10}), the probability that the equation $Ae_0 \oplus Ce_0 \oplus A'' e_0 \oplus C'' e_0 = 0$ holds is biased. Thus, we have

$$LP_R((e_0, e_0, 0, 0) \to (e_0, e_0)) = 2^{-20}.$$

Similarly, we obtain the following equations for $i, j, k, l \in \{0, 1, 2, 3, 4\}$

$$LP_R((e_i, e_j, 0, 0) \to (e_k, e_l)) = 2^{-20}.$$

5   Linear Probability of RC6 with Weak Keys

In this section, we consider the linear probability of $2r$-round RC6 with Type II weak keys by taking "multiple paths". In general, the linear characteristic probability depends on the key-value. Here, for simplicity, we consider RC6$\oplus$. We assume that key is randomly distributed. In the case that the least significant five bits of extended-key related to linear approximation is fixed (especially, in the case of weak-keys), we can calculate the precise linear probability for each linear approximation. We will discuss how to calculate it in Section 7.

Table 2. 4-round multiple linear path of RC6⊕

round	output mask of (A, C)		
0		(e_i, e_j)	
	↙	...	↘
2	(e_0, e_0)	....	(e_4, e_4)
	↘	...	↙
4		(e_k, e_l)	
LP	2^{-40}	...	2^{-40}

Let us consider 4-round RC6⊕. If we set the input mask for (A, C) being (e_i, e_j) $(i, j \in \{0, ..., 4\})$ and its output mask being (e_k, e_l) $(k, l \in \{0, ..., 4\})$, then for any $i, j, s, t \in \{0, ..., 4\}$ the following holds:

$$|Bias_R((e_i, e_j, 0, 0) \to (e_s, e_t))| = 2^{-10}.$$

Thus we can show that the absolute value of linear characteristic per path of 4-round RC6⊕ is 2^{-40} by the piling up lemma in average of the key.

Since there exist at least $25 = 5^2$ linear characteristic paths such that input mask (e_i, e_j) and output mask (e_k, e_l) are equal but any other intermediate masks are different from the input mask, we can calculate linear characteristic over multiple paths. (See Table 2.)

$$LP_{RC6\oplus_{(4)}}((e_0, e_0, 0, 0) \to (e_0, e_0, 0, 0))$$
$$= \sum_{s,t} LP_R((e_0, e_0) \to (e_s, e_t)) LP_R((e_s, e_t) \to (e_0, e_0))$$
$$= 25 \cdot (2^{-40}) = 2^{-35.356}$$

Moreover, the linear probability of $2r$-round RC6⊕ is derived as follows.

$$LP_{RC6\oplus_{(2r)}}((e_i, e_j, 0, 0) \to (e_k, e_l, 0, 0))$$
$$\geq 2^{-20}(25 \cdot (2^{-20}))^{r-1}$$
$$= 2^{-20-15.356(r-1)}$$

It is easy to consider the linear approximation of $(2r + 1)$-round RC6⊕ from the linear approximation of $2r$-round RC6⊕ obtained above. (See Table 3.)

Next, we consider Type II weak-keys of RC6. It is easy to see that if $lsb_4(K) = 0$ then arithmetic addition of some fixed 32-bit value K (say, $Y = X + K$ mod 2^{32}) does not cause any carry-over in the least significant 5 bits. In this case, the equation $LP_{add_K}(e_i \to e_i) = 1$ always holds for $i \in \{0, ..., 4\}$. Such keys can be generated with probability 2^{-4} if K is randomly distributed.

This implies that the linear probability of RC6 with weak key of Type II is independent of keys. Thus, we can say that the resistance of RC6 with such keys against multiple linear attack is reduced to the one of RC6⊕ against multiple linear attack. In this sense, we can regard such keys as weak ones. (For example, some weak keys of 3-round RC6 are characterized as ones with least significant four bits each of $S[0], S[1], S[4], S[5]$ is 0. So, the fraction of such weak keys is 2^{-16}.)

Table 3. The linear probability of RC6 with Type II weak keys (or RC6$\oplus$)

round	input mask	output mask	linear probability ($\log_2$)	fraction of weak keys of RC6
3	$(0,0,e_i,e_j)$	$(e_k,e_l,0,0)$	-20.000	2^{-16}
5	$(0,0,e_i,e_j)$	$(e_k,e_l,0,0)$	-35.356	2^{-24}
7	$(0,0,e_i,e_j)$	$(e_k,e_l,0,0)$	-50.712	2^{-32}
9	$(0,0,e_i,e_j)$	$(e_k,e_l,0,0)$	-66.068	2^{-40}
11	$(0,0,e_i,e_j)$	$(e_k,e_l,0,0)$	-81.424	2^{-48}
13	$(0,0,e_i,e_j)$	$(e_k,e_l,0,0)$	-96.780	2^{-56}
15	$(0,0,e_i,e_j)$	$(e_k,e_l,0,0)$	-112.136	2^{-64}
17	$(0,0,e_i,e_j)$	$(e_k,e_l,0,0)$	-127.492	2^{-72}

Note that i,j,k,l range over $\{0,...,4\}$.

Table 4. The linear probability of 2-round RC6

Input mask $(\Gamma A, \Gamma C)$		$(e_i, e_j), i, j \in \{0,...,4\}$		
Output mask $(\Gamma A', \Gamma C')$	(e_0,e_0)	(e_k,e_0)	(e_0,e_l)	(e_k,e_l)
Linear prob. of addition	1	2^{-2}	2^{-2}	2^{-4}
Linear prob. of 1-round RC6	2^{-20}	2^{-22}	2^{-22}	2^{-24}

Note that $k \neq 0, l \neq 0$.

6   Linear Probability of RC6

There are several researches about success probability of linear approximation for arithmetic addition $Y = X + K$ on the assumption that K is randomly chosen but fixed ([7,10,15,4]). In this paper, we consider linear approximation only of the form $Xe_i \oplus Ye_i = 0$, which is a relation between a 1-bit of input and a 1-bit of output. Let us see it more precisely.

It is well known that the expectation (w.r.t. keys) of the bias of linear equations satisfies that $LP_{add_K}(e_0 \to e_0) = 1$, and $LP_{add_K}(e_i \to e_i) = 2^{-2}, (i \neq 0)$ on the average of K. By utilizing these equations, it is easy to calculate the linear probability of 2-round RC6 with the key addition. We note that the linear probability of key addition can be obtained only from output masks. The linear probability of 2-round RC6 follows from the linear probability of addition, the linear probability of 1-round RC6$\oplus$, obtained in the previous section, and the piling up lemma. (See Table 4.)

Any output mask $(\Gamma A', \Gamma C')$ corresponds with one of $(e_0, e_0), (e_k, e_0), (e_0, e_l)$ and (e_k, e_l). The number of output masks of each type is $1, 4, 4$ and 16, respectively. Thus we have the linear probability of 4-round RC6 over multiple paths such that the input mask is of the form $(e_i, e_j, 0, 0)$ and the output mask is of the form $(e_0, e_0, 0, 0)$ as follows.

$$LP_{RC6_{(4)}}((e_i, e_j, 0, 0) \to (e_0, e_0, 0, 0))$$
$$= 2^{-20}(2^{-20} + 4 \cdot 2^{-22} + 4 \cdot 2^{-22} + 16 \cdot 2^{-24})$$
$$= 2^{-38}$$

Table 5. The linear probability of RC6

round	input mask (A, C, B, D)	output mask (A',C',B',D')	linear probability $(\log_2)$
3	$(0, 0, e_i, e_j)$	$(e_k, e_l, 0, 0)$	$-20 - 2\mu(\{i, j, k, l\})$
5	$(0, 0, e_i, e_j)$	$(e_k, e_l, 0, 0)$	$-38 - 2\mu(\{i, j, k, l\})$
7	$(0, 0, e_i, e_j)$	$(e_k, e_l, 0, 0)$	$-56 - 2\mu(\{i, j, k, l\})$
9	$(0, 0, e_i, e_j)$	$(e_k, e_l, 0, 0)$	$-74 - 2\mu(\{i, j, k, l\})$
11	$(0, 0, e_i, e_j)$	$(e_k, e_l, 0, 0)$	$-92 - 2\mu(\{i, j, k, l\})$
13	$(0, 0, e_i, e_j)$	$(e_k, e_l, 0, 0)$	$-110 - 2\mu(\{i, j, k, l\})$

Note that $i, j, k, l \in \{0, ..., 4\}$, $\mu(X) = \#\{x \neq 0 | x \in X\}$.

Similarly, we have the linear probability of $2r$-round RC6 over the same multiple paths as follows.

$$LP_{RC6_{(2r)}}((e_i, e_j, 0, 0) \to (e_0, e_0, 0, 0)) \geq 2^{-20-18(r-1)}$$

Furthermore, we have the linear probability of $2r$-round RC6 over multiple paths of the other types.

$$LP_{RC6_{(2r)}}((e_i, e_j, 0, 0) \to (e_k, e_0, 0, 0)) \geq 2^{-22-18(r-1)}$$
$$LP_{RC6_{(2r)}}((e_i, e_j, 0, 0) \to (e_0, e_l, 0, 0)) \geq 2^{-22-18(r-1)}$$
$$LP_{RC6_{(2r)}}((e_i, e_j, 0, 0) \to (e_k, e_l, 0, 0)) \geq 2^{-24-18(r-1)}$$

By utilizing the linear approximation of $2r$-round RC6, it is not hard to consider the linear approximation of $2r + 1$-round RC6, We note that the linear probability is affected by the extended-key that is added to input data B and D to the first round. Namely, we have to take into account that the linear probability depends on the bit position of the input mask. We illustrate an estimation of the linear probability of reduced-round RC6 in Table 5.

7   Linear Probability of a Fixed Key

In this section, we give a way to calculate the more precise linear probability of the linear approximation $Ae_i \oplus Ce_j \oplus A'e_k \oplus C'e_l = 0$ for RC6 with any fixed key. In Section 5 and Section 6, we calculated the linear probability in average of keys.

On the other hand, especially as in the case of Type I weak keys, that is the least significant five bits of the extended key is 0, then, by keeping the sign of the bias of linear equation in mind when summing the bias of linear equation, we can generalize the piling up lemma to calculate the bias of linear equation more precisely.

Now, we consider the linear probability of RC6 with Type I weak key. For a simpler exposition, we calculate the linear probability of 4-round RC6 with the weak keys such that the input and the output mask are both $(e_0, e_0, 0, 0)$. We show the bias of each linear characteristic in Table 6. For example, the linear characteristic for paths from the 0th round through the second round to the fourth round trace $(e_0, e_0) \to (e_1, e_1) \to (e_0, e_0)$ is $(-2^{-10}) \cdot (-2^{-10}) = +2^{-20}$.

Table 6. The bias of linear characteristic path such that the input and output masks are both $(e_0, e_0, 0, 0)$ (4-round RC6 with Type I weak key)

Input mask	00 00 00 00 00 00 00 00 00 00 00 00 00 00 00 00 00 00 00 00 00 00 00 00 00
1st R. Bias $(\times 2^{-10})$	1 -1 1 -1 1 -1 1 -1 1 -1 1 -1 1 -1 1 -1 1 -1 1 -1 1 -1 1 1
2nd R. mask	00 01 02 03 04 10 11 12 13 14 20 21 22 23 24 30 31 32 33 34 40 41 42 43 44
3rd R. Bias $(\times 2^{-10})$	1 -1 1 -1 1 -1 1 1 1 -1 1 1 1 1 -1 -1 1 1 1 1 1 1 -1 1 1 1
Output mask	00 00 00 00 00 00 00 00 00 00 00 00 00 00 00 00 00 00 00 00 00 00 00 00 00
Total Bias $(\times 2^{-20})$	1 1 1 1 1 1 1 -1 -1 1 1 -1 1 -1 1 1 -1 -1 1 -1 1 1 1 -1 1

Table 7. Matrix M of the bias $(\times 2^{-10})$ of linear equations for 2-round RC6 with weak key

Input (i,j)	Output Mask (k,l)																								
	00	01	02	03	04	10	11	12	13	14	20	21	22	23	24	30	31	32	33	34	40	41	42	43	44
00	1	-1	1	-1	1	-1	1	-1	1	-1	1	-1	1	-1	1	-1	1	-1	1	-1	1	-1	1	-1	1
01	-1	1	-1	1	-1	-1	1	-1	1	-1	1	-1	1	-1	1	-1	1	-1	1	-1	1	-1	1	-1	-1
02	1	-1	1	-1	1	1	-1	1	-1	1	1	-1	1	-1	1	1	-1	1	-1	1	1	-1	1	-1	-1
03	-1	1	-1	1	-1	-1	1	-1	1	-1	1	-1	1	-1	1	1	-1	1	-1	1	-1	1	-1	1	1
04	1	-1	1	-1	1	1	-1	1	-1	1	1	-1	1	-1	1	1	-1	1	-1	1	1	-1	1	-1	1
10	-1	-1	1	1	-1	1	1	-1	-1	1	-1	-1	1	1	-1	1	1	-1	-1	1	-1	-1	1	1	-1
11	1	-1	-1	1	1	-1	1	1	-1	-1	1	-1	-1	1	1	-1	1	1	-1	-1	1	1	1	1	-1
12	1	1	-1	-1	1	-1	-1	1	1	-1	-1	1	1	-1	-1	1	1	-1	-1	1	1	1	1	1	-1
13	-1	1	1	-1	-1	1	1	-1	-1	1	1	-1	-1	1	1	-1	-1	1	1	-1	1	1	1	1	1
14	-1	-1	1	1	-1	1	1	-1	-1	1	1	1	-1	-1	1	1	1	-1	-1	1	1	1	-1	-1	1
20	1	1	1	1	-1	-1	-1	-1	-1	1	1	1	1	1	-1	-1	-1	-1	-1	1	1	1	1	1	-1
21	1	-1	-1	-1	-1	1	-1	-1	-1	-1	-1	1	1	1	1	-1	1	1	1	1	1	-1	-1	-1	-1
22	1	1	-1	-1	1	1	-1	-1	1	-1	-1	1	1	1	-1	-1	1	1	1	-1	-1	1	1	1	1
23	1	1	1	-1	-1	1	1	-1	-1	1	-1	-1	1	1	-1	-1	1	1	-1	-1	1	1	1	-1	1
24	1	1	1	1	-1	1	1	1	1	-1	-1	-1	-1	-1	1	-1	-1	-1	-1	1	-1	-1	-1	-1	1
30	-1	-1	-1	-1	1	1	1	1	-1	-1	-1	-1	1	1	1	1	1	1	-1	-1	-1	-1	-1	-1	-1
31	-1	1	1	1	1	1	-1	-1	-1	-1	-1	-1	-1	-1	-1	1	1	1	1	-1	1	1	1	1	1
32	1	-1	-1	-1	-1	1	1	1	-1	-1	1	1	1	1	-1	1	1	1	-1	1	1	1	1	1	1
33	1	1	1	-1	-1	1	1	-1	-1	1	1	1	-1	-1	-1	-1	1	1	-1	-1	1	1	1	-1	1
34	1	1	1	1	-1	1	1	1	1	-1	1	1	1	1	-1	-1	-1	-1	-1	1	-1	-1	-1	-1	1
40	1	1	1	1	1	-1	-1	-1	-1	-1	1	1	1	1	1	-1	-1	-1	-1	-1	1	1	1	1	1
41	-1	1	1	1	1	-1	1	1	1	1	-1	-1	-1	-1	1	-1	-1	-1	-1	1	1	1	1	1	1
42	1	1	-1	-1	-1	1	1	-1	-1	1	1	1	-1	-1	-1	1	1	-1	-1	-1	1	1	1	1	1
43	1	1	1	-1	-1	1	1	-1	-1	1	1	1	-1	-1	1	1	1	-1	-1	-1	-1	1	1	1	1
44	1	1	1	1	-1	1	1	1	1	-1	1	1	1	1	-1	1	1	1	1	-1	-1	-1	-1	-1	1

The number of linear characteristic for paths through $(e_s, e_t, 0, 0)$ in the second round is totally $5^2 = 25$. Among them, there are 17 positive $(= 2^{-10})$ bias of linear characteristics and 8 negative $(= -2^{-10})$ bias of linear characteristics. By taking account of the sign of linear characteristic of each path, we obtain the linear characteristic and the linear probability as follows. (The validity of this observation is demonstrated by computer experiments.)

$$Bias_{RC6(\text{Type I weak key})}((e_0, e_0, 0, 0) \to (e_0, e_0, 0, 0))$$
$$= (17 - 8) \cdot (2^{-10}2^{-10}) = 2^{-16.83}$$
$$LP_{RC6(\text{Type I weak key})}((e_0, e_0, 0, 0) \to (e_0, e_0, 0, 0)) = 2^{-33.66}$$

We note that linear probability obtained here is much higher than the linear probability of RC6 with average keys (2^{-38}) and of RC6$\oplus$ $(2^{-35.356})$.

Next, we generalize the above method to calculate precisely the linear probability of the input mask and the output mask pattern $(e_i, e_j, 0, 0) \to (e_k, e_l, 0, 0)$ for $2r$-round RC6 with Type I weak key.

For $i, j, k, l \in \{0, ..., 4\}$, let $m = (k - i)(\mod 32), n = (l - j)(\mod 32)$. We consider the 25×25 matrix $M = (a_{(ij)(kl)})$ such that $a_{(ij)(kl)} = (-1)^{n \cdot e_i \oplus m \cdot e_j}$. (See Table 7.) Then, the bias of linear equation of $2r$-round RC6 with Type I weak key can be calculated as follows.

Table 8. The bias $(\times 2^{-4})$ of linear equation for addition$(+)$

lsb_5	k_4	k_3	k_2	k_1	k_0	lsb_5	k_4	k_3	k_2	k_1	k_0	lsb_5	k_4	k_3	k_2	k_1	k_0	lsb_5	k_4	k_3	k_2	k_1	k_0
0	16	16	16	16	16	8	0	-16	16	16	16	16	-16	16	16	16	16	24	0	-16	16	16	16
1	14	12	8	0	-16	9	-2	-12	8	0	-16	17	-14	12	8	0	-16	25	2	-12	8	0	-16
2	12	8	0	-16	16	10	-4	-8	0	-16	16	18	-12	8	0	-16	16	26	4	-8	0	-16	16
3	10	4	-8	0	-16	11	-6	-4	-8	0	-16	19	-10	4	-8	0	-16	27	6	-4	-8	0	-16
4	8	0	-16	16	16	12	-8	0	-16	16	16	20	-8	0	-16	16	16	28	8	0	-16	16	16
5	6	-4	-8	0	-16	13	-10	4	-8	0	-16	21	-6	-4	-8	0	-16	29	10	4	-8	0	-16
6	4	-8	0	-16	16	14	-12	8	0	-16	16	22	-4	-8	0	-16	16	30	12	8	0	-16	16
7	2	-12	8	0	-16	15	-14	12	8	0	-16	23	-2	-12	8	0	-16	31	14	12	8	0	-16

Table 9. The maximum linear probability of RC6 with Type I weak key

Rounds	Input Mask (i, j)	Output Mask (k, l)	Bias	LP	comment
3	$(*,*)$	$(*,*)$	$1 \cdot 2^{-10}$	2^{-20}	1 in 2^{20}
5	$(3,8)$	$(2,2)$	$21 \cdot 2^{-20}$	$2^{-31.214}$	1 in 2^{30}
7	$(1,1)$	$(3,3)$	$101 \cdot 2^{-30}$	$2^{-46.682}$	1 in 2^{40}
9	$(1,1)$	$(0,0)$	$633 \cdot 2^{-40}$	$2^{-61.386}$	1 in 2^{50}
11	$(1,1)$	$(2,2)$	$4449 \cdot 2^{-50}$	$2^{-75.760}$	1 in 2^{60}
13	$(2,2)$	$(2,2)$	$24798 \cdot 2^{-60}$	$2^{-90.804}$	1 in 2^{70}
15	$(2,2)$	$(0,0)$	$134645 \cdot 2^{-70}$	$2^{-105.922}$	1 in 2^{80}
17	$(2,2)$	$(0,0)$	$942657 \cdot 2^{-80}$	$2^{-120.306}$	1 in 2^{90}

$$\Phi^{(r)} = 2^{-10r} M^r,$$
$$Bias_{RC6(\text{Type I weak key})(2r)}((e_i, e_j, 0, 0) \to (e_k, e_l, 0, 0)) = \Phi^{(r)}{}_{(ij)(kl)},$$

where M^r means the exponentiation of the integer matrix M.

In case of RC6 with arbitrary key, for the least significant five bits of extended key $lsb_5(S[4i - 2])$, $lsb_5(S[4i - 1])$, we calculate $(k_{i4}, k_{i3}, k_{i2}, k_{i1}, k_{i0})$ and $(h_{i4}, h_{i3}, h_{i2}, h_{i1}, h_{i0})$ by using Table 8, and also calculate the following matrix.

$$K_i = diag(k_{i0}h_{i0}, k_{i0}h_{i1}, ..., k_{i4}h_{i4}),$$

where $diag(a_0, a_1, ...)$ is the 25×25 matrix whose diagonal elements are $a_0, a_1, ...$ and other elements are all 0. For example, when $lsb_5(S[2]) = 4$ and $lsb_5(S[3]) = 26$, then K_1 is calculated as follows.

$$K_1 = diag(2^{-3}, -2^{-2}, 0, -2^{-1}, 2^{-1}, 0, 0, 0, 0, 0, -2^{-2}, 2^{-1}, ..., -1, 1)$$

Then, the bias of linear equation of $2r$-round RC6 can be calculated by using a "Generalized Piling up lemma" of matrix representation as follows.

$$Bias_{RC6(2r)}((e_i, e_j, 0, 0) \to (e_k, e_l, 0, 0))$$
$$= \Psi^{(r)}[K]_{(ij)(kl)}$$
$$\Psi^{(r)}[K] = \prod_{i=1}^{r}(2^{-10}M \cdot K_i)$$

Table 9 shows linear masks that take the maximum linear probability of RC6 with Type I weak key in the elements of 25×25 matrix calculated as above. Now, we can get the maximal linear probability of $(2r + 1)$-round RC6 with the weak keys by combining the discussion in the case of $2r$-round RC6 and one-round addition to the input side.

Table 10. Linear approximations of $2r$-round RC6 for multiple linear approximation

linear approximation	linear probability	number
$Be_0 \oplus De_0 \oplus A'e_0 \oplus C'e_0 = 0$	$2^{-20-18(r-1)}$	1
$Be_i \oplus De_0 \oplus A'e_0 \oplus C'e_0 = 0$	$2^{-22-18(r-1)}$	4
$Be_0 \oplus De_j \oplus A'e_0 \oplus C'e_0 = 0$	$2^{-22-18(r-1)}$	4
$Be_0 \oplus De_0 \oplus A'e_k \oplus C'e_0 = 0$	$2^{-22-18(r-1)}$	4
$Be_0 \oplus De_0 \oplus A'e_0 \oplus C'e_l = 0$	$2^{-22-18(r-1)}$	4

8   Multiple Linear Approximation of RC6

"Multiple Linear Approximation", which is proposed by Kaliski and Robshaw, is a technique to enable to attack ciphers using less amount of data. This technique is quite effective if there exist several linear approximations that have almost maximum linear probability.

Let ϵ_i be the bias of linear equation $L_i : X\Gamma_{X_i} \oplus Y\Gamma_{Y_i} = 0$, $(i = 1, ..., n)$ with respect to $Y = F(X)$. Then we define *weight* according to ϵ_i as being $w_i = \epsilon_i/(\epsilon_0 + ... + \epsilon_n)$. Let N be the number of known plaintexts and N_i the number of known plaintexts that satisfy linear approximation L_i. Then by utilizing the difference between $w_i N_i$ and $N/2$ it is not hard to distinguish a cipher from random permutations. The necessary number N of known plaintexts to distinguish a cipher from random permutations is $C/(\sum_{i=1}^n \epsilon_i^2)$, where C is a parameter which determines the success probability (e.g., $C = 4$ implies that the success probability is 95%).

By careful consideration of multiple linear approximation, we can see that it is sufficient for estimating necessary number of plaintexts to break a cipher that we get linear approximations whose linear equations are linearly independent. Recall the linear approximations which are discussed in the previous section. The linear approximations we should consider are all of the form $Be_i \oplus De_j \oplus A'e_k \oplus C'e_l = 0$. It is not difficult to see that there are at most 17 linearly independent linear approximations. We utilize 17 linear approximations (shown in Table 10), which are linearly independent and whose linear probabilities are comparatively high, in order to improve the efficiency of breaking RC6.

We estimate the necessary number N of plaintexts to distinguish RC6 from random permutations by applying linear approximations shown in Table 10 to the technique of multiple linear approximation. We note that the coefficients in the equations below are introduced in order to increase the success probability up to 95%.

$$N = 4 \cdot (1/(2^{-20-18(r-1)} + 16 \cdot 2^{-22-18(r-1)}))$$
$$= 4 \cdot 1/((1 + 2^{-2}) \cdot 2^{-18r}) = 2^{3.68+18r}$$

Similarly, we estimate the necessary number N of plaintexts to distinguish RC6$\oplus$ (or RC6 with Type II weak key) from random permutations by seeing Table 3.

$$N = 4/(17 \cdot 2^{-20-15.356(r-1)}) = 2^{2.556+(15.356)r}$$

Table 11. Distinguishing attack of RC6

round	RC6	RC6(Type II weak key)		RC6(Type I weak key)	
3	$2^{21.68}$	$2^{17.912}$	(1 in 2^{16})	$2^{17.912}$	(1 in 2^{20})
5	$2^{39.68}$	$2^{33.268}$	(1 in 2^{24})	$2^{30.1261}$	(1 in 2^{30})
7	$2^{57.68}$	$2^{48.624}$	(1 in 2^{32})	$2^{45.712}$	(1 in 2^{40})
9	$2^{75.68}$	$2^{63.980}$	(1 in 2^{40})	$2^{60.260}$	(1 in 2^{50})
11	$2^{93.68}$	$2^{79.336}$	(1 in 2^{48})	$2^{75.111}$	(1 in 2^{60})
13	$2^{111.68}$	$2^{94.692}$	(1 in 2^{56})	$2^{90.061}$	(1 in 2^{70})
15	$2^{129.68}$	$2^{110.048}$	(1 in 2^{63})	$2^{104.701}$	(1 in 2^{80})
17	$2^{147.68}$	$2^{125.404}$	(1 in 2^{70})	$2^{119.423}$	(1 in 2^{90})
19	$2^{165.68}$	$2^{140.760}$	(1 in 2^{78})	$2^{134.227}$	(1 in 2^{100})

Moreover, in case of RC6 with Type I weak key, we can pick up linearly independent 17 linear approximations according to the estimation for the linear probability of each input-output masks in Section 8.

Our attacking method, described in this section, is a known plaintext attack. It means that we do not restrict the form of inputs and thus that we can make full use of inputs, that is, the number 2^{128} of plaintexts. Therefore, we can say that 13-round RC6 is distinguishable from random permutations and also that 15-round RC6⊕, 17-round RC6 with weak keys whose fraction is 2^{-90} is distinguishable from random permutations. We summarize these results in Table 11.

9   Key Recovery of RC6

In this section, we consider the key recovery of $(2r + 2)$-round RC6 by utilizing the distinguishability result of $(2r + 1)$-round RC6. We adopt a typical method "1-round elimination attack" as same as the method by Knudsen and Meier [13] for key recovery in the following: apply multiple linear approximation to RC6 through the 2nd round to the $2r + 2$th round, search exhaustively for the extended key $S[0], S[1]$ (64 bits in total), and use them and the value in the position before key addition of the first round to find a target key. (See Figure 3.)

Since the necessary data size for distinguishing attacker to success the attack is $4p^{-1}$ with probability 95%, (which is calculated using the linear approximation with the linear probability p), we claim by our experience that we need the number $4np^{-1}$ of known plaintexts and the number $4p^{-1}2^n$ of computation of the round-function for the number $n(= 64)$ of the target key bits with probability 95% by the one-round elimination method. Thus, we can summarize the necessary data size and complexity to find the target extended key by the one-round elimination method in Table 12.

Thus we conclude that the 64-bit target extended key of 14-round RC6 can be recovered with probability 95% by Multiple Linear Attack with the number $2^{119.68}$ of known plaintexts and the number $2^{185.68}$ of computation of the round-function. Also that the 64-bit target extended key of 18-round RC6 with weak key, (the fraction is 2^{-90}), can be recovered with probability 95% by Multiple

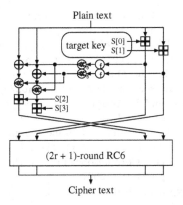

Fig. 3. 1-round elimination attack

Table 12. Key recovery of RC6

Rounds	Target	#Texts	Complexity	Comments
4	RC6	$2^{29.68}$	$2^{95.68}$	
6	RC6	$2^{47.68}$	$2^{113.68}$	
8	RC6	$2^{65.68}$	$2^{131.68}$	
10	RC6	$2^{83.68}$	$2^{149.68}$	
12	RC6	$2^{101.68}$	$2^{167.68}$	
14	RC6	$2^{119.68}$	$2^{185.68}$	
16	RC6 weak key	$2^{118.048}$	$2^{184.048}$	1 in 2^{64} (Type II)
18	RC6 weak key	$2^{127.423}$	$2^{193.423}$	1 in 2^{90} (Type I)

Linear Attack with the number $2^{127.423}$ of known plaintexts and the number of 2^{64} of memory, and the number $2^{193.423}$ of computation of the round-function.

References

1. A. Biryukov and E. Kushilevitz. Improved cryptanalysis of RC5. EUROCRYPT'98, LNCS 1403, pp.85–99, 1998.
2. J. Borst, B. Preneel, and J. Vandewalle. Linear cryptanalysis of RC5 and RC6. FSE'99, LNCS 1636, pp.16–30, 1999.
3. S. Contini, R.L. Rivest, M.J.B. Robshaw, and Y.L. Yin. The security of the RC6 block cipher. v.1.0, August 20, 1998.
   Available at http://www.rsasecurity.com/rsalabs/rc6/.
4. S. Contini, R.L. Rivest, M.J.B. Robshaw, and Y.L. Yin. Improved analysis of some simplified variants of RC6. FSE'99, LNCS 1636, pp.1–15, 1999.
5. H. Gilbert, H. Handschuh, A. Joux and S. Vaudenay, A Statistical Attack on RC6. FSE 2000, LNCS 1978, pp.64–74, 2001.
6. M.H. Heys. Linearly weak keys of RC5. *IEE Electronic Letters*, Vol.33, pp.836–838, 1997.
7. B.S. Kaliski Jr. and M.J.B. Robshaw. Linear cryptanalysis using multiple approximations. CRYPTO'94, LNCS 839, pp.26–39, 1994.

8. B.S. Kaliski Jr. and M.J.B. Robshaw. Linear cryptanalysis using multiple approximations and FEAL. FSE'94, LNCS 1008, pp.249–264, 1995.
9. B.S. Kaliski Jr. and Y.L. Yin. On differential and linear cryptanalysis of the RC5 encryption algorithm. CRYPTO'95, LNCS 963, pp.171–184, 1995.
10. B.S. Kaliski Jr. and Y.L. Yin. On the security of the RC5 encryption algorithm. Available at http://www.rsasecurity.com/rsalabs/rc6/.
11. J. Kelsey, B. Schneier, and D. Wagner. Mod n cryptanalysis, with applications against RC5P and M6. FSE'99, LNCS 1363, pp.139–155, 1999.
12. L.R. Knudsen and M.J.B. Robshaw. Non-linear approximations in linear cryptanalysis. EUROCRYPT'96, LNCS 1070, pp.224–236, 1996.
13. L.R. Knudsen and W. Meier. Correlations in RC6 with a reduced number of rounds. FSE 2000, LNCS 1978, pp.94–108, 2001.
14. M. Matsui. Linear cryptanalysis method for DES cipher. EUROCRYPT'93, LNCS 765, pp.386–397, 1993.
15. S. Moriai, K. Aoki and K. Ohta. Key-dependency of linear probability of RC5. *IEICE Transactions on Fundamentals of Electronics, Communications and Computer Sciences*, Vol.E80-A, No.1, 1997.
16. R.L. Rivest. The RC5 encryption algorithm. FSE'94, LNCS 1008, pp.86–96, 1995.
17. R.L. Rivest, M.J.B. Robshaw, R. Sidney and Y.L. Yin. The RC6 block cipher. v1.1, August 20, 1998. Available at http://www.rsasecurity.com/rsalabs/rc6/.
18. K. Nyberg. Linear approximation of block ciphers. EUROCRYPT'94, LNCS 950, pp.439–444, 1994.
19. A.A. Selcuk. New results in linear cryptanalysis of RC5. FSE'98, LNCS 1372, pp.1–16, 1998.

On the Security of CAMELLIA against the Square Attack

Yongjin Yeom, Sangwoo Park, and Iljun Kim

National Security Research Institute
161 Gajeong-dong, Yuseong-gu, Daejeon, 305-350 Korea
{yjyeom, psw}@etri.re.kr, ijkim@dingo.etri.re.kr

Abstract. Camellia is a 128 bit block cipher proposed by NTT and Mitsubishi. We discuss the security of Camellia against the square attack. We find a 4 round distinguisher and construct a basic square attack. We can attack 5 round Camellia by guessing one byte subkey and using 2^{16} chosen plaintexts. Cosidering the key schdule, we may extend this attack up to 9 round Camellia including the first FL/FL^{-1} function layer.

1 Introduction

Camellia[5] is a 128-bit block cipher which was announced by NTT and Mitsubishi in 2000. It has the modified Feistel structure with irregular rounds, so called the FL/FL^{-1} function layer. The round function is based on that of the block cipher E2[13] by NTT whereas the FL/FL^{-1} layer comes from MISTY[18] by Mitsubishi. Camellia was submitted to the standardization and the evaluation projects such as ISO/IEC JTC 1/SC 27, CRYPTREC, and NESSIE. Recently, Camellia was selected as an algorithm for the second phase of the NESSIE project.

Currently, the most efficient methods analyzing Camellia are truncated differential cryptanalysis and higher order differential attack. Kanda and Matsumoto[12] studied the security against truncated differential cryptanalysis from the designer's standpoint. They found the upper bound of the best bytewise characteristic probability and proved that Camellia with more than 11 rounds are secure against truncated differential cryptanalysis. Most analyses on Camellia consider simplified version without FL/FL^{-1} function layers. For instance, S. Lee et al.[15] attacked eight round Camellia using truncated differential cryptanalysis. M. Sugita et al.[19] found a nontrivial 9 round bytewise characteristics and a seven round impossible differential for Camellia. Kawabata and Kaneko[11] showed that Camellia can be attacked by higher order differential attack up to 10 rounds. Some other analyses can be found in [5].

The square attack was a dedicated attack on the block cipher SQUARE[6] and applied to block ciphers of the SPN structure such as Rijndael[7,8,16], CRYPTON[9], and Hierocrypt[3]. In order to apply the square attack on the Feistel structure, Lucks[17] introduced the saturation attack, as a variation of the square attack. He analyzed the Twofish algorithm of the modified Feistel

J. Daemen and V. Rijmen (Eds.): FSE 2002, LNCS 2365, pp. 89–99, 2002.
© Springer-Verlag Berlin Heidelberg 2002

structure. Recently, Y. He and S. Qing[10] showed that six round Camellia are breakable by square attack.

In this paper, we apply the square attack to Camellia including FL/FL^{-1} function layers. We suggest the basic attack which breaks 5 round Camellia using 2^{16} chosen plaintexts and 2^8 key guessings. Also, the key schedule is considered so that the square attack on 256 bit Camellia is faster than exhaustive key search up to 9 rounds.

Section 2 briefly describes the structure of Camellia. A basic attack based on a 4 round distinguisher is given in Section 3 and extensions of the basic attack up to 9 round Camellia is proposed in Section 4.

2   Description of the Camellia

Camellia has a 128 bit block size and supports 128, 192 and 256 bit keys. The design of Camellia is based on the Feistel structure and its number of rounds is 18(128 bit key) or 24(192, 256 bit key). The FL/FL^{-1} function layer is inserted at every 6 rounds in order to thwart future unknown attacks. Before the first round and after the last round, there are pre- and post-whitening layers which use bitwise exclusive-or operations with 128 bit subkeys, respectively.

One round substitution and permutation structure is adopted as the round function F. Let $X_L^{(r)}$ and $X_R^{(r)}$ be the left and the right halves of the r round inputs, respectively, and $k^{(r)}$ be the r round subkey. Then the Feistel structure of Camellia can be written as

$$X_L^{(r+1)} = X_R^{(r)} \oplus F(X_L^{(r)}, k^{(r)}),$$
$$X_R^{(r+1)} = X_L^{(r)}.$$

In the following substitution S, the four types of S-boxes s_1, s_2, s_3, and s_4 are used. Each of them is affinely equivalent to an inversion over $GF(2^8)$. Actually, s_2, s_3, s_4 are variations of s_1. The only property of S-boxes used for the square attack is that they are one-to-one functions. The substitution function $S : \{0,1\}^{64} \rightarrow \{0,1\}^{64}$ which consists of S-boxes is also a one-to-one function defined by

$$(x_1, \ldots, x_8) \xrightarrow{S} (s_1(x_1), s_2(x_2), s_3(x_3), s_4(x_4), s_2(x_5), s_3(x_6), s_4(x_7), s_1(x_8)).$$

The permutation function $P : \{0,1\}^{64} \rightarrow \{0,1\}^{64}$ maps $(z_1, \ldots, z_8)$ to $(z_1', \ldots, z_8')$ defined by

$$z_1' = z_1 \oplus z_3 \oplus z_4 \oplus z_6 \oplus z_7 \oplus z_8,$$
$$z_2' = z_1 \oplus z_2 \oplus z_4 \oplus z_5 \oplus z_7 \oplus z_8,$$
$$z_3' = z_1 \oplus z_2 \oplus z_3 \oplus z_5 \oplus z_6 \oplus z_8,$$
$$z_4' = z_2 \oplus z_3 \oplus z_4 \oplus z_5 \oplus z_6 \oplus z_7,$$
$$z_5' = z_1 \oplus z_2 \oplus z_6 \oplus z_7 \oplus z_8,$$

$$z_6' = z_2 \oplus z_3 \oplus z_5 \oplus z_7 \oplus z_8,$$
$$z_7' = z_3 \oplus z_4 \oplus z_5 \oplus z_6 \oplus z_8,$$
$$z_8' = z_1 \oplus z_4 \oplus z_5 \oplus z_6 \oplus z_7.$$

We can also express the function P in the matrix form:

$$
\begin{pmatrix} z_8' \\ z_7' \\ z_6' \\ z_5' \\ z_4' \\ z_3' \\ z_2' \\ z_1' \end{pmatrix}
=
\begin{pmatrix}
0\,1\,1\,1\,1\,0\,0\,1 \\
1\,0\,1\,1\,1\,1\,0\,0 \\
1\,1\,0\,1\,0\,1\,1\,0 \\
1\,1\,1\,0\,0\,0\,1\,1 \\
0\,1\,1\,1\,1\,1\,1\,0 \\
1\,0\,1\,1\,0\,1\,1\,1 \\
1\,1\,0\,1\,1\,0\,1\,1 \\
1\,1\,1\,0\,1\,1\,0\,1
\end{pmatrix}
\begin{pmatrix} z_8 \\ z_7 \\ z_6 \\ z_5 \\ z_4 \\ z_3 \\ z_2 \\ z_1 \end{pmatrix}.
$$

The round function $F : \{0,1\}^{64} \times \{0,1\}^{64} \to \{0,1\}^{64}$ is defined as a composition of S and P functions as follows:

$$(X, k) \xrightarrow{F} P(S(X \oplus k)).$$

At every 6 rounds the functions FL and FL^{-1} are inserted. We denote bitwise-and, bitwise-or operations by $\cap$, $\cup$ and a n bit rotation by $\lll n$. The left 64 bit half (X_L, X_R) is mapped to (Y_L, Y_R) by the function FL.

$$(X_L \| X_R, kl_L \| kl_R) \xrightarrow{FL} (Y_L, Y_R),$$

where

$$Y_R = ((X_L \cap kl_L) \lll 1) \oplus X_R,$$
$$Y_L = (Y_R \cup kl_R) \oplus X_L.$$

and the inverse FL^{-1} of FL is used for the right half as follows:

$$(Y_L \| Y_R, kl_L \| kl_R) \xrightarrow{FL^{-1}} (X_L, X_R),$$

where

$$X_L = (Y_R \cup kl_R) \oplus Y_L,$$
$$X_R = ((X_L \cap kl_L) \lll 1) \oplus Y_R.$$

The key schedule of Camellia will be briefly considered in Section 4.

3   Basic Square Attack

The concept of the Λ-set, which was introduced by Daemen et al. [6], plays an important role in the square attack.

Let $\mathcal{F}$ be a collection of state bytes $X = (x_1, x_2, \ldots, x_n)$ where x_i is the i-th byte of X. If the i-th bytes of elements in $\mathcal{F}$ are different one another, the i-th byte is called an 'active' byte. Likewise, the j-th byte is 'passive' (or fixed), if the j-th bytes of states in $\mathcal{F}$ have the same value.

A collection $\mathcal{F}$ of 256 state bytes is called a Λ-set, if every byte of $\mathcal{F}$ is either active or passive. More precisely, if X and Y are arbitrary elements of a Λ-set $\mathcal{F}$, then

$$\begin{cases} x_i \neq y_i, & \text{if the } i\text{-th byte is active,} \\ x_i = y_i, & \text{otherwise,} \end{cases}$$

where x_i and y_i are the i-th byte of X and Y, respectively. Note that an arbitrary collection $\mathcal{F}$ has non-active and non-passive bytes in general. The i-th byte in a collection $\mathcal{F}$ is called balanced, if $\bigoplus_{X \in \mathcal{F}} x_i = 0$.

The main operations of the Camellia are bitwise exclusive-or(XOR) and substitution using one-to-one 8×8 S-boxes s_i. If an active(passive) byte of a Λ-set is used as an input of S-boxes s_i, then the output is also active(passive). But the output of s_i is not necessarily balanced when its input is balanced.

Some properties of XOR operation can be summarized as shown in Table 1.

Table 1. Some properties of XOR operation

XOR($\oplus$)	active byte	passive byte	balanced byte
active byte	balanced byte	active byte	balanced byte
passive byte	active byte	passive byte	balanced byte
balanced byte	balanced byte	balanced byte	balanced byte

3.1 Four Round Distinguishers

Let $X_L^{(r)}$, $X_R^{(r)}$ be the left and the right inputs of the r-th round. Then we can construct a 4 round distinguisher as follows:

Choose

$$X_L^{(1)} = (\alpha_1, \alpha_2, \ldots, \alpha_8), \quad X_R^{(1)} = (A, \beta_2, \ldots, \beta_8)$$

as a Λ-set $\mathcal{F}$ of 256 input plaintexts, where α_i, β_j are constants and A is an active bytes of $\mathcal{F}$. Because $X_L^{(1)}$ is passive, the output of the first round function F is also passive. Thus, the input of the 2nd round can be written of the form

$$X_L^{(2)} = (A, \gamma_2, \ldots, \gamma_8), \quad X_R^{(2)} = (\alpha_1, \alpha_2, \ldots, \alpha_8),$$

where γ_i are constants. In the 2nd round, the input $X_L^{(2)}$ for F is transformed as follows:

$$(A, \gamma_2, \ldots, \gamma_8) \xrightarrow{F} (B, C, D, \delta_4', E, \delta_6', \delta_7', F),$$

where B, C, D, E, and F are active. Thus, we have

$$X_L^{(3)} = (B, C, D, \delta_4, E, \delta_6, \delta_7, F), \quad X_R^{(3)} = (A, \gamma_2, \ldots, \gamma_8)$$

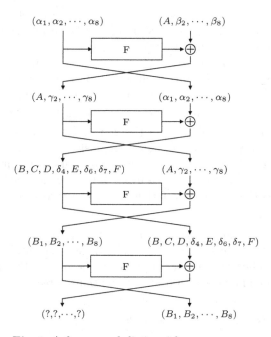

Fig. 1. A four round distinguisher

as an input for the 3rd round. Applying the 3rd round function to $X_L^{(3)}$, we expect that each state byte in the left half of the input for the 4th round is balanced. This implies that all bytes in the right half of the 4th round output are balanced. Thus, we obtain a 4 round distinguisher.

Note that only 2 round functions are effectively activated in this 4 round distinguisher. This corresponds to the 2 round distinguisher for the SPN structure. If we change the position of the active byte in $X_R^{(1)}$, we obtain 8 different distinguishers.

3.2   Five Round Square Attack

From the above distinguisher, we can construct a basic square attack on the 5 round Camellia without pre- and post-whitenings.

Step 1. Guess the 1st byte k of the first round key.
Step 2. As input plaintexts, choose a Λ-set $\mathcal{F}$ of the form

$$\mathcal{F} = \{(X_L(i), X_R(i))|0 \le i \le 255\}, \tag{1}$$

where for arbitrarily chosen constants α_i, β_j,

$$X_L(i) = (i, \alpha_2, \ldots, \alpha_8),$$
$$X_R(i) = (s_1(i \oplus k), s_1(i \oplus k), s_1(i \oplus k), \beta_4, s_1(i \oplus k), \beta_6, \beta_7, s_1(i \oplus k)).$$

Step 3. If k is a correct key, we can expect the left half of the 2nd round inputs consists of constant states. For example, the 1st output byte z_1' of the 1st round function is $z_1'(i) = z_1 \oplus z_3 \oplus z_4 \oplus z_6 \oplus z_7 \oplus z_8$, where $z_1 = s_1(i \oplus k)$ and $z_3, z_4, \ldots, z_8$ are constants solely depending upon $\alpha_2, \ldots, \alpha_8$. Taking exclusive-or of $z_1'(i)$ and the 1st byte $s_1(i \oplus k)$ of $X_R(i)$, we have a constant byte which is independent of i. Using the same argument, we can show that each byte of $X_L^{(2)}$ is a constant.

Step 4. The right half $X_R^{(2)}$ of the 2nd round input is identical to $X_L^{(1)}$. Thus we can use the 4 round distinguisher previously mentioned.

Step 5. Let C_L, C_R be corresponding outputs of the input Λ-set $\mathcal{F}$ which is chosen in Step 2. If all bytes of C_R are balanced, then we can accept k as the correct key. Otherwise, go to Step 1 and guess another key and repeat.

For this 5 round attack, we use 2^8 times 5 round encryptions in every key guessing. A wrong key can pass the balance test with a probability 2^{-64}, i.e. negligible. Thus, the number of plaintexts needed for this attack is $2^8 \times 2^8 = 2^{16}$, and the same number of 5 round encryptions is required.

3.3   Six Round Square Attack

We can extend this basic attack to 6 round Camellia by adding a round at the beginning. The key idea for 6 round attack is to choose a collection of plaintexts whose 1 round output is a Λ-set $\mathcal{F}$ as described in (1). To do this, we assume additional 5 bytes of the first round key.

Let $k_1^{(2)}$ be the first byte of the second round key and $k_i^{(1)}$ the i-th byte of the 1st round key. Suppose that we guess $k_1^{(1)}$, $k_2^{(1)}$, $k_3^{(1)}$, $k_5^{(1)}$, $k_8^{(1)}$, and $k_1^{(2)}$, correctly. Then we can find a set $\mathcal{F}^{(1)}$ of plaintexts so that the second round input is a Λ-set $\mathcal{F}^{(2)}$ of the form

$$\mathcal{F}^{(2)} = \left\{ \left(X_L^{(2)}(i), X_R^{(2)}(i) \right) \,\middle|\, 0 \le i \le 255 \right\}, \tag{2}$$

where

$$X_L^{(2)}(i) = (i, \alpha_2, \ldots, \alpha_8),$$
$$X_R^{(2)}(i) = (s(i), s(i), s(i), \beta_4, s(i), \beta_6, \beta_7, s(i)),$$
$$s(i) = s_1(i \oplus k_1^{(2)}).$$

It is easy to see that the left half $X_L^{(1)}(i)$ of an input Λ-set

$$\mathcal{F}^{(1)} = \left\{ \left(X_L^{(1)}(i), X_R^{(1)}(i) \right) \,\middle|\, 0 \le i \le 255 \right\} \tag{3}$$

should be exactly equal to $X_R^{(2)}(i)$ for each i and the right half $X_R^{(1)}(i)$ of that can be determined by the subkeys $k_1^{(1)}$, $k_2^{(1)}$, $k_3^{(1)}$, $k_5^{(1)}$, and $k_8^{(1)}$. For example, the 1st output byte z_1' of the 1st round function can be written as

$$z_1'(i) = s_1(s(i) \oplus k_1^{(1)}) \oplus s_3(s(i) \oplus k_3^{(1)}) \oplus s_4(\beta_4 \oplus k_4^{(1)}) \oplus s_3(\beta_6 \oplus k_6^{(1)})$$
$$\oplus s_4(\beta_7 \oplus k_7^{(1)}) \oplus s_1(s(i) \oplus k_8^{(1)})$$
$$= s_1(s(i) \oplus k_1^{(1)}) \oplus s_3(s(i) \oplus k_3^{(1)}) \oplus s_1(s(i) \oplus k_8^{(1)}) \oplus \beta,$$

where $\beta = s_4(\beta_4 \oplus k_4^{(1)}) \oplus s_3(\beta_6 \oplus k_6^{(1)}) \oplus s_4(\beta_7 \oplus k_7^{(1)})$ is independent of i. Thus we can choose

$$i \oplus s_1(s(i) \oplus k_1^{(1)}) \oplus s_3(s(i) \oplus k_3^{(1)}) \oplus s_1(s(i) \oplus k_8^{(1)})$$

as the first byte S_1 of $X_R^{(1)}(i)$ so that the first byte of $X_L^{(2)}(i)$ is active. Similarly, remaining bytes of $X_R^{(1)}(i)$ can be calculated.

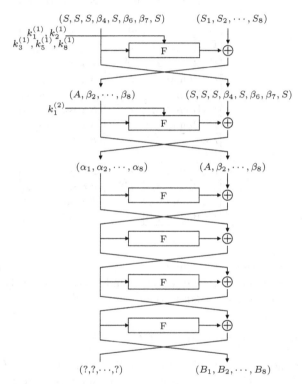

Fig. 2. A square attack on 6 round Camellia

For each Λ-set $\mathcal{F}^{(1)}$ determined by the 6 byte subkeys guessing, we can check the balance of the right half of the 6th round output. In this check, a wrong key can be accepted with a probability 2^{-64}. Thus, the 6 round attack requires $2^8 \times 2^{48}$ plaintexts and 2^{48} subkeys guessing.

By adding a round at the end, we obtain another square attack(See [10]) on 6 round Camellia. But it makes the attack exceeding 6 rounds with the FL/FL^{-1} layer much harder.

4   Key Schedule and Extension of the Basic Attack

4.1   Key Schedule of Camellia

To extend the basic attack on over 6 round Camellia with FL/FL^{-1} function layer, we consider the key schedule. The round keys are bitwise rotations of K_L, K_R, K_A, and K_B which are calculated from the master keys K_L and K_R. In this calculation, they use the reduced rounds of Camellia with a constant key. We do not describe the details here(See [1]). Table 2 shows how to select 1–10 round keys from K_L, K_R, K_A, and K_B.

Table 2. Subkeys for 192/256-bit secret key

	subkey	value
F (Round 1)	$k^{(1)}$	$(K_B \lll_0)_{L(64)}$
F (Round 2)	$k^{(2)}$	$(K_B \lll_0)_{R(64)}$
F (Round 3)	$k^{(3)}$	$(K_R \lll_{15})_{L(64)}$
F (Round 4)	$k^{(4)}$	$(K_R \lll_{15})_{R(64)}$
F (Round 5)	$k^{(5)}$	$(K_A \lll_{15})_{L(64)}$
F (Round 6)	$k^{(6)}$	$(K_A \lll_{15})_{R(64)}$
FL	$kl_{1(64)}$	$(K_R \lll_{30})_{L(64)}$
FL^{-1}	$kl_{2(64)}$	$(K_R \lll_{30})_{R(64)}$
F (Round 7)	$k^{(7)}$	$(K_B \lll_{30})_{L(64)}$
F (Round 8)	$k^{(8)}$	$(K_B \lll_{30})_{R(64)}$
F (Round 9)	$k^{(9)}$	$(K_L \lll_{45})_{L(64)}$
F (Round 10)	$k^{(10)}$	$(K_L \lll_{45})_{R(64)}$

Note that seven and eight round keys $k^{(7)}$ and $k^{(8)}$ are nothing but 30 bit rotations of the first and the second round keys, respectively. This property will be used to attack more than 6 rounds of Camellia.

4.2   An Observation on the FL/FL^{-1} Layer

Consider the reduced model of 6 round Camellia with the FL/FL^{-1} layer. As mentioned previously, if we assume 6 byte subkeys correctly, every byte of the right half of the 6th round outputs is balanced. By guessing additional 7 bits of subkey kl_2, we can partially invert FL/FL^{-1} layer.

Let (C_L, C_R) be an output of the FL^{-1} function. Then we can determine the leftmost 7 bits of the input $y_{R,4}$ for the FL^{-1} function from two bytes $x_{R,4}$ and $x_{L,4}$ using the relation

$$y_{R,4} = x_{R,4} \oplus (x_{L,4} \cap kl_{2L,4}) \lll 1,$$

where $y_{R,4}$, $x_{R,4}$, $x_{L,4}$ and $kl_{2L,4}$ are the fourth bytes of Y_R, X_R, X_L and kl_{2L}, respectively.

4.3   Square Attacks on 256 bit Camellia up to 7, 8, and 9 Rounds

Now, we consider 256 bit Camellia and its key schedule. We can extend the previous observation on the FL/FL^{-1} layer to a 7 round square attack. We should assume 7 byte subkeys $k_1^{(7)}, \ldots, k_7^{(7)}$ out of the seventh round key $k^{(7)}$ to determine two byte outputs $x_{R,4}$ and $x_{L,4}$ of FL^{-1} function. But the 7th round key $k^{(7)}$ is nothing but the 30 bit left rotation of the first round key $k^{(1)}$. In fact, we only guess additional 18 bits. Thus, the number of bits we need to guess for 7 round Camellia is $73 = 58(\text{round } 1, 7) + 8(\text{round } 2) + 7(FL^{-1} \text{ layer})$. When we apply the square attack to 7 round Camellia, we can check only 7 bits of the 6th round outputs. Thus, a wrong key can pass the test with a probability 2^{-7}. We need 11 Λ-sets as input plaintexts to eliminate a wrong key. The algorithm to attack 7 round Camellia can be summarized as follows:

Step 1. Guess 6 byte subkeys $k_1^{(1)}$, $k_2^{(1)}$, $k_3^{(1)}$, $k_5^{(1)}$, $k_8^{(1)}$, and $k_1^{(2)}$ of the first and the second round.

Step 2. Prepare 11 Λ-sets as plaintexts so that inputs of the third round are of the form

$$X_L^{(3)} = (\alpha_1, \alpha_2, \ldots, \alpha_8), \quad X_R^{(3)} = (A, \beta_2, \ldots, \beta_8).$$

Note that the only byte A of them is active. Thus, we expect the right half of the 6th round outputs is balanced, if key guessing is correct.

Step 3. Partially decrypt outputs and test the balance of them.

   3.1. Guess additional 25 bit subkeys for FL^{-1} and the 7th round.

   3.2. Decrypt ciphertexts and determine 7 bits of the right half of the 6th round outputs.

   3.3. Check if this 7 bits are balanced for all 11 Λ-sets.

   3.4. If so, accept 73 bit subkeys as a correct key.

   3.5. Otherwise, discard 25 bit subkeys guessed in Step 3.1 and choose another 25 bits. If all possible 25 bits are checked, go to Step 1 and repeat Step 2 and Step 3.

For each subkey candidate, we need to encrypt 11 Λ-sets, which costs

$$2^{48}(\text{subkeys}) \times 2^8(\Lambda\text{-set size}) \times 11(\text{the number of } \Lambda\text{-sets})$$

encryptions of 7 round. Also, one round decryptions and partial invertings of FL^{-1} function are needed for them with the computational complexity

$$2^{73}(\text{subkeys}) \times 2^8(\Lambda\text{-set size}) \times 11(\text{the number of } \Lambda\text{-sets}).$$

Thus, total amount of cipher execution is approximately $2^{81.7}$ encryptions.

With helpful comments of anonymous referees, this attack could be improved as follows: for a given 6 byte subkeys of the first and the second rounds, first prepare 4 Λ-sets and see whether 2^{25} subkeys for the FL^{-1} and the 7th round pass balanced tests. With probability $1/8$, one of the subkeys can pass these tests. In that case, the remaining 7 Λ-sets can be exercised. This procedure reduced the plaintext cost to $(4 + 7/8)2^{56}$. Also, the time complexity can be reduced to $2^{80.2}$ encryptions.

By assuming all round keys in round 8 and 9, we construct an attack algorithm on 9 round Camellia. One byte of the 8th round key is already guessed in the second round. Therefore, 193 bits of subkey guessing is needed to attack 9 round Camellia. It is of course infeasible but faster than exhaustive key search.

5   Conclusion

We have discussed the security of Camellia against the square attack. We have treated the reduced round Camellia without pre- and post-whitenings including the FL/FL^{-1} layers. The key schedule has been considered to reduce the number of subkey guess and how to treat the FL/FL^{-1} function layers has been presented.

Table 3. Summary of attacks on 256 bit Camellia

Rounds	FL/FL^{-1}	Methods	Plaintexts	Time	Comments
5	N/A	Square Attack	$2^{10.3}$	2^{48}	He & Qing[10](Pre-Whitening)
5	N/A	Square Attack	2^{16}	2^{16}	This paper
6	N/A	Higher Order DC	2^{17}	$2^{19.4}$	Kawabata & Kaneko[11]
6	N/A	Square Attack	$2^{11.7}$	2^{112}	He & Qing[10](Pre-Whitening)
6	N/A	Square Attack	2^{56}	2^{56}	This paper
7	×	Higher Order DC	2^{19}	$2^{61.2}$	Kawabata & Kaneko[11]
7	×	Truncated DC	$2^{82.6}$	192	S. Lee et al.[15]
7	○	Square Attack	$2^{58.3}$	$2^{80.2}$	This paper
8	×	Higher Order DC	2^{20}	2^{126}	Kawabata & Kaneko[11]
8	×	Truncated DC	$2^{83.6}$	$2^{55.6}$	S. Lee et al.[15]
8	○	Square Attack	$2^{59.7}$	$2^{137.6}$	This paper
9	×	Higher Order DC	2^{21}	$2^{190.8}$	Kawabata & Kaneko[11]
9	○	Square Attack	$2^{60.5}$	$2^{202.2}$	This paper
10	×	Higher Order DC	2^{21}	$2^{254.7}$	Kawabata & Kaneko[11]

Table 3 summarizes attacks on 256 bit Camellia by the number of rounds. Time complexities in the table is the number of encryptions.

Up to 9 rounds, the square attack is a faster way to attack Camellia than the brute force key search.

Acknowledgment. We would like to thank anonymous referees for their helpful comments and suggestions. As mentioned at the end of Section 4, we could reduce the plaintext requirement as well as time complexity according to the anonymous advice.

References

1. K. Aoki, T. Ichikawa, M. Kanda, M. Matsui, S. Moriai, J. Nakajima and T. Tokita. *Camellia: A 128–Bit Block Cipher Suitable for Multiple Platforms.* Proceedings of Selected Areas in Cryptography (to appear in LNCS by Springer-Verlag) (2000) 41–54.
2. K. Aoki, T. Ichikawa, M. Kanda, M. Matsui, S. Moriai, J. Nakajima and T. Tokita. *Camellia – A 128-bit Block Cipher.* Technical Report of IEICE, ISEC2000–6 (2000).
3. P. Barreto, V. Rijmen, J. Nakahara Jr., B. Preneel, J. Vandewalle and H. Kim. *Improved Square Attacks against Reduced–Round Hierocrypt.* Proceedings of Fast Software Encryption (to appear in LNCS by Springer-Verlag) (2001) 173–182.
4. E. Biham. *Cryptanalysis of Ladder-DES.* Fast Software Encryption, LNCS 1267, Springer-Verlag (1997) 134–138.
5. Camellia Home Page. http://info.isl.ntt.co.jp/camellia/.
6. J. Daemen, L. R. Knudsen and V. Rijmen. *The Block Cipher SQUARE.* Fast Software Encryption, LNCS 1267, Springer-Verlag (1997) 149–165.
7. J. Daemen and V. Rijmen. *AES Proposal: Rijndael (Document version 2).* AES Submission (1999).
8. N. Ferguson, J. Kelsey, S. Lucks, B. Schneier, M. Stay, D. Wagner and D. Whiting. *Improved Cryptanalysis of Rijndael.* Fast Software Encryption, LNCS 1978, Springer-Verlag (2000) 213–230.
9. C. D'Halluin, G. Bijnens, V. Rijmen and B. Preneel. *Attack on the Six Rounds of CRYPTON.* Fast Software Encryption, LNCS 1636, Springer-Verlag (1999) 46–59.
10. Y. He and S. Qing, *Square Attack on Reduced Camellia Cipher.* ICICS 2001, LNCS 2229, Springer-Verlag (2001) 238–245.
11. T. Kawabata and T. Kaneko. *A Study on Higher Order Differential Attack of Camellia.* the 2nd open NESSIE workshop (2001).
12. M. Kanda and T. Matsumoto. *Security of Camellia against Truncated Differential Cryptanalysis.* Proceedings of Fast Software Encryption (to appear in LNCS by Springer-Verlag) (2001) 298–312.
13. K. Kanda, S. Moriai, K. Aoki, H. Ueda, M. Ohkubo, Y. Takashima, K. Ohta and T. Matsumoto. *A New 128-bit Block Cipher E2.* Technical Report ISEC98-12, The Institute of Electronics, Information and Communication Engineers. (1998).
14. L. R. Knudsen. *Analysis of Camellia.* a contribution for ISO/IEC JTC1 SC27. http://info.isl.ntt.co.jp/camellia/Publications/knudsen.ps (2000).
15. S. Lee, S. Hong, S. Lee, J. Lim and S. Yoon. *Truncated Differential Cryptanalysis of Camellia.* ICISC 2001, (to appear in LNCS by Springer-Verlag) (2001).
16. S. Lucks. *Attacking Seven Rounds of Rijndael under 192–bit and 256–bit Keys.* Proceedings of 3rd AES Conference (2000).
17. S. Lucks. *The Saturation Attack – a Bait for Twofish.* Proceedings of Fast Software Encryption (to appear in LNCS by Springer-Verlag) (2001) 1–15.
18. M. Matsui. *New Block Encryption Algorithm MISTY.* Fast Software Encryption, LNCS 1267, Springer-Verlag (1997) 54–68.
19. M. Sugita, K. Kobara and H. Imai, *Security of Reduced Version of the Block Cipher Camellia against Truncated and Impossible Differential Cryptanalysis.* ASIACRYPT 2001, LNCS 2248, Springer-Verlag (2001) 193–207.
20. Y. L. Yin. *A Note on the Block Cipher Camellia.* a contribution for ISO/IEC JTC1 SC27. http://info.isl.ntt.co.jp/camellia/Publications/yiqun.ps (2000).

Saturation Attacks on Reduced Round Skipjack

Kyungdeok Hwang[1], Wonil Lee[1], Sungjae Lee[2], Sangjin Lee[1], and Jongin Lim[1]

[1] Center for Information and Security Technologies(CIST),
Korea University, Anam Dong, Sungbuk Gu,
Seoul, KOREA
{kdhwang, wonil, sangjin, jilim}@cist.korea.ac.kr
[2] Korea Information Security Agency(KISA)
sjlee@kisa.or.kr

Abstract. This paper describes saturation attacks on reduced-round versions of Skipjack. To begin with, we will show how to construct a 16-round distinguisher which distinguishes 16 rounds of Skipjack from a random permutation. The distinguisher is used to attack on 18(5∼22) and 23(5∼27) rounds of Skipjack. We can also construct a 20-round distinguisher based on the 16-round distinguisher. This distinguisher is used to attack on 22(1∼22) and 27(1∼27) rounds of Skipjack. The 80-bit user key of 27 rounds of Skipjack can be recovered with 2^{50} chosen plaintexts and $3 \cdot 2^{75}$ encryption times.

1 Introduction

In April 1993, the Clinton administration announced a proposed encryption technology that, according to the announcement "will bring the Federal Government together with industry in a voluntary program to improve the security and privacy of telephone communication while meeting the legitimate needs of law enforcement." Subsequently, in July 1993, a more formal announcement appeared in the Federal Register as a request for comments on a proposed Federal Information Processing Standard. The overall approach was initially referred to as Clipper, whereas the specific encryption algorithm is known as Skipjack.

Skipjack is a 64-bit block cipher and was first made public by the NSA in 1998[9,10]. After the publication, several approaches to analysis of Skipjack have been made. The first analysis by Biham et al. [1] studied some of the detailed properties of G and in particular some of the properties of the substitution table S. This provided the first description of some differential and linear cryptanalytic attacks on reduced-round versions of Skipjack. They[2,3] also considered the role of truncated differentials in Skipjack and some variants. Biham et al.[5] presented impossible differential attacks that are faster than exhaustive search for the user key if Skipjack is reduced by at least one round. So far, the attacks[5] are the best known attacks on Skipjack. Knudsen et al.[6] also published a range of attacks on reduced-round variants of Skipjack. They concentrate on the role of truncated differentials and demonstrated the effectiveness of boomerang attacks on Skipjack. But they could not improve on the impossible differential attacks

J. Daemen and V. Rijmen (Eds.): FSE 2002, LNCS 2365, pp. 100–111, 2002.

on the 31 rounds of Skipjack. In addition, most recently Granboulan[7] found several flaws in the differential cryptanlysis of Knudsen et al.

In this paper, we describe saturation attacks on reduced-round versions of Skipjack. Saturation attack[8] is based on the idea of choosing a set of $k \times 2^w$ plaintexts such that each of the 2^w inputs for a w-bit permutation occurs exactly k times. The saturation attack exploits the fact that if the input set for the w-bit permutation is saturated then the output set of the permutation is saturated.

It should be emphasized that our attacks do not improve on the impossible differential attacks[5]. But this paper shows how to apply saturation attack to Skipjack for the first time.

The paper is organized as follows: In Section 2, preliminaries to the text of this paper is presented. The description of Skipjack is briefly given in Section 3. Section 4 explains how to construct a 16- and 20-round distinguisher. In Section 5, we show how to use the 16-round distinguisher to attack on 18(5~22) and 23(5~27) rounds of Skipjack. Moreover, using the 20-round distinguisher we also describe attacks on 22(1~22) and 27(1~27) rounds of Skipjack. Finally, in Section 6 we summarize this paper.

2 Preliminaries

We denote by I_n the set of all n-bit data. In this paper, a word always means a 16-bit data. We denote by $(\alpha^i, \beta^i, \gamma^i, \delta^i)$ an input data of the round i, where each of the Greek small letters is a constant word. By the notation, $(\alpha^{i+1}, \beta^{i+1}, \gamma^{i+1}, \delta^{i+1})$ means an output data of the round i. We denote by $(\mathbf{A}^i, \beta^i, \gamma^i, \delta^i)$ a set of input data of the round i, where $\mathbf{A}^i$ is a subset of I_{16} and each of the Greek small letters is a constant word (i.e.,$(\mathbf{A}^i, \beta^i, \gamma^i, \delta^i) = \{(\alpha^i, \beta^i, \gamma^i, \delta^i) \in I_{16}^4 | \alpha^i \in \mathbf{A}^i\}$). In a similar way, we can also define $(\mathbf{A}^i, \mathbf{B}^i, \gamma^i, \delta^i)$, $(\mathbf{A}^i, \mathbf{B}^i, \mathbf{C}^i, \delta^i)$, etc.

We will use the notion of a multiset to define a "saturated set". A multiset with $k \cdot 2^w$ entries in I_w is said to be "k-saturated " if every value in I_w is found exactly k times in the multiset. If $k = 1$, a saturated multiset is 1-saturated as I_w. From now on, "1-saturated" is shortly said to be "saturated". A set $M(\subseteq I_w)$ is said to be "balanced" if the following equation holds :

$$\bigoplus_{x_i \in M} x_i = 0.$$

Note that if M is a k-saturated multiset then M is balanced.

3 Skipjack

Skipjack[9] is a 64-bit iterated block cipher with 32 rounds of two types, called Rule A and Rule B. Each round is described in the form of a linear feedback shift register with additional non-linear keyed G permutation. Encryption with Skipjack consists of first applying eight rounds of Rule A, then eight rounds of

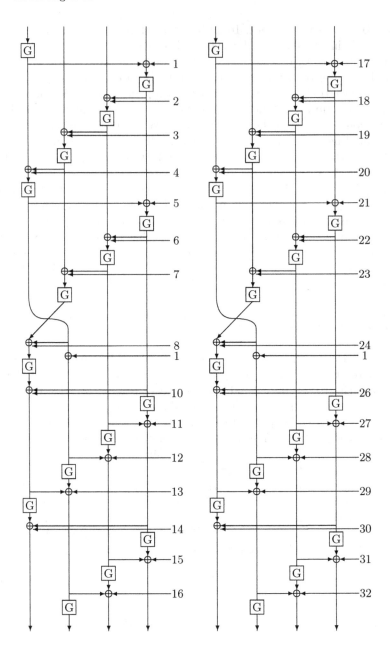

Fig. 1. Skipjack

Rule B, once again eight rounds of Rule A and finally eight rounds of Rule B. The original definitions of Rule A and Rule B are given in table 1 where w_i is a word, *counter* is the round number and G is a four-round Feistel permutation

Table 1. Rule A and B

Rule A	Rule B
$w_1^{k+1} = G^k(w_1^k) \oplus w_4^k \oplus counter^k$	$w_1^{k+1} = w_4^k$
$w_2^{k+1} = G^k(w_1^k)$	$w_2^{k+1} = G^k(w_1^k)$
$w_3^{k+1} = w_2^k$	$w_3^{k+1} = w_1^k \oplus w_2^k \oplus counter^k$
$w_4^{k+1} = w_3^k$	$w_4^{k+1} = w_3^k$

whose F is defined as an 8×8-bit S box, and each round of G is keyed by eight
bits of the key. The key scheduling of Skipjack takes a 10-byte key, and uses four
of them at a time to key each G permutation. The first four bytes are used to
key the first G permutation, and each additional G permutation is keyed by the
next four bytes cyclically, with a cycle of five rounds.

The description of table 1 becomes simpler if we unroll the rounds, and keep
the four words in the shift register stationary. Figure 1 describes this representation of Skipjack. In this paper, the existence of the *counter* is ignored since it
has no cryptanalytic significance in our attack.

4 Distinguishers

It is well known that a good block cipher behaves like a random permutation.
In this section, we describe distinguishers for Skipjack. In other words, given a
well-chosen set of plaintexts, we will find properties in the corresponding set of
ciphertexts, which are unlikely in the case of a random permutation. This holds
for reduced-round versions of Skipjack under arbitrary keys. In the following, we
describe how to construct a 16- and 20-round distinguisher.

4.1 A 16-Round Distinguisher

We describe how to construct a 16-round($5\sim20$) distinguisher. Using this distinguisher, we will show that Skipjack reduced from 32 to 18 rounds and to 23
rounds can be broken by an attack which is faster than exhaustive search.

We concentrate on the 16 rounds of Skipjack starting from round 5 and
ending at round 20 (i.e., without the first four rounds and the last twelve rounds).
For the sake of clarity, we use the original round numbers of the full Skipjack,
i.e., from 5 to 20, rather than from 1 to 16. The 16-round distinguisher is shown
in Figure 2.

Consider a set of 2^{16} plaintexts $(\alpha^5, \mathbf{B}^5, \gamma^5, \delta^5)$ where α^5, γ^5 and δ^5 are
three arbitrary constant words and $\mathbf{B}^5$ is saturated. Then the corresponding
data set after round 7 is $(\alpha^8, \mathbf{B}^8, \gamma^8, \delta^8)$ where α^8, γ^8 and δ^8 are new constant
words and $\mathbf{B}^8$ is saturated. In addition, the corresponding data set after round
9 is $(\mathbf{A}^{10}, \beta^{10}, \gamma^{10}, \delta^{10})$ where $\mathbf{A}^{10}$ is saturated, and β^{10}, γ^{10} and δ^{10} are new
constant words. We also observe that the set of output data of round 13 is
$(\mathbf{A}^{14}, \mathbf{B}^{14}, \gamma^{14}, \delta^{14})$ where $\mathbf{A}^{14}, \mathbf{B}^{14}$ are saturated and the set of output data

of round 17 is $(\mathbf{A}^{18}, \mathbf{B}^{18},\ \mathbf{C}^{18}, \mathbf{D}^{18})$ where $\mathbf{A}^{18}$, $\mathbf{B}^{18}$, $\mathbf{C}^{18}$ and $\mathbf{D}^{18}$ are all saturated. Moreover, the set of output data of round 18 is $(\mathbf{A}^{19}, \mathbf{B}^{19}, \mathbf{C}^{19}, \mathbf{D}^{19})$ where $\mathbf{A}^{19}$, $\mathbf{B}^{19}$ and $\mathbf{D}^{19}$ are saturated, but $\mathbf{C}^{19}$ is generally not saturated since $\mathbf{C}^{19} = \{\gamma^{18} \oplus \delta^{19} | \gamma^{18} \in \mathbf{C}^{18}, \delta^{19} \in \mathbf{D}^{19}\}$ holds. However note that $\mathbf{C}^{19}$ is balanced.

The corresponding data set after round 19 is $(\mathbf{A}^{20}, \mathbf{B}^{20}, \mathbf{C}^{20}, \mathbf{D}^{20})$ where $\mathbf{A}^{20}$ and $\mathbf{D}^{20}$ are saturated, but neither $\mathbf{B}^{20}$ nor $\mathbf{C}^{20}$ is generally saturated. This fact is denoted by $(\mathbf{A}^{20}, ?, ?, \mathbf{D}^{20})$ in Figure 2. The corresponding data set after round 20 which is the last round in the distinguisher is $(\mathbf{A}^{21}, \mathbf{B}^{21}, \mathbf{C}^{21}, \mathbf{D}^{21})$ where $\mathbf{D}^{21}$ is saturated but $\mathbf{A}^{21}$, $\mathbf{B}^{21}$ and $\mathbf{C}^{21}$ are generally not saturated. This fact is denoted by $(?, ?, ?, \mathbf{D}^{21})$ in figure 2. Note that $\mathbf{A}^{20}$, $\mathbf{D}^{21}$ is balanced.

As a result, given any set of 2^{16} plaintexts $(\alpha^5, \mathbf{B}^5, \gamma^5, \delta^5)$ where α^5, γ^5 and δ^5 are three arbitrary constant words and $\mathbf{B}^5$ is saturated, $\mathbf{A}^{20}$ and $\mathbf{D}^{21}$ are saturated and therefore balanced with probability 1. On the other hand, the probability that a random permutation satisfies the property is 2^{-32}. Therefore we can distinguish Skipjack from a random permutation with high probability.

The reason that a saturation attack works on the reduced-round versions of Skipjack can be explained by the fact that $\mathbf{A}^{20}$ and $\mathbf{D}^{21}$ are always balanced in this distinguisher.

4.2   An Extension to 20-Round

We show how to extend the distinguisher from 16 to 20-round. Using the 20-round(1~20) distinguihser, we will show that Skipjack reduced from 32 to 22 rounds and to 27 rounds can be broken by an attack which is faster than exhaustive search.

We concentrate on the 20 rounds of Skipjack starting from round 1 and ending at round 20 (i.e., without the last twelve rounds). The 20-round distinguisher is shown in Figure 3.

The result of this subsection will now be briefly summarized : Given any set of 2^{48} plaintexts $(\mathbf{A}^1, \mathbf{B}^1, \gamma^1, \mathbf{D}^1)$ where $\mathbf{A}^1$, $\mathbf{B}^1$ and $\mathbf{D}^1$ are saturated and γ^1 is a constant word, $\mathbf{A}^{20}$ and $\mathbf{D}^{21}$ are balanced with probability 1. On the other hand, the probability that a random permutation satisfies the property is 2^{-32}.

This result is derived from using the 16-round distinguisher. For the specific explanation of the result, we will partition the set of 2^{48} plaintexts into 2^{32} subsets with 2^{16} elements in the following paragraph. If we perform it, then each of the 2^{32} subsets turns into the form of the input set of the 16-round distinguisher after round 4. Accordingly, for each of the 2^{32} subsets, $\mathbf{A}^{20}$ and $\mathbf{D}^{21}$ are saturated and therefore balanced by the property of the 16-round distinguisher. For that reason, given the set of 2^{48} plaintexts $(\mathbf{A}^1, \mathbf{B}^1, \gamma^1, \mathbf{D}^1)$, $\mathbf{A}^{20}$ and $\mathbf{D}^{21}$ are always 2^{32}-saturated and therefore balanced.

Now we present that the set of the 2^{48} plaintexts can be partitioned. Let 2^{48} plaintexts $(\mathbf{A}^1, \mathbf{B}^1, \gamma^1, \mathbf{D}^1)$ be given as is stated above. We will concentrate on the data set after round 4 to partition the set of the 2^{48} plaintexts. Figure 3 describes this situation. At first, let $\alpha^5, \delta^5 \in I_{16}$ be fixed with any constant words and $\mathbf{B}^5 = I_{16}$(saturated set). Then γ^5 is determined by the equation $\mathrm{G}(\gamma^1 \oplus$

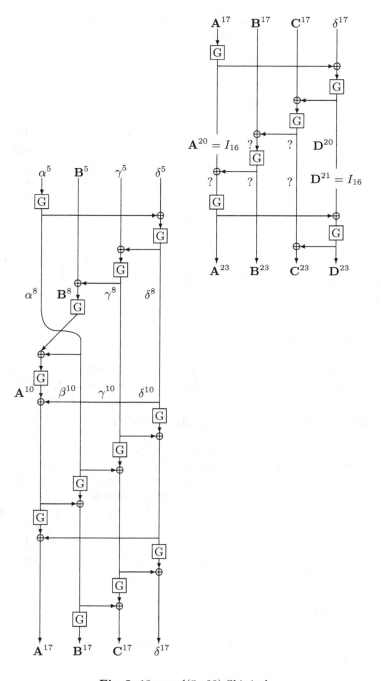

Fig. 2. 18-round(5∼22) Skipjack

$\delta^5) = \gamma^5$. In addition, for each $\beta_i^5 \in \mathbf{B}^5$, the corresponding tuple $(\alpha_i^1, \beta_i^1, \delta_i^1)$ is determined by the following equations.

$$\alpha_i^1 = \mathbf{G}^{-1}(\alpha^5 \oplus \beta_i^5)$$
$$\beta_i^1 = \mathbf{G}^{-1}(\beta_i^5) \oplus \gamma^5$$
$$\delta_i^1 = (\alpha^5 \oplus \beta_i^5) \oplus \mathbf{G}^{-1}(\delta^5)$$

If $i \neq j$ then $(\alpha_i^1, \beta_i^1, \delta_i^1) \neq (\alpha_j^1, \beta_j^1, \delta_j^1)$ holds since G is permutation. Therefore, if α^5 and $\delta^5 \in I_{16}$ are fixed and $\mathbf{B}^5$ is saturated then the number of the corresponding set $\{(\alpha_i^1, \beta_i^1, \gamma^1, \delta_i^1) | 0 \leq i \leq 2^{16} - 1\}$ is 2^{16}. Conversely, if the set $\{(\alpha_i^1, \beta_i^1, \gamma^1, \delta_i^1) | 0 \leq i \leq 2^{16} - 1\}$ is given, the corresponding set after round 4 is $(\alpha^5, \mathbf{B}^5, \gamma^5, \delta^5)$, where $\mathbf{B}^5$ is saturated, i.e., the input form of the 16-round distinguisher which is presented in section 4.1.

What's more, note that the fixed pair (α^5, δ^5) can be any one of 2^{32} elements. So, there are 2^{32} disjoint subsets of $\{(\alpha_i^1, \beta_i^1, \gamma^1, \delta_i^1) | 0 \leq i \leq 2^{16} - 1\}_{0 \leq j \leq 2^{32}-1}$. By the explanation, we can easily obtain the fact that $(\mathbf{A}^1, \mathbf{B}^1, \gamma^1, \mathbf{D}^1)$ can be partitioned into the 2^{32} disjoint subsets of $\{(\alpha_i^1, \beta_i^1, \gamma^1, \delta_i^1) | 0 \leq i \leq 2^{16} - 1\}_{0 \leq j \leq 2^{32}-1}$.

5   Saturation Attack

In this section, we use the distinguishers which is presented in section 4.1 and 4.2 to recover the user keys of the reduced-round versions of Skipjack.

5.1   Attack with the 16-Round Distinguisher

Attack on 18-round(5~22) Skipjack. It will be shown that we can recover K_{21} and K_{22} of the 18-round Skipjack(which is describe in Figure 2) using the 16-round distinguisher, where K_{21} and K_{22} are the subkeys of the round 21 and 22, respectively. Note that the attack is a chosen plaintext attack.

Let a set of 2^{16} plaintexts, $\mathsf{P} = (\alpha^5, \mathbf{B}^5, \gamma^5, \delta^5)(= \{(\alpha^5, \beta_i^5, \gamma^5, \delta^5) | 0 \leq i \leq 2^{16} - 1\})$ be chosen as required for the 16-round disginguisher. And ask for the corresponding set of ciphertexts, $\mathsf{C} = (\mathbf{A}^{23}, \mathbf{B}^{23}, \mathbf{C}^{23}, \mathbf{D}^{23})(= \{(\alpha_i^{23}, \beta_i^{23}, \gamma_i^{23}, \delta_i^{23}) | 0 \leq i \leq 2^{16} - 1\})$. Then the following equations hold with probability 1 by the property of the 16-round distinguisher.

$$\bigoplus_{0 \leq i \leq 2^{16}-1} \alpha_i^{20} = \bigoplus_{0 \leq i \leq 2^{16}-1} \mathbf{G}_{K_{21}}^{-1}(\alpha_i^{23}) \oplus \beta_i^{23} = 0, \quad K_{21} \in I_{32} \tag{1}$$

$$\bigoplus_{0 \leq i \leq 2^{16}-1} \delta_i^{21} = \bigoplus_{0 \leq i \leq 2^{16}-1} \alpha_i^{23} \oplus \mathbf{G}_{K_{22}}^{-1}(\delta_i^{23}) = 0, \quad K_{22} \in I_{32} \tag{2}$$

If K_{21} is the right subkey, it always satisfies the equation (1). While on the other, arbitrary subkey satisfies (1) with probability 2^{-16}. Since the length of the subkey is 32-bit, the number of subkeys satisfying (1) is about 2^{16}. Thus we need the other set of plaintexts to find the right subkey. For the about 2^{16}

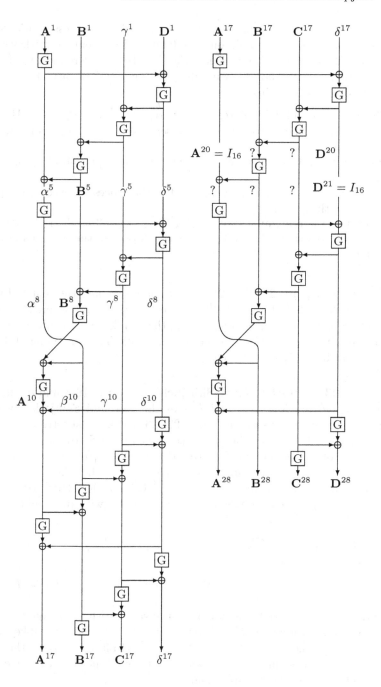

Fig. 3. 27-round(1∼27) Skipjack

remaining candidate subkeys, if we perform the above process again with the other set of plaintexts, the only one candidate subkey remains with very high probability. Hence the subkey is almost the right key of the round 21. Also using the same way, we can find K_{22} with the equation (2). This attack can now be summarized in the following :

1. Choose two sets of 2^{16} plaintexts, $\mathsf{P}_1 = (\alpha_1^5, \mathbf{B}^5, \gamma_1^5, \delta_1^5)$, $\mathsf{P}_2 = (\alpha_2^5, \mathbf{B}^5, \gamma_2^5, \delta_2^5)$ as required for the 16-round distinguihser. Ask for the corresponding sets of ciphertexts, C_1 and C_2.
2. For each candidate subkey of K_{21}, calculate the equation (1) using P_1 and C_1.
3. For the remaining candidate subkeys after the process **2**, execute the process **2** again with P_2 and C_2.
4. Determine the remaining subkey after the process **3** as the right key.
5. Using the equation (2), find the right key K_{22} as in the previous process **2**, **3** and **4**.

The attack requires $2 \cdot 2^{16} = 2^{17}$ chosen plaintexts and the required work is about $2^{44} \simeq 2(2^{16} \cdot 2^{32} \cdot 2^{-5} + 2^{16} \cdot 2^{16} \cdot 2^{-5})$ encryption times where 2^{-5} means a G operation of Skipjack encryption. Using the simple key schedule of Skipjack, we can directly find 64 bits of all the user key bits. The remaining 16 bits can also be found by exhaustive search.

Attack on 23-round(5~27) of Skipjack. Using the 16-round distinguisher we can also recover K_{22}, K_{26} and K_{27} of the 23-round Skipjack where K_{22}, K_{26} and K_{27} are subkeys of the round 22, 26 and 27, respectively. Note that $K_{22} = K_{27}$ holds because of the simple key schedule of Skipjack.

Let a set of 2^{16} plaintexts, $\mathsf{P} = (\alpha^5, \mathbf{B}^5, \gamma^5, \delta^5)(= \{(\alpha^5, \beta_i^5, \gamma^5, \delta^5)|0 \le i \le 2^{16}-1\})$ be chosen as required for the 16-round distinguisher. And ask for the corresponding set of ciphertexts, $\mathsf{C} = (\mathbf{A}^{28}, \mathbf{B}^{28}, \mathbf{C}^{28}, \mathbf{D}^{28})(= \{(\alpha_i^{28}, \beta_i^{28}, \gamma_i^{28}, \delta_i^{28})| 0 \le i \le 2^{16}-1\})$. Then the following equation holds with probability 1 by the property of the 16-round distinguisher.

$$\bigoplus_{0 \le i \le 2^{16}-1} \delta_i^{21} = \bigoplus_{0 \le i \le 2^{16}-1} \beta_i^{28} \oplus G_{K_{22}}^{-1}(G_{K_{26}}^{-1}(G_{K_{27}}^{-1}(\gamma_i^{28}) \oplus \delta_i^{28})) = 0 \qquad (3)$$

$$K_{22}, K_{26}, \text{ and } K_{27} \in I_{32} \quad (K_{22} = K_{27})$$

If (K_{22}, K_{26}) is the right subkey pair, it always satisfies the equation (3). Of course, K_{27} is determined by K_{22}. The probability that arbitrary subkey satisfies (3) is 2^{-16}. Since the length of the subkey pair is 64-bit at this time, the number of subkey pairs satisfying (3) is about 2^{48}. Thus we need other three sets of plaintexts to find the right subkey pair.

So, the attack requires $4 \cdot 2^{16} = 2^{18}$ chosen plaintexts and about $3 \cdot 2^{75} \simeq 2^{16} \cdot 2^{64} \cdot \frac{3}{2^5} + 2^{16} \cdot 2^{48} \cdot \frac{3}{2^5} + 2^{16} \cdot 2^{32} \cdot \frac{3}{2^5} + 2^{16} \cdot 2^{16} \cdot \frac{3}{2^5}$ encryption times. We can find 64 bits of all the user key bits using the key schedule of Skipjack. The remaining 16 bits can also be found by exhaustive search.

5.2   Attack with the 20-Round Distinguisher

Attack on 22-Round(1~22) of Skipjack. If we use the 20-round distinguisher for attacking the 22-round Skipjack, we are able to recover K_{21} and K_{22} where K_{21} and K_{22} are the subkeys of the round 21 and 22, respectively.

Let a set of 2^{48} plaintexts, $\mathsf{P} = (\mathbf{A}^1, \mathbf{B}^1, \gamma^1, \mathbf{D}^1)(= \{(\alpha_i^1, \beta_i^1, \gamma^1, \delta_i^1)|0 \le i \le 2^{48}-1\})$ be chosen as required for the 20-round distinguisher. And ask for the corresponding set of ciphertexts, $\mathsf{C} = (\mathbf{A}^{23}, \mathbf{B}^{23}, \mathbf{C}^{23}, \mathbf{D}^{23})(= \{(\alpha_i^{23}, \beta_i^{23}, \gamma_i^{23}, \delta_i^{23})| 0 \le i \le 2^{48} - 1\})$. Then the following equations always hold by the property of the 20-round distinguisher. And subkey finding method is the same as in the attack on the 18-round(5~22) Skipjack.

$$\bigoplus_{0 \le i \le 2^{48}-1} \alpha_i^{20} = \bigoplus_{0 \le i \le 2^{48}-1} \mathrm{G}_{K_{21}}^{-1}(\alpha_i^{23}) \oplus \beta_i^{23} = 0, \quad K_{21} \in I_{32} \tag{4}$$

$$\bigoplus_{0 \le i \le 2^{48}-1} \delta_i^{21} = \bigoplus_{0 \le i \le 2^{48}-1} \alpha_i^{23} \oplus \mathrm{G}_{K_{22}}^{-1}(\delta_i^{23}) = 0, \quad K_{22} \in I_{32} \tag{5}$$

Since the length of a word is 16-bit and the number of the set C is 2^{48}, for each 16-bit data channel, there must be words appeared repeatedly in the channel. For each possible word w, let the number of repetition that the word w appears in the channel be denoted by num. Then the following property holds.

$$\bigoplus_{num} w \triangleq \underbrace{w \oplus \cdots \oplus w}_{num} = \begin{cases} 0 & \text{if } num \text{ is even} \\ w & \text{if } num \text{ is odd} \end{cases} \tag{6}$$

In principle, for each candidate subkey, we need 2^{48} operations of G^{-1} function to calculate the equation (4). But using the property (6) we can reduce the complexity of the subkey finding method in the following way.

To begin with, for each possible α_i^{23}, we examine the number of repetition that the word α_i^{23} appears in the data channel. Then by this result and the property (6), we can calculate $\bigoplus_{0 \le i \le 2^{48}-1} \mathrm{G}_{K_{21}}^{-1}(\alpha_i^{23})$ with at most 2^{16} operations of G^{-1}. The summing up of this attack can be shown in the following :

1. Choose two sets of 2^{48} plaintexts, $\mathsf{P}_1 = (\mathbf{A}^1, \mathbf{B}^1, \gamma_1^1, \mathbf{D}^1)$, $\mathsf{P}_2 = (\mathbf{A}^1, \mathbf{B}^1, \gamma_2^1, \mathbf{D}^1)$ as required for the 20-round distinguisher. Ask for the corresponding sets of ciphertexts, C_1 and C_2.
2. For each candidate subkey of K_{21}, evaluate the equation (4) using P_1 and C_1 (At this time, use the property (6) to reduce the complexity of the calculation of the equation (4)).
3. For the remaining candidate subkeys after the process **2**, perform the process **2** again with P_2 and C_2.
4. Determine the remaining subkey after the process **3** as right key.
5. Using the equation (5), find the right key K_{22} as in the previous process **2, 3** and **4**.

The attack requires $2 \cdot 2^{48} = 2^{49}$ chosen plaintexts and about $2(2^{16} \cdot 2^{32} \cdot 2^{-5} + 2^{16} \cdot 2^{16} \cdot 2^{-5}) \simeq 2^{44}$ encryption times. We can directly find 64 bits of all the user key bits using the key schedule. The remaining 16 bits can be found by exhaustive search.

Attack on 27-round(1~27) of Skipjack. Using the 20-round distinguisher we can also recover K_{22}, K_{26} and K_{27} of the 27-round Skipjack(which is describe in Figure 3) where K_{22}, K_{26} and K_{27} are subkeys of the round 22, 26 and 27, respectively. Note that $K_{22} = K_{27}$ holds.

Let a set of 2^{48} plaintexts, $\mathsf{P} = (\mathbf{A}^1, \mathbf{B}^1, \gamma^1, \mathbf{D}^1)(= \{(\alpha_i^1, \beta_i^1, \gamma^1, \delta_i^1)|0 \le i \le 2^{48}-1\})$ be chosen as required for the 20-round distinguisher. And ask for the corresponding set of ciphertexts, $\mathsf{C} = (\mathbf{A}^{28}, \mathbf{B}^{28}, \mathbf{C}^{28}, \mathbf{D}^{28})(= \{(\alpha_i^{28}, \beta_i^{28}, \gamma_i^{28}, \delta_i^{28})| 0 \le i \le 2^{48} - 1\})$. Then the following equation always holds by the property of the 20-round distinguisher. Subkey finding method is also the same as in the attack on the 23-round(5~ 27) Skipjack.

$$\bigoplus_{0 \le i \le 2^{48}-1} \delta_i^{21} = \bigoplus_{0 \le i \le 2^{48}-1} \beta_i^{28} \oplus \mathrm{G}_{K_{22}}^{-1}(\mathrm{G}_{K_{26}}^{-1}(\mathrm{G}_{K_{27}}^{-1}(\gamma_i^{28}) \oplus \delta_i^{28})) = 0 \qquad (7)$$

$$K_{22}, K_{26}, \text{ and } K_{27} \in I_{32} \quad (K_{22} = K_{27})$$

The attack requires $4 \cdot 2^{48} = 2^{50}$ chosen plaintexts and about $2^{16} \cdot 2^{64} \cdot \frac{3}{2^5} + 2^{16} \cdot 2^{48} \cdot \frac{3}{2^5} + 2^{16} \cdot 2^{32} \cdot \frac{3}{2^5} + 2^{16} \cdot 2^{16} \cdot \frac{3}{2^5} \simeq 3 \cdot 2^{75}$ encryption times. Using the key schedule, we can find 64 bits of the user key. The remaining 16 bits can be found by exhaustive search.

Table 2. Complexities of Saturation Attacks Against Reduced-Round Skipjack

rounds	plaintexts	running time
18(5~22)	2^{17}	2^{44}
22(5~26)	2^{18}	2^{76}
23(5~27)	2^{18}	$3 \cdot 2^{75}$
22(1~22)	2^{49}	2^{44}
26(1~27)	2^{50}	2^{76}
27(1~27)	2^{50}	$3 \cdot 2^{75}$

6   Conclusion

In this paper we have described saturation attacks on reduced-round versions of Skipjack. We have showed how to construct a 16-round distinguisher. The distinguisher can be used to attack on 18(5~22) and 23(5~27) rounds of Skipjack.

We could also construct a 20-round distinguisher based on the 16-round distinguisher. This distinguisher can be used to attack on 22(1~22) and 27(1~27) rounds of Skipjack. The complexities of these attacks are summarized in table 2. It should be emphasized that our attacks do not improve on the impossible differential attacks[5]. But this paper shows how to apply saturation attack to Skipjack for the first time.

Acknowledgment. We would like to thank Seokhie Hong and Jaechul Sung for many helpful discussions.

References

1. E. Biham, A. Biryukov, O. Dunkelmann, E. Richardson and A. Shamir, *Initial Observations on the Skipjack Encryption Algorithm*, June 25, 1998. Available at http://www.cs.technion.ac.il/ ~biham/Reports/Skipjack/.
2. E. Biham, A. Biryukov, O. Dunkelmann, E. Richardson and A. Shamir, *Cryptanalysis of Skipjack-3XOR in 2^{20} time and using 2^9 chosen plaintexts*, July 2, 1998. Available at http:// www.cs.technion.ac.il/~biham/Reports/Skipjack/.
3. E. Biham, A. Biryukov, O. Dunkelmann, E. Richardson and A. Shamir, *Cryptanalysis of Skipjack-4XOR*, June 30, 1998. Available at http://www.cs. technion.ac.il/~biham/Reports/Skipjack/.
4. E. Biham, A. Biryukov, and A. Shamir, *Initial Observations on the Skipjack: Cryptanalysis of Skipjack-3XOR*, SAC'98, 1998. Available at http://www.cs. technion.ac.il/~biham/Reports/ Skipjack/.
5. Eli Biham, Alex Biryukov, and Adi Shamir, *Cryptanalysis of Skipjack reduced to 31 rounds using impossible differentials*, EUROCRYPT'99, LNCS 1592, Springer-Verlag, 1999 pp. 12–23. Available at
   http://www.cs.technion.ac.il/~biham/Reports/ Skipjack/.
6. Lars R. Knudsen, M.J.B. Robshaw, and David Wagner, *Truncated differentials and Skipjack*, CRYPTO'99, LNCS 1666, Springer-Verlag, August 1999 pp. 165–180.
7. Louis Granboulan, *Flaws in differential Cryptanalysis of Skipjack*, Fast Software Encryption Workshop 2001, LNCS 1039, Springer-Verlag April, 2001, pp. 341–346.
8. S. Lucks, *The Saturation Attack – a Bait for Twofish*, Fast Software Encryption Workshop 2001, LNCS 1039, Springer-Verlag, 2001, pp. 189–203
9. National Institute of Standards and Technology, *Skipjack and KEA Algorithm Specifications, version 2.0*, Available at http://crsc.nist.gov/encryption/skipjack-kea.htm.
10. National Institute of Standards and Technology, *NSA Releases Fortezza Algorithms*, Available at http://crsc.nist.gov/encryption/encryption/nsa-press.pdf.

Integral Cryptanalysis
(Extended Abstract)

Lars Knudsen[1]* and David Wagner[2]

[1] Dept. of Mathematics, DTU, Building 303, DK-2800 Lyngby, Denmark
lars@ramkilde.com
[2] University of California Berkeley, Soda Hall, Berkeley, CA 94720, USA
daw@cs.berkeley.edu

Abstract. This paper considers a cryptanalytic approach called integral cryptanalysis. It can be seen as a dual to differential cryptanalysis and applies to ciphers not vulnerable to differential attacks. The method is particularly applicable to block ciphers which use bijective components only.

Keywords: Cryptanalysis, block ciphers, integrals, MISTY.

1 Introduction

The last three decades have seen considerable progress in understanding the basic operating principles of block ciphers. One of the most significant advances was the introduction in 1990 of differential cryptanalysis [3]. In differential cryptanalysis, one considers the propagation of differences between (pairs of) values.

In this paper, we consider a cryptanalytic technique which considers the propagation of sums of (many) values. This approach can thus be seen as a dual to differential cryptanalysis [3]. A number of these ideas have been exploited before in specific scenarios, but in this paper we unify and extend previous work in a single consistent framework, and we propose the name *integral cryptanalysis* for this set of techniques.

Integrals have a number of interesting features. They are especially well-suited to analysis of ciphers with primarily bijective components. Moreover, they exploit the simultaneous relationship between many encryptions, in contrast to differential cryptanalysis where one considers only pairs of encryptions. Consequently, integrals apply to a number of ciphers not vulnerable to differential cryptanalysis. These features have made integrals an increasingly popular tool in recent cryptanalysis work, and this motivates our systematic study of integrals.

We begin by formulating integral cryptanalysis in a general group-theoretic setting and develop a consistent notation for expressing integral attacks. We also

---

* Part of this author's work was done while visiting University of California San Diego on leave from the Department of Informatics, University of Bergen, Norway supported by the Norwegian Research Council

J. Daemen and V. Rijmen (Eds.): FSE 2002, LNCS 2365, pp. 112–127, 2002.

Table 1. Summary of some of our cryptanalytic results. For MISTY, all results are key-recovery attacks of the full cipher (including the *FL* functions). "Gen. Feistel" are key-recovery attacks of the generalised Feistel networks [26] with 64-bit blocks and bijective 8-bit S-boxes. All attacks use chosen plaintexts.

Cipher	(rounds)	Complexity [Data]	[Time]	Comments
MISTY1	(4)	2^{20}	2^{89}	see [19] (previously known)
MISTY1	(4)	$2^{22.25}$	2^{45}	see [20] (previously known)
MISTY1	(4)	2^{38}	2^{62}	see [19] (previously known)
MISTY1	(4)	25	2^{27}	integrals (new)
MISTY1	(5)	2^{34}	2^{48}	integrals (new)
MISTY2	(5)	2^{20}	2^{89}	see [19] (previously known)
MISTY2	(5)	2^{38}	2^{62}	see [19] (previously known)
MISTY2	(4)	9	2^{55}	integrals (new)
MISTY2	(6)	2^{34}	2^{71}	integrals (new)
Gen. Feistel	(13)	$2^{9.6}$	2^{32}	basic integral (new)
Gen. Feistel	(14)	$2^{10.6}$	2^{56}	basic integral (new)
Gen. Feistel	(14)	2^{16}	2^{24}	second-order integral (new)
Gen. Feistel	(15)	$2^{17.6}$	2^{40}	second-order integral (new)
Gen. Feistel	(16)	$2^{18.6}$	2^{64}	second-order integral (new)
Gen. Feistel	(16)	$2^{33.6}$	2^{56}	fourth-order integral (new)
Gen. Feistel	(17)	$2^{34.6}$	2^{80}	fourth-order integral (new)
Gen. Feistel	(17)	$2^{49.6}$	2^{72}	sixth-order integral (new)

introduce an important extension to previous work, the *higher-order integral attack*. See Section 2.

In the main body of the paper, we first explain the well-known attacks on Square, Rijndael, Crypton (see Section 3), using our new concepts and notation, then we apply these techniques to a number of other ciphers: MISTY (Section 4), Nyberg's generalized Feistel networks (Section 5), see Table 1 for a summary of some of these results. Many of these attacks illustrate our new notion of higher-order integrals and its utility for cryptanalysis. Finally, we discuss how to extend these techniques to non-word-oriented ciphers (Section 6) and how to combine integrals with interpolation attacks (Section 7), we draw attention to some related work (Section 8), and we conclude the paper (Section 9). Due to page constraints we did not include our results on Skipjack. These can be found in the full version of this paper.

In this paper, a time complexity of n means that the time of an attack corresponds to performing n encryptions of the underlying block cipher.

2   Fundamentals of Integral Cryptanalysis

Let $(G, +)$ be a finite abelian group of order k. Consider the product group $G^n = G \times \ldots \times G$, that is, the group with elements of the form $v = (v_1, \ldots, v_n)$ where $v_i \in G$. The addition of G^n is defined component-wise, so that $u + v = w$ holds for $u, v, w \in G^n$ just when $u_i + v_i = w_i$ for all i.

Let S be a multiset of vectors. An integral over S is defined as the sum of all vectors in S. In other words, the integral is $\int S = \sum_{v \in S} v$, where the summation is defined in terms of the group operation for G^n. (For a multiplicative group this would usually be called a "product" instead.)

In integral cryptanalysis, n will represent the number of words in the plaintext and ciphertexts, and m denotes the number of plaintexts and ciphertexts considered (at a time). Typically, $m = k$ (recall that k is defined as the order of G, i.e., $k = |G|$), the vectors $v \in S$ represent the plaintext and ciphertexts, and $G = GF(2^s)$ or $G = Z/kZ$. In an attack, one tries to predict the values in the integrals after a certain number of rounds of encryption. For this purpose it is advantageous to distinguish between the three cases: where all ith words are equal; are all different; or sum to a certain value predicted in advance. Let $S \subseteq G^n$ be as before, and consider some fixed index i. We consider these cases.

$$v_i = c \qquad \text{for all } v \in S \tag{1}$$
$$\{v_i : v \in S\} = G \tag{2}$$
$$\sum_{v \in S} v_i = c' \tag{3}$$

where $c, c' \in G$ represent some known values that are fixed in advance.

Let us consider the typical case where $m = k$, that is, the number of vectors in the set S equals the number of elements in the considered group. If all ith words are equal then clearly the ith word in the integral will take the value of the neutral element of G (Lagrange's theorem). Furthermore, there is a result from group theory which allows us to predict the integral in the case when all ith words are different; it allows us to characterize the sum of all elements in G [11, Problem 2.1, p. 116].

Theorem 1. *Let $(G, +)$ be a finite abelian additive group, and let $H = \{g \in G : g + g = 0\}$ be the subgroup of elements of order 1 or 2. Write $s(G)$ for the sum $\sum_{g \in G} g$ of all the elements of G. Then $s(G) = \sum_{h \in H} h$. Moreover $s(G) \in H$, i.e., $s(G) + s(G) = 0$.*

Thus for $G = GF(2^s)$ we get $s(G) = 0$ and for Z/mZ we get $s(Z/mZ) = m/2$ when m is even or 0 when m is odd. There is an analogue for multiplicatively written groups.

Theorem 2. *Let $(G, *)$ be a finite abelian multiplicative group and let $H = \{g \in G : g * g = 1\}$ be the subgroup of elements of order 1 or 2. Write $p(G)$ for the product $\prod_{g \in G} g$ of all of the elements of G. Then $p(G) = \prod_{h \in H} h$. Moreover $p(G) \in H$, i.e., $p(G) * p(G) = 1$.*

For example, when $G = (Z/pZ)^*$ where p is prime, $p(G) = -1$ (Wilson's theorem).

Thus, in all the above three cases (1), (2), and (3) we have a tool to predict the value of the sum of all words.

In differential cryptanalysis over a group G, one typically considers differences defined in terms of the subtraction or division, e.g. $dx = x' - x$ for an additive group or $dx = x' \cdot x^{-1}$ for a multiplicative group. We claim that the right operation for integrals is addition or multiplication.

Suppose the cipher computes $w_j = u_j + v_j$ where u_j, v_j, w_j are intermediate values. Suppose also the integral predicts that the words u_j and v_j are of the forms (1), (2), or (3). What can we say about the words w_j? We can at least say that $\sum w_j = \sum u_j + \sum v_j$, where the sum is taken over some set of encryptions. Thus, if the sum of the words u_j and v_j are known, the sum of the words w_j can be determined. Moreover, if the words u_j are all equal and the words v_j are all different, then the words w_j are all different, and so on.

A good cipher also contains nonlinear components, or nonlinear S-boxes. Assume that at some point in the cipher the function f is applied to a word, i.e., $v_j = f(u_j)$. Clearly, if the words u_j are all equal (of the form (1)), then so are the words v_j. Also, if f is a permutation (bijection) and if the words u_j are all different (of the form (2)), then so are the words v_j.

By analogy to higher-order differentials (see next section), we define higher-order integrals. Consider a set $\tilde{S} = S_1 \cup \ldots \cup S_s$ made up of s sets of vectors, where each S_i forms an integral. Then clearly, if one can determine the sum of the elements of S_i for each i, then one can also determine the sum of all vectors in $\tilde{S}$. Suppose the words in a cipher can take m values each. Consider a set of m vectors (representing a set of plaintexts) which differ only in one particular word. The sum over the vectors of this set is called a first-order integral. Consider next a set of m^d vectors which differ in d components, such that each of the m^d possible values for the d-tuple of values from these components occurs exactly once. The sum of this set is called an dth-order integral.

Let us introduce the following symbols for words in an integral. For a first-order integral, the symbol '$\mathcal{C}$' (for "Constant") in the ith entry, means that the values of all ith words in the collection of texts are equal. The symbol '$\mathcal{A}$' (for "All") means that all words in the collection of texts are different, and the symbol '$\mathcal{S}$' (for "Sum") means that the sum of all ith words can be predicted. Finally, we will write '?' when the sum of words can not be predicted.

For dth-order integrals we use $\mathcal{C}$ and ? as before, and we use the notation $\mathcal{A}^d$ to denote that the corresponding component participates in a dth-order integral. If we assume that one word can take m different values, then $\mathcal{A}^d$ means that in the integral the particular word takes all values exactly m^{d-1} times. We shall use $\mathcal{A}$ as a short notation for $\mathcal{A}^1$. To express further the interdependencies between particular words we introduce the following notation: The terms $\mathcal{A}_i^d$ mean that in the integral the string concatenation of all words with subscript i take the m^d values exactly once.

Integrals can be probabilistic just like differentials [3]. However, all integrals for the specific ciphers given in this paper are of probability one.

Comparison with other Concepts

First we note that integrals are somewhat similar to truncated differentials [16, 15,18]. In the latter, one often is only interested in whether the words in a pair are equal or different [2]. Thus integrals restricted to pairs of texts with only the values 0 and $\mathcal{A}$ coincide with such truncated differentials. Integrals, though, can also represent texts with the value $\mathcal{S}$; truncated differentials cannot, which may make integrals a more powerful tool in some cases.

Also, integrals are somewhat similar to higher-order differentials. Let $(G, +)$ be an Abelian group. For a function $f : G \to G$ the first-order derivative [21] at the point a is defined as $f_a(x) = f(x + a) - f(x)$. This is the definition of a differential or characteristic that is traditionally used in cryptanalysis. One can extend the definition of differentials to higher orders. One defines [21] the ith-order derivative of f at the point $a_1, \ldots, a_i$ as follows:

$$f_{a_1,\ldots,a_i}(x) = f_{a_i}(f_{a_1,\ldots,a_{i-1}}(x)). \tag{4}$$

As an example, a third-order derivative is:

$$\begin{aligned} f_{a,b,c}(x) = f(x + a + b + c) - f(x + a + b) - f(x + b + c) - f(x + a + c) + \\ f(x + a) + f(x + b) + f(x + c) - f(x) \end{aligned}$$

Thus for general groups the higher-order derivatives (or higher-order differentials) are not the same as integrals, since in an integral one would consider the sum of all elements in a set. In groups of characteristic two, an sth-order differential is the exclusive-or of all 2^s different words, and therefore also an integral. But where for integrals one distinguishes between the three cases (1), (2), and (3), for higher-order differentials only the value of (4) is used.

Higher-order differentials have traditionally been used in cryptanalytic attacks on ciphers which consist of subfunctions with a low algebraic degree. For groups with characteristic two it holds that an sth-order differential of a function of algebraic degree s is a constant (and consequently an $(s + 1)$st-order differential of a function of algebraic degree s is zero).

To sum up, in some cases integrals contain both truncated and higher-order differentials, but there are cases where integrals can be specified for more rounds than either of the other two. On the other hand, in contrast to truncated and higher-order differentials, integrals do not seem to apply as well to ciphers using non-bijective S-boxes/subcomponents.

In the remainder of this paper we give examples of integrals for a variety of ciphers.

3   Square, Rijndael, and Crypton

In *FSE'97* an integral attack was given on the block cipher Square [5]. This attack can be applied also to the ciphers Rijndael [7,9] and Crypton [8]. All three ciphers are 128-bit block ciphers operating on bytes. The sixteen bytes are arranged in a 4×4 matrix. One round of the ciphers consists of the addition of a subkey, a substitution of each byte, and a linear transformation, MixColumn, which modifies the four bytes in a column in the matrix. In the following we shall apply integral cryptanalysis to Rijndael only. Due to the similarity between these

Table 2. A 3-round (first-order) integral for Rijndael, where $\mathcal{S} = 0$

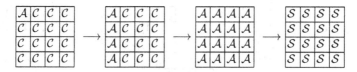

three ciphers, the attack applied to the other two ciphers is quite similar to the attack on Rijndael [8].

Consider a collection of 256 texts, which have different values in one byte and equal values in all other bytes. Then it follows that after two rounds of encryption the texts take all 256 values in each of the sixteen bytes, and that after three rounds of encryption the sum of the 256 bytes in each position is zero [5]. Also, note that there are 16 such integrals since the position of the non-constant byte in the plaintexts can be in any of the sixteen bytes. The integral is illustrated in Table 2. This integral can be used to attack four rounds of Rijndael (or Square or Crypton) with small complexity (note that the final round is special and does not include MixColumn) counting over one key byte at a time. Simply guess a key byte and compute byte-wise backwards to check if the sum of all 256 values is zero.

The attack can be extended to five rounds using the same integral over the first three rounds. Guess one key byte in the fifth round and four in the fourth round, in total five key bytes at a time. The attack can be further extended to six rounds using the same integral as above but used now from the second round and onwards. Here one chooses a collection of 2^{32} plaintexts, such that for each guess of four key bytes in the first round, one can find a collection of 256 ciphertexts after one round of encryption which form an integral. Guess further one key byte in the sixth round and four in the fifth round, in total nine bytes.

The following observation was made which led to an improvement in the running time of the attack [9]. Instead of guessing four key bytes in the first round, one uses all 2^{32} texts in the analysis. The main observation is that the 2^{32} plaintexts together form 2^{24} copies of the above integrals (starting in the second round). Since the text in each integral sums to zero in any byte after the fourth round, so does the sum of all 2^{32} texts. This attack finds less key bits than the original Square attack of *FSE'97* [5], but the running time is greatly improved. (The improvement is in the key-search part of the attack). Table 3 depicts this four-round fourth-order integral.

4 MISTY

Integrals can be used to attack some reduced-round variants of Matsui's MISTY1 and MISTY2 [24]. We refer to the MISTY specifications [24] for the description of these ciphers and for the notation used in the following.

In earlier work, Sakurai and Zheng noted the following property of the MISTY2 round function [28]. Let $F(x, y)$ denote the left half of the output of

Table 3. A four-round fourth-order integral for Rijndael with 2^{32} texts.

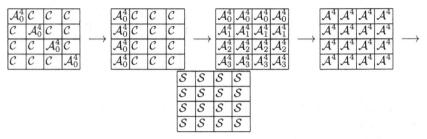

three rounds of MISTY2 on plaintext $\langle x, y \rangle$. They observe that F has the form $F(x, y) = f(x) \oplus g(y)$, where f and g are some key-dependent bijective mappings. Consequently, if we pick sets S, T each containing two arbitrary 32-bit values, then we will have

$$\sum_{\langle x,y \rangle \in S \times T} F(x, y) = 0. \tag{5}$$

We note that this may be viewed as a three-round integral for MISTY2.

This provides an efficient chosen-plaintext attack on four rounds of MISTY2. Choosing S', T' with $|S'| = |T'| = 3$ gives us four independent ways to choose S, T with $S \subset S'$, $T \subset T'$, $|S| = |T| = 2$, and thus this use of 'structures' yields four independent integrals of the above form. For each integral, we guess KO_{44}, KI_{43}, KO_{43}, and $KI_{421} \oplus \text{truncate}(KI_{422})$ (55 bits in all), and peel off enough of the last round to check that the 17th through 23rd bits of the input to the last round XOR to zero (a 7-bit condition on each of the four integrals). Guesses that survive this filtering phase can be further tested by guessing KI_{422} and checking an additional 9-bit condition. In this way we expect that all incorrect guesses will be eliminated, and then the remainder of the key may be recovered easily. In summary, this breaks four rounds of MISTY2 with work comparable to 2^{55} trial encryptions and just 9 chosen plaintexts. A known-plaintext variant would need 2^{33} texts and comparable work.

Also, there is an attack on six rounds of MISTY2, which works as follows. Consider the integral $\langle \mathcal{A}, \mathcal{C} \rangle$. After four rounds we have $\langle \mathcal{S}, \mathcal{S} \rangle$, where $\mathcal{S} = 0$. Note that texts with sum to zero at the input to the function FL, also sum to zero after FL, in other words, there is a probability one integral throught the FL function. After five rounds we have $\langle \mathcal{S}, ? \rangle$. Using Kühn's [19, page 328] alternative MISTY description, one sees that in a 6-round version, one can compute backwards from the ciphertext through $FO6$ to the 16 rightmost bits (which has the value $\mathcal{S}$ in the integral) by guessing at most 50 key bits. Note here that the key AKO_{63} in Kühns representation need not be guessed. It can be moved to the end of the first round of $FO6$ (in the right half) by exoring it to AKO_{64} and to AKO_{65}. With one structure of 2^{32} chosen plaintexts, 2^{34} of the 2^{50} possible values of the target key bits would be left suggested, thus with four structures one can expect only one suggested and correct value of 50 key bits. In total the attack needs 2^{34} chosen plaintext and has a time complexity of 2^{80}.

The attack on six rounds of MISTY2 can be further improved. Consider the second FI-function in $FO6$ (in Kühns reprensentation). The nine-bit key AKI_{ij} can be moved up before the "truncate step" if it is added to the left half (truncated to 7 bits) of the output of FI. Then in $FO6$ it should be added to the seven most significant bits of both halves of the output of $FO6$. In this version of an attack one would count on only 41 key bits. The disadvantage is that one can test on only seven data bits. So with one structure 2^{34} keys are left suggested. One can now either check on a few other structures or introduce the "remaining" 9 key bits from the before mentioned attack and run that attack. As an example, run the improved attack on two structures, which leaves 2^{27} out of 2^{41} possible values of the target key bits. Then run the first attack, with nine additional unknown key bits, which leaves 2^{18} out of 2^{50} possible values. With four structures, in total 2^{34} chosen plaintexts, the time complexity of the attack is 2^{71}.

We can also attack five rounds of MISTY1 using a related idea. There is a four-round integral $\langle \mathcal{C}, \mathcal{A} \rangle \rightarrow \langle ?, \mathcal{S} \rangle$. We collect four instances of this integral with 2^{34} chosen texts and apply a 1-R attack [3]. Note that $FO5$ has the same structure as three MISTY2 rounds, so it has a Sakurai-Zheng property [28]. In other words, we can write bits 1–7 of the right half of the block just before applying $FO5$ as a function $f_{KO_{51}}(C) \oplus g_{KO_{52}}(C) \oplus k'$ of the ciphertext C, for some functions f, g and some key-dependent constant k'. Our integral predicts that this value will sum to zero when summed over each integral of 2^{32} ciphertexts, or equivalently,

$$\sum_i f_{KO_{51}}(C_i) = \sum_i g_{KO_{52}}(C_i).$$

This gives a 7-bit condition for each integral, so taken together our four integrals will yield a 28-bit condition. We note that one can use a meet-in-the-middle technique to find solutions to this equation efficiently: we enumerate all 2^{16} possibilities for $\sum_i f_{KO_{51}}(C_i)$, then merge this list with the 2^{16} possibilities for $\sum_i g_{KO_{52}}(C_i)$, and their intersection yields candidates for KO_{51} and KO_{52}. Then further key material can be recovered by using guesses at KI_{512} and KI_{522} to check a 16-bit condition on each integral, and so on. These ideas allow us to break five rounds of MISTY1 with 2^{34} chosen plaintexts and work comparable to 2^{48} trial encryptions. Many tradeoffs between the time and data complexities are possible.

There is also an attack on four rounds of MISTY1 (without $FL5, FL6$) with very low data complexity. We apply the Sakurai-Zheng property twice: once to predict the sum of four outputs of $FO2$, and then a second time in the 1-R analysis to recover key material from $FO4$. We choose 25 plaintexts whose left halves are fixed and whose right halves range over the values $\langle x, y \rangle \in S' \times S'$, for some set S' containing five arbitrary 16-bit values. This choice ensures that we find 16 quadruples S, T of inputs to $FO2$ that each satisfy the conditions of Equation 5 (having survived $FL2$ without disruption, thanks to the choice of plaintexts). Thus, the XOR of the left half of the output of $FO2$ over each such quadruple will be zero. This propagates to the input of the fourth round undisturbed by $FL3$ at each bit position where KL_{32} has a one bit. We obtain a 7-bit condition on each quadruple of ciphertexts,

$$\text{truncate}(KL_{32}) \wedge \sum_i f_{KO_{41}}(C_i) = \text{truncate}(KL_{32}) \wedge \sum_i g_{KO_{42}}(C_i).$$

Guessing truncate(KL_{32}) and applying meet-in-the-middle techniques will typically let us find KO_{41} and KO_{42} with about 2^{30} simple steps of computation. The attack can be continued as before by guessing KI_{412}, KI_{422}, and the rest of KL_{32}. We expect that these techniques will give an attack on four rounds of MISTY1 that, for most keys, uses about 25 chosen plaintexts and takes time comparable to 2^{27} trial encryptions.

5   Generalised Feistel Networks

Nyberg has proposed a *generalised Feistel network* with block size $2nd$ bits [26]. We briefly describe the construction here. Let $X_0, \ldots, X_{2n-1}$ be the inputs to one round of the cipher. Given n S-boxes $F_0, \ldots, F_{n-1}$, where $F_i : \{0,1\}^d \to \{0,1\}^d$, and n round keys $K_0, \ldots, K_{n-1}$, the output of the round $Z_0, \ldots, Z_{2n-1}$ is defined as follows:

$$Y_i = X_i \oplus F_i(K_i \oplus X_{2n-1-i}), \text{ for } i = 0, \ldots, n-1$$
$$Y_i = X_i \text{ for } i = n, \ldots, 2n-1$$
$$Z_i = Y_{i-1} \text{ for } i = 0, \ldots, 2n-1,$$

where all indices are computed modulo $2n$. The integrals and attacks to follow are independent of the key schedule. Therefore, it is assumed that all round keys are independent and chosen uniformly at random.

As a special example, Nyberg considers the cases where the S-boxes are bijective. In this case the probabilities of differentials can be upper bounded to p^{2n} where p is the probability of a non-trivial differential through the S-boxes. With $n = 4$ and $d = 8$ there are S-boxes for which $p = 2^{-6}$ and the probabilities of all differentials over 12 rounds are bounded by 2^{-48}. Also, the probabilities of linear hulls over 12 rounds can be bounded by 2^{-48} [26].

For the above network with $n = 4$, $d = 8$, and bijective S-boxes, there exists an integral of probability one over eleven rounds using only 256 texts. Consider Table 4. It follows that for the integral where all first words are different, and where all other words are held constant, the sum of the first words after eleven rounds of encryption is zero. Thus, for a 12-round version of this cipher it is trivial to find eight bits of the key in the last round using the above integral by simply computing backwards from the ciphertexts to the outputs of the eleventh round. Also, a 13-round version can be attacked using the integral by computing backwards from the ciphertext to the eleventh round output by guessing only three key bytes. In this case, the attack must be repeated a few times to be

Table 4. An 11-round integral with 256 texts for the generalized Feistel cipher with $n = 4$, $d = 8$ and using bijective S-boxes. In the integral $\mathcal{S} = 0$.

Ciphertexts after round								
	$\mathcal{A}$	$\mathcal{C}$	$\mathcal{C}$	$\mathcal{C}$	$\mathcal{C}$	$\mathcal{C}$	$\mathcal{C}$	$\mathcal{C}$
1	$\mathcal{C}$	$\mathcal{A}$	$\mathcal{C}$	$\mathcal{C}$	$\mathcal{C}$	$\mathcal{C}$	$\mathcal{C}$	$\mathcal{C}$
2	$\mathcal{C}$	$\mathcal{C}$	$\mathcal{A}$	$\mathcal{C}$	$\mathcal{C}$	$\mathcal{C}$	$\mathcal{C}$	$\mathcal{C}$
3	$\mathcal{C}$	$\mathcal{C}$	$\mathcal{C}$	$\mathcal{A}$	$\mathcal{C}$	$\mathcal{C}$	$\mathcal{C}$	$\mathcal{C}$
4	$\mathcal{C}$	$\mathcal{C}$	$\mathcal{C}$	$\mathcal{C}$	$\mathcal{A}$	$\mathcal{C}$	$\mathcal{C}$	$\mathcal{C}$
5	$\mathcal{C}$	$\mathcal{C}$	$\mathcal{C}$	$\mathcal{C}$	$\mathcal{A}$	$\mathcal{A}$	$\mathcal{C}$	$\mathcal{C}$
6	$\mathcal{C}$	$\mathcal{C}$	$\mathcal{C}$	$\mathcal{A}$	$\mathcal{A}$	$\mathcal{A}$	$\mathcal{A}$	$\mathcal{C}$
7	$\mathcal{C}$	$\mathcal{C}$	$\mathcal{A}$	$\mathcal{A}$	$\mathcal{S}$	$\mathcal{A}$	$\mathcal{A}$	$\mathcal{A}$
8	$\mathcal{A}$	$\mathcal{A}$	$\mathcal{A}$	$\mathcal{S}$	?	$\mathcal{S}$	$\mathcal{A}$	$\mathcal{A}$
9	$\mathcal{A}$	$\mathcal{S}$	$\mathcal{S}$	?	?	?	$\mathcal{S}$	$\mathcal{A}$
10	$\mathcal{A}$	$\mathcal{S}$	?	?	?	?	?	$\mathcal{S}$
11	$\mathcal{S}$	?	?	?	?	?	?	?

able to uniquely determine the secret keys. This attack would run in total time approximately 2^{32} using $3 \cdot 2^8$ chosen texts. A 14-round and a 15-round version can be attacked by guessing a total of six respectively ten key bytes in a straight forward extension of the previous attack. These attacks would run in total times approximately 2^{56} respectively 2^{88} using $6 \cdot 2^8$ respectively $10 \cdot 2^8$ chosen texts. However, in these cases it is advantageous to use an integral of higher order. Consider the 13-round second-order integral in Table 5.

It follows by a closer look at the structure of the cipher that the values of the 16 bits of the first and second words after one round of encryption are a permutation of the 16-bit values of the first and eighth words of the plaintexts. Therefore by choosing 2^{16} plaintexts different only in the first and eighth words (counting from the left) one gets a collection of 2^8 integrals of the form in Table 4 this time starting from the second round. Therefore, one would expect to be able to determine the sum of the first words after 12 rounds. However, this integral goes one round further. To see this, consider Table 5. The question is why after nine rounds of encryption the fifth words sum to zero (the $\mathcal{S}$ after nine rounds of encryption in Table 5). It follows by simple observations that the 16-bit value $(x \mid y)$ consisting of the fifth (x) and sixth words (y) after five rounds of encryption is a permutation of the 16 varying bits in the plaintexts. Furthermore, the fourth word after seven rounds of encryption has the form $g_1(x)$ and the fifth word after seven rounds of encryption has the form $g_2(x) + g_3(y)$, for some bijective key-dependent functions g_i. Therefore, these 16 bits are a permutation of x and y and therefore also a permutation of the 16 varying bits in the plaintexts. This is illustrated in the integral where the two words are assigned the symbol $\mathcal{A}_4^2$. It then follows that the fifth words after eight rounds of encryption take all possible values equally many times. By similar arguments, it follows that the fourth and seventh words after rounds of encryption map one-to-one to the 16 varying bits in the plaintexts; therefore the fourth words after 8 rounds of encryption take all possible values equally many times. The

Table 5. A 13-round integral with 2^{16} texts for the generalised Feistel cipher with $n = 4$, $d = 8$ and using bijective S-boxes. In the integral $\mathcal{S} = 0$.

Ciphertexts after round	$\mathcal{A}_0^2$	$\mathcal{C}$	$\mathcal{C}$	$\mathcal{C}$	$\mathcal{C}$	$\mathcal{C}$	$\mathcal{C}$	$\mathcal{A}_0^2$
1	$\mathcal{A}_0^2$	$\mathcal{A}_0^2$	$\mathcal{C}$	$\mathcal{C}$	$\mathcal{C}$	$\mathcal{C}$	$\mathcal{C}$	$\mathcal{C}$
2	$\mathcal{C}$	$\mathcal{A}_0^2$	$\mathcal{A}_0^2$	$\mathcal{C}$	$\mathcal{C}$	$\mathcal{C}$	$\mathcal{C}$	$\mathcal{C}$
3	$\mathcal{C}$	$\mathcal{C}$	$\mathcal{A}_0^2$	$\mathcal{A}_0^2$	$\mathcal{C}$	$\mathcal{C}$	$\mathcal{C}$	$\mathcal{C}$
4	$\mathcal{C}$	$\mathcal{C}$	$\mathcal{C}$	$\mathcal{A}_0^2$	$\mathcal{A}_0^2$	$\mathcal{C}$	$\mathcal{C}$	$\mathcal{C}$
5	$\mathcal{C}$	$\mathcal{C}$	$\mathcal{C}$	$\mathcal{C}$	$\mathcal{A}_0^2$	$\mathcal{A}_0^2$	$\mathcal{C}$	$\mathcal{C}$
6	$\mathcal{C}$	$\mathcal{C}$	$\mathcal{C}$	$\mathcal{A}_1^2$	$\mathcal{A}_0^2$	$\mathcal{A}_1^2$	$\mathcal{A}_0^2$	$\mathcal{C}$
7	$\mathcal{C}$	$\mathcal{C}$	$\mathcal{A}_3^2$	$\mathcal{A}_4^2$	$\mathcal{A}_4^2$	$\mathcal{A}_3^2$	$\mathcal{A}_2^2$	$\mathcal{A}_2^2$
8	$\mathcal{A}_2^2$	$\mathcal{A}_5^2$	$\mathcal{A}_5^2$	$\mathcal{A}_3^2$	$\mathcal{A}_4^2$	$\mathcal{A}_4^2$	$\mathcal{A}_3^2$	$\mathcal{A}_2^2$
9	$\mathcal{A}_2^2$	$\mathcal{A}_2^2$	$\mathcal{A}^2$	$\mathcal{A}^2$	$\mathcal{S}$	$\mathcal{A}_4^2$	$\mathcal{A}_4^2$	$\mathcal{A}^2$
10	$\mathcal{A}^2$	$\mathcal{S}$	$\mathcal{S}$	$\mathcal{S}$	?	$\mathcal{S}$	$\mathcal{A}_4^2$	$\mathcal{A}_4^2$
11	$\mathcal{A}_4^2$	$\mathcal{S}$	?	?	?	?	$\mathcal{S}$	$\mathcal{A}_4^2$
12	$\mathcal{A}_4^2$	$\mathcal{A}_4^2$	?	?	?	?	?	$\mathcal{S}$
13	$\mathcal{S}$	?	?	?	?	?	?	?

Table 6. A 14-round integral with 2^{32} texts for the generalised Feistel cipher with $n = 4$, $d = 8$ and using bijective S-boxes. In the integral $\mathcal{S} = 0$.

Ciphertexts after round	$\mathcal{A}^4$	$\mathcal{A}^4$	$\mathcal{C}$	$\mathcal{C}$	$\mathcal{C}$	$\mathcal{C}$	$\mathcal{A}^4$	$\mathcal{A}^4$
	...	...	...	...	...	...	...	...
14	$\mathcal{S}$	?	?	?	?	?	?	?

fact that both the fourth and fifth words after eight rounds of encryption take all possible values equally many times explains why the sum of the fifth words after nine rounds of encryption is zero. Finally we note that there are other ways of specifying interdependencies of the words in the integral of Table 5. As an example the symbols after six rounds of encryption could also be specified as

$$\mathcal{C}, \mathcal{C}, \mathcal{C}, \mathcal{A}_1^2, \mathcal{A}_1^2, \mathcal{A}_0^2, \mathcal{A}_0^2, \mathcal{C}.$$

One can find eight key bits of a 14-round version using the integral by simply computing backwards from the ciphertexts to the outputs of the thirteenth round. The time complexity of this attack is approximately 2^{24} using 2^{16} texts. A 15-round and a 16-round version can be attacked by guessing a total of three respectively six key bytes in a straight forward extension of the previous attack. These attacks would run in total times approximately 2^{40} respectively 2^{64} using $3 \cdot 2^{16}$ respectively $6 \cdot 2^{16}$ chosen texts. Clearly, the attacks using the second-order integral are much faster than the attacks using a first-order integral, but on the down side they require more chosen plaintexts.

Let us go one step further and consider the fourth-order integral of Table 6. This integral contains 2^{16} copies of the second-order integral of Table 5 but starting here from the second round and onwards. Therefore, one can determine (at least) the sum of the first words after 14 rounds of encryption. Using this integral there are attacks on a 15-round, 16-round and a 17-round version which run in total times approximately 2^{40}, 2^{56} respectively 2^{80} using 2^{32}, $3 \cdot 2^{32}$ respectively $6 \cdot 2^{32}$ chosen texts.

There exists a sixth-order integral over 15 rounds with a total of 2^{48} chosen texts. This would enable an attack on a 17-round version of total time complexity approximately 2^{72} using $3 \cdot 2^{48}$ chosen texts.

Finally we note that there are impossible differentials for the above ciphers. With $n = 4$ we have detected a 14-round differential of probability zero. A set of plaintexts which differ only in the first word will never result in ciphertexts (after 14 rounds of encryption) different in only the fourth words. The differential can be used to distinguish a 14-round version of the generalised Feistel network from a randomly chosen permutation using about 2^{50} chosen texts. For comparison the integral of Table 6 can be used to distinguish the cipher from a randomly chosen permutation using only 2^{32} chosen texts with good advantage.

6 DES

So far we have only considered round functions that break the block into several independent words and then operate only in a word-oriented fashion. However, this restriction is not always satisfied: in some ciphers—for example, DES—the round function operates on individual bits (not words) and the inputs to the S-boxes are correlated. In this more general case, our previous techniques for constructing integrals may not apply.

In this section we consider more general round functions. In particular, we show that the existence of integrals is not limited to word-oriented ciphers or to S-boxes whose inputs are independent. Since DES is a classic example where neighboring S-boxes in the same round are fed related inputs and where the round function works at the bit level, we will use the DES round function as a concrete example of how to build integrals for more general S-box networks.

As a starting example to illustrate the possibility of finding integrals on the DES round function F, we give a simple integral. Let the inputs to F take on values of the form $u_z = \langle z, z, z, z \rangle$, where z varies over all values in $\{0,1\}^8$. Then we claim that $\sum_z F(u_z) = 0$, i.e., the XOR of the corresponding 2^8 outputs of the F function will be zero. This fact will imply that the above structure of 2^8 texts yields an integral for one round of DES.

The proof of the claim requires a bit of knowledge about the form of the DES F function. Recall that the DES round function takes the form $F = P \circ S \circ E$ where $E : \{0,1\}^{32} \to \{0,1\}^{48}$ expands its input by duplicating some input bits, where $S : \{0,1\}^{48} \to \{0,1\}^{32}$ is composed of eight parallel S-boxes

$$S(x_1, \ldots, x_{48}) = \langle S_1(x_1, \ldots, x_6), \ldots, S_8(x_{43}, \ldots, x_{48}) \rangle$$

where each S-box has a corresponding map $S_i : \{0,1\}^6 \to \{0,1\}^4$, and where the bit-permutation P is irrelevant to our discussion. Also, the expansion function E

ensures that the inputs $\langle v_1, \ldots, v_6 \rangle, \langle w_1, \ldots, w_6 \rangle$ to any two consecutive S-boxes satisfy $v_5 = w_1$ and $v_6 = w_2$.

Now we can see why the above integral works. When the input to the F function is $\langle z, z, z, z \rangle$, the odd S-boxes receive $\langle z_8, z_1, \ldots, z_5 \rangle$ as input, and the even S-boxes receive $\langle z_4, \ldots, z_8, z_1 \rangle$. Note that each S-box takes on all 2^4 possible output values exactly four times if its input takes on each possible 6-bit input value exactly once. Consequently, if we focus on any one S-box, we see that its output will take on all 4-bit values exactly four times as we range over all possible choices of z, which means that these outputs will XOR to zero. Since this is true for each S-box, and since the P bit-permutation is linear with respect to XOR, we see that $\sum_z F(u_z) = 0$ (where the addition operation is the exclusive-or). This gives a simple integral for the F function containing 2^8 inputs.

There are more complicated integrals that use fewer input texts. For example, if we consider F-function inputs of the form

$$u = \langle d, e, f, a, b, e, f, c, d, a, b, e, f, c, d, a, b, e, f, c, d, a, b, e, f, c, d, a, b, e, f, c \rangle$$

where the 6-bit value $\langle a, b, c, d, e, f \rangle$ varies over all 2^6 possibilities, we find that $\sum F(u) = 0$. (The input to S_1 is $\langle c, d, e, f, a, b \rangle$ and hence takes on all possibilities exactly once; the input to S_2 is $\langle a, b, e, f, c, d \rangle$; and in general, the input to each S-box is a permutation of the 6 bits a, b, c, d, e, f.)

In fact, there even exist integrals containing only 2^5 inputs. We use the following property of the S-boxes: $S_i(w_1, \ldots, w_6)$ is a bijective function of $\langle w_2, \ldots, w_5 \rangle$ when w_1, w_6 are held fixed. With this observation, we consider inputs of the form

$$u = \langle a, b, c, d, e, a, b, c, d, a, b, e, c, a, b, d, e, a, b, c, d, a, e, b, c, d, e, a, b, c, d, e \rangle$$

where the 5-bit value $\langle a, b, c, d, e \rangle$ ranges over all 2^5 possibilities. This choice ensures that each S-box has an input pattern of the form $\langle i, j, k, l, m, i \rangle$ (where i, j, k, l, m represent some re-ordering of the bits a, b, c, d, e), and then the XOR of the corresponding 2^5 outputs will be zero, as required. We leave it as an open question to determine whether there exist integrals for the DES F function that use a smaller number of inputs.

We stress that we do not know of any way to use these integrals to mount an attack on more than a few rounds of DES. Thus, the main interest of these observations is likely to be in their motivational value: they show that it may be possible to find integrals even on fairly complicated round functions.

7   Integral-Interpolation Attacks

An interesting property of integrals is that they can be combined with interpolation attacks [13]. Consider a cipher whose first half may be covered by an integral and whose second half may be approximated using a low-degree polynomial. Suppose that we have a set of chosen plaintext/ciphertext pairs (P_i, C_i) following the integral, and let Z_i denote the corresponding intermediate values predicted by the integral. Assuming that the integral ends with an S, we have

$\sum_i Z_i = 0$. Suppose we can write Z_i as a polynomial function of the ciphertext, so that $Z_i = p(C_i)$ for some low-degree polynomial $p(x) = a_d x^d + \ldots + a_1 x + a_0$ with $d = \deg p$. Then we can conclude that

$$0 = \sum_i Z_i = \sum_i p(C_i) = \sum_i \sum_{j=0}^{d} a_j C_i^j = \sum_{j=0}^{d} \tau_j \cdot a_j \qquad \text{where } \tau_j = \sum_i C_i^j. \quad (6)$$

Note that the τ_j's are known, since the ciphertexts are. Treating the coefficients a_j as formal unknowns, we thus see that Eqn. 6 gives us a single linear relation on the $d + 1$ variables $a_0, \ldots, a_d$.

If we repeat the above experiment $d + 2$ times, obtaining $d + 2$ sets of texts following the integral, we will have $d + 2$ linear equations in $d + 1$ unknowns. Applying Gaussian elimination, we will find a linear relationship that the ciphertexts must obey when this block cipher is used.

In other words, this allows a distinguishing attack on the underlying block cipher. When the first half of the cipher can be covered by an integral containing 2^s plaintexts, and when the second half can be expressed as a polynomial of degree d, the complexity of the attack will be approximately $d \cdot 2^s$ chosen texts and $d^2 \cdot 2^s + d^3$ work. It is an open question whether these techniques may be effectively extended to apply where we have a probabilistic polynomial relation [14] or rational polynomial relation [13] for the last half of the cipher.

Although we do not know of any concrete examples where this combination yields improved attacks, we conjecture that the opportunity to combine attack techniques in this way may be of interest.

8   Related Work

The attack techniques we exploit here were first introduced in [5], but under a different name: these techniques were previously described as "the Square attack", instead of "integrals." The name "integrals" has since been proposed independently by both Knudsen [17] and Yu, Zhang, and Xiao [12] to describe this general class of attacks. Also, in [6] the attack was described in terms of "lambda-sets" and applied also to reduced-round versions of the ciphers SHARK [27] and SAFER K [23].

Since their introduction, integrals have been used to cryptanalyse reduced-round versions of Square [5], SAFER K [18], SAFER+ [12], Crypton [8], Rijndael [9], Twofish [22], Hierocrypt [1], IDEA [25], and Camellia [10]. We have shown here additional examples of applications of integrals. Thus, this class of techniques seems to be of broad interest.

Recently Biryukov and Shamir applied a variant of integral cryptanalysis to an SP-network with secret S-boxes and secret linear transformations [4]. They called their technique the *multi-set* attack, where one distinguishes between whether all values in a multi-set are equal, are all different, all occur an even number of times, and where the exclusive-or sum of all values is zero. Thus, there is some resemblence to our definition of integrals and higher-order integrals.

9   Conclusions

In this paper we studied integral cryptanalysis, an attack which applies particularly well to block ciphers that use bijective components. The basic integral attack was introduced some years ago, but without a specific name attached to it. We argued that integral cryptanalysis is the obvious name for the attacks. A powerful extension, the higher-order integral, was given. These new attacks were applied to a range of ciphers. Also, a possible combination of integral cryptanalysis and the interpolation attacks was outlined. We believe that attacks based on higher-order integrals will find many applications in the future.

Acknowledgements. We thank Ulrich Kühn for many helpful comments.

References

1. P. Barreto, V. Rijmen, J. Nakahara Jr., B. Preneel, J. Vandewalle, and H.Y. Kim. "Improved SQUARE attacks against reduced-round HIEROCRYPT". *Fast Software Encryption 2001*, Springer-Verlag, to appear.
2. E. Biham, A. Biryukov, A. Shamir, "Cryptanalysis of Skipjack reduced to 31 rounds using impossible differentials," In J. Stern, editor, *Advances in Cryptology: EUROCRYPT'99, LNCS 1592*, pp. 12–23. Springer Verlag, 1999.
3. E. Biham, A. Shamir, *Differential Cryptanalysis of the Data Encryption Standard*, Springer-Verlag, 1993.
4. A. Biryukov, A. Shamir, "Structural Cryptanalysis of SASAS, " *Advances in Cryptology - EUROCRYPT 2001*, LNCS 2045, Springer-Verlag, pp. 394–405, 2001.
5. J. Daemen, L. Knudsen, and V. Rijmen. The block cipher Square. In E. Biham, editor, *Fast Software Encryption, Fourth International Workshop, Haifa, Israel, January 1997, LNCS 1267*, pages 149–165. Springer Verlag, 1997.
6. J. Daemen, L.R. Knudsen, and V. Rijmen, "Linear Frameworks for Block Ciphers," Designs, Codes and Cryptography, Volume 22, No 1, 2001, pp. 65-87.
7. J. Daemen, V. Rijmen, "AES Proposal: Rijndael," *AES Round 1 Technical Evaluation CD-1: Documentation*, National Institute of Standards and Technology, Aug 1998.
8. C. D'Halluin, G. Bijnens, V. Rijmen, and B. Preneel. Attack on Six Rounds of Crypton. In L. Knudsen, editor, *Fast Software Encryption, Sixth International Workshop, Rome, Italy, March 1999, LNCS 1636*, pages 46–59. Springer Verlag, 1999.
9. N. Ferguson, J. Kelsey, B. Schneier, M. Stay, D. Wagner, and D. Whiting. Improved cryptanalysis of Rijndael. In B. Schneier, editor, *Fast Software Encryption, 7th International Workshop, FSE 2000, New York, USA, April 2000, LNCS 1978*, pages 213–230. Springer Verlag, 2001.
10. Y. He, S. Qing, "Square Attack on Reduced Camellia Cipher", *ICICS 2001*, LNCS 2229, Springer-Verlag.
11. I.N. Herstein, *Topics in Algebra*, 2nd ed., John Wiley & Sons, 1975.
12. Y. Hu, Y. Zhang, and G. Xiao, "Integral cryptanalysis of SAFER+", *Electronics Letters*, vol.35, (no.17), IEE, 19 Aug. 1999, p.1458-1459.

13. T. Jakobsen and L. Knudsen. The interpolation attack on block ciphers. In E. Biham, editor, *Fast Software Encryption, Fourth International Workshop, Haifa, Israel, January 1997, LNCS 1267*, pages 28–40. Springer Verlag, 1997.

14. T. Jakobsen, Cryptanalysis of block ciphers with probabilistic non-linear relations of low degree. In H. Krawczyk, editor, *Advances in Cryptology: CRYPTO'98, LNCS 1462*, pages 212–222. Springer Verlag, 1998.

15. L.R. Knudsen and T. Berson. Truncated differentials of SAFER. In Gollmann D., editor, *Fast Software Encryption, Third International Workshop, Cambridge, UK, February 1996, LNCS 1039*, pages 15–26. Springer Verlag, 1995.

16. L.R. Knudsen. Truncated and higher order differentials. In B. Preneel, editor, *Fast Software Encryption - Second International Workshop, Leuven, Belgium, LNCS 1008*, pages 196–211. Springer Verlag, 1995.

17. L.R. Knudsen, "Block Ciphers: State of the Art". Copies of transparencies for lecture at the International Course on State of the Art and Evolution of Computer Security and Industrial Cryptography, Katholieke Universiteit Leuven, Belgium, June, 1997.

18. L.R. Knudsen, "A Detailed Analysis of SAFER K", *Journal of Cryptology*, vol.3, no.4, Springer-Verlag, 2000, pp.417–436.

19. U. Kühn. Cryptanalysis of reduced-round MISTY. In B. Pfitzmann, editor, *Advances in Cryptology - EUROCRYPT'2001, LNCS 2045*, pages 325–339. Springer Verlag, 2001.

20. U. Kühn, "Improved Cryptanalysis of MISTY1," These proceedings.

21. X. Lai, "Higher Order Derivations and Differential Cryptanalysis," *Communications and Cryptography: Two Sides of One Tapestry*, Kluwer Academic Publishers, 1994, pp. 227–233.

22. S. Lucks, "The Saturation Attack—a Bait for Twofish", *Fast Software Encryption 2001*, Springer-Verlag, to appear.

23. J.L. Massey. SAFER K-64: A byte-oriented block-ciphering algorithm. In R. Anderson, editor, *Fast Software Encryption - Proc. Cambridge Security Workshop, Cambridge, U.K.,* LNCS 809, pages 1–17. Springer Verlag, 1994.

24. M. Matsui. New block encryption algorithm MISTY. In E. Biham, editor, *Fast Software Encryption, Fourth International Workshop, Haifa, Israel, January 1997,* LNCS 1267, pages 54–68. Springer Verlag, 1997.

25. J. Nakahara Jr., P.S.L.M. Barreto, B. Preneel, J. Vandewalle, H.Y. Kim, "SQUARE Attacks Against Reduced-Round PES and IDEA Block Ciphers", IACR Cryptology ePrint Archive, Report 2001/068, 2001.

26. K. Nyberg. Generalized Feistel networks. In Kwangjo Kim and Tsutomu Matsumoto, editors, *Advances in Cryptology - ASIACRYPT'96, LNCS 1163*, pages 91–104. Springer Verlag, 1996.

27. V. Rijmen, J. Daemen, B. Preneel, A. Bosselaers, and E. De Win. The cipher SHARK. In Gollmann D., editor, *Fast Software Encryption, Third International Workshop, Cambridge, UK, February 1996, LNCS 1039*, pages 99–112. Springer Verlag, 1996.

28. K. Sakurai and Y. Zheng, "On Non-Pseudorandomness from Block Ciphers with Provable Immunity against Linear Cryptanalysis", *IEICE Transactions on Fundamentals of Electronics, Communications and Computer Science*, Vol. E80-A, No.1, 1997, pp.19-24.

Improved Upper Bounds of Differential and Linear Characteristic Probability for Camellia

Taizo Shirai, Shoji Kanamaru, and George Abe

Sony Corporation
7-35 Kitashinagawa 6-chome, Shinagawa-ku, Tokyo, 141-0001 Japan
{Taizo.Shirai, Shoji.Kanamaru, George.Abe}@jp.sony.com

Abstract. We discuss the security of the block cipher Camellia against differential attack and linear attack. The security of Camellia against these attacks has been evaluated by upper bounds of maximum differential characteristic probability (MDCP) and maximum linear characteristic probability (MLCP) calculated by the least numbers of active S-boxes which are found by a search method[2]. However, we found some truncated differential paths generated by the method have wrong properties. We show a new evaluation method for truncated differential and linear paths to discard such wrong paths by using linear equations systems and sets of nonzero conditions. By applying this technique to Camellia, we found tighter upper bounds of MDCP and MLCP for reduced-round Camellia. As a result, 10-round Camellia without FL/FL^{-1} has no differential and linear characteristic with probability higher than 2^{-128}.

1 Introduction

Camellia[1] is a block cipher which was suggested as a candidate for the NESSIE project[14] and recently selected for the 2nd phase of the project, and also suggested as a candidate for the CRYPTREC project in Japan[15]. The security of Camellia has been studied by many researchers [2,4,5,7,13]. Among them, designers of Camellia tried to evaluate its security against differential attack[3] and linear attack[11] by showing the upper bounds of maximum differential characteristic probability (MDCP) and maximum linear characteristic probability (MLCP) for each reduced round Camellia. Since both of maximum differential probability and maximum linear probability of Camellia's S-boxes are 2^{-6}, the MDCP and MLCP are upper-bounded by $(2^{-6})^d$ and $(2^{-6})^l$, respectively, where d is the least number of differentially active S-boxes, and l is the least number of linearly active S-boxes. The designers modified a search method for truncated differential probability[9,10,12] to count the least number of active S-boxes and then searched d and l [2]. The search algorithm works fast because it treats one byte differences at once by using truncated differences [8]. They also showed that 12-round Camellia and 11-round Camellia without FL/FL^{-1} has no differential or linear characteristic with probability higher than 2^{-128}.

But at the FSE2001, Kanda pointed out that the search algorithm using truncated differential theory is not enough to evaluate the security of Camellia

J. Daemen and V. Rijmen (Eds.): FSE 2002, LNCS 2365, pp. 128–142, 2002.
© Springer-Verlag Berlin Heidelberg 2002

because it has SPN-type round function [5]. In SPS-type round function like E2[6], each split data in F-functions are substituted by the second substitution layer before output. But SPN-type round function doesn't have such a second substitution layer. Therefore, if a truncated differential is applied to SPN round function, a strong relation appears between each byte in the output of the F-function. By approximating these relationships, he showed the upper bounds of truncated differential probability of Camellia.

In this paper, we investigate these relationships more strictly. And we show a new evaluation method exploiting the relations all over the cipher to count the least number of active S-box more strictly. Then we apply the proposed evaluation method to Camellia, and we show that the upper bounds of MDCP and MLCP for Camellia are tighter than ever. We reveal that 10-rounds Camellia without FL/FL^{-1} has no differential or linear characteristic with probability higher than 2^{-128}.

This paper is organized as follows: In Section 2 we give the description of Camellia and the SPN-type round function treated in this paper. In Section 3 we use an example to show a contradiction in the truncated differential path generated by Camellia's evaluation method. In Section 4 we present a new method to evaluate the validity of truncated differential paths. In Section 5 we apply the proposed method to Camellia in practice, and show the revised upper bound of MDCP and MLCP for Camellia. Section 6 summarizes our conclusion.

2   Preliminaries

2.1   Description of SPN Round Function and Camellia

Camellia [1] is a Feistel network block cipher with block size of 128 bits and applicable to 128, 192 and 256 key bits. The round numbers are determined by a key length, 18 rounds for 128 bits key, 24 rounds for 192 and 256 bits keys. The F-function of Camellia is composed of a so-called SPN(Substitution Permutation Network) type structure. Fig 1 shows the i-th round of a Feistel cipher with SPN round function to be treated in this paper.

The block size is $2 \times m \times n$ bits. $m \times n$ bits of key and $m \times n$ bits of data are inputted to the F-function in each round. The input data of F-function is split into m pieces of n bits data, represented by $X_i = (X_i[1], X_i[2], \ldots, X_i[m])$. After the key adding operation, each data is inputted to non-linear bijective function $S : \{0,1\}^n \to \{0,1\}^n, (1 \leq j \leq m)$. Then the output of S-functions $S_i = (S_i[1], S_i[2], \ldots, S_i[m])$ are inputted to linear transformation layer $P : \{0,1\}^{mn} \to \{0,1\}^{mn}$. $Y_i = (Y_i[1], Y_i[2], \ldots, Y_i[m])$ represents an output vector of the P-function at the i-th round. Y_i is also an output of the F-function at the i-th round.

In general, any linear transformation P can be represented by a matrix form. In this paper, we specially concentrate on a linear transformation P which can be represented by a square matrix of order m over $GF(2^n)$ produced by some irreducible polynomial f defined in the cipher. Let $P_{mat} = (a_{ij})$ be a matrix

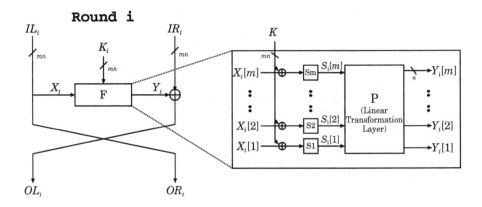

Fig. 1. Feistel Network with SPN round function

of order m over $GF(2^n)$ corresponding to the P-function. By using P_{mat}, the relation between S_i and Y_i can be described as follows:

$$\begin{pmatrix} Y[1] \\ Y[2] \\ \vdots \\ Y[m] \end{pmatrix} = \begin{pmatrix} a_{11} & a_{12} & \cdots & a_{1m} \\ a_{21} & a_{22} & \cdots & a_{2m} \\ \vdots & \vdots & \ddots & \vdots \\ a_{m1} & a_{m2} & \cdots & a_{mm} \end{pmatrix} \begin{pmatrix} S[1] \\ S[2] \\ \vdots \\ S[m] \end{pmatrix}$$

In the context of the matrix multiplication, each n bits of data are comprehended as a bit representation of an element in $GF(2^n)$. Throughout this paper, we treat n bits of data as a bit representation of an element in $GF(2^n)$. Thus the exclusive-or operation $\oplus$ and add operation $+$ of n bits of data are same.

In the case of Camellia, the round function can be viewed as $n = 8, m = 8$ SPN-type F-function, and the P-function is represented by the following matrix:

$$P_{Camellia} = \begin{pmatrix} 1 0 1 1 0 1 1 1 \\ 1 1 0 1 1 0 1 1 \\ 1 1 1 0 1 1 0 1 \\ 0 1 1 1 1 1 1 0 \\ 1 1 0 0 0 1 1 1 \\ 0 1 1 0 1 0 1 1 \\ 0 0 1 1 1 1 0 1 \\ 1 0 0 1 1 1 1 0 \end{pmatrix}$$

All elements in Camellia's matrix can be represented by only two elements, 0 and 1, in $GF(2^8)$.

Camellia also has a key scheduling algorithm, a key whitening layer and key-dependent linear functions FL and FL^{-1} which are inserted every 6 rounds. Details of these are described in [1]. In this paper, we assume that round keys are independent and uniformly random.

2.2   Definitions

We use the following definitions in this paper.

Definition 1. (Active S-box)
*An S-box which has non-zero input difference is called **differentially active**
S-box, and an S-box which has non-zero output linear mask is called **linearly**
active S-box.*

Definition 2. (χ function)
*For any difference $\Delta X \in \{0,1\}^n$, a function $\chi : \{0,1\}^n \to \{0,1\}$ is defined as
follows:*

$$\chi(\Delta X) = \begin{cases} 0 \ if \ \Delta X = \{0\}^n \\ 1 \ if \ \Delta X \neq \{0\}^n \end{cases}$$

*For any differential vector $\Delta X = (\Delta X[1], \Delta X[2], \ldots, \Delta X[m])$, $\Delta X[i] \in \{0,1\}^n$
truncated difference of ΔX is defined as*

$$\delta X = \chi(\Delta X) = (\chi(\Delta X[1]), \chi(\Delta X[2]), \ldots, \chi(\Delta X[m]))$$

Definition 3. (truncated differential probability of F-function) *[5] Let
$\delta X, \delta Y$ be input truncated difference and output truncated difference of F-
function , respectively. Then the truncated differential probability $p_F(\delta X \to \delta Y)$
is defined as follows:*

$$p_F(\delta X \to \delta Y) = \frac{\displaystyle\sum_{\chi(\Delta X)=\delta X} \sum_{\chi(\Delta Y)=\delta Y} \Pr_{X \in (\{0,1\}^n)^m} [F(X) \oplus F(X \oplus \Delta X) = \Delta Y]}{\#\{\Delta X | \chi(\Delta X) = \delta X\}}$$

where $\#\{A\}$ denotes the number of elements in set A.

Specially, in the case of SPN-type round function defined above and the output difference of the S-function is assumed to be uniform[9,10,12], the truncated
differential probability $p'_F(\delta X \to \delta Y)$ can be defined approximately as follows:

$$p'_F(\delta X \to \delta Y) = \frac{\#\{\Delta S | \chi(\Delta S) = \delta X, \chi(P(^t\Delta S)) = \delta Y\}}{\#\{\Delta S | \chi(\Delta S) = \delta X\}}$$

3   Wrong Truncated Differential Paths

The algorithm used in Camellia's evaluation counts the least number of active
S-boxes for any round of Camellia[2]. Since the algorithm has exploited an approximation at the XOR operation in each round function, some of truncated
differential paths generated by the algorithm cannot exist in reality.

Before proposing a new judgment method for the existence of truncated
differential paths, we show an example of such wrong truncated differential
paths and consider contradictions between truncated differences.

A truncated differential path is defined as follows:

Definition 4. A truncated differential path

For an r round Feistel block cipher, let δPR be a truncated difference of the right half of plain text and δCR be a truncated difference of the right half of cipher text. For each i-th round, let $\delta X_i, \delta Y_i$ be input truncated difference and output truncated difference of F-function, respectively. Then the truncated differential path TP is represented by:

$$TP = (\delta PR, \delta X_1, \delta Y_1, \ldots, \delta X_r, \delta Y_r, \delta CR) \in (\{0,1\}^m)^{2r+2}$$

The truncated difference of left half of plain text δPL and cipher text δCL can be represented by $\delta X_1, \delta X_r$, respectively.

For the *i*-th round, shown as Fig.1, let $\delta IL_i, \delta IR_i, (2 \le i \le r)$ be the left and right input truncated differences, respectively. And let $\delta OL_i, \delta OR_i, (1 \le i \le r-1)$ be left and right output truncated differences, respectively. Then there are following relations between these truncated differences and truncated differential path: $\delta IR_i = \delta X_{i-1}, \delta IL_i = \delta OR_i = \delta X_i, \delta OL_i = \delta X_{i+1}$.

3.1   Algorithm to Find Truncated Differential Paths

We show an outline of an algorithm used in the designer's evaluation to count the least number of active S-boxes of Camellia without FL/FL^{-1}[2]. This algorithm is modified version of Matsui's algorithm[9,10] which is originally developed to estimates truncated differential probabilities of E2[6].

Algorithm

INPUT: a round number N

OUTPUT: a table of the least number of active S-boxes for all pattern of output truncated difference $\delta CR, \delta CL$ for N round Camellia without FL/FL^{-1}.

1. Make a table (F-table) of truncated differential probability of F-function for all δX and δY. The number of entries in the F-table is 2^{16}.
2. Using the F-table, make a table (R-table) of truncated differential probability of the round function for all $\delta IL, \delta IR$ and δCL. The number of entries in the R-table is 2^{24}.
3. (Inductive step) Using a table of the least number of active S-boxes for $N-1$ round of Camellia and R-table, calculate the least number of the active S-boxes for N round's output $\delta CL, \delta CR$ and store it in a table. The number of entries in the table is 2^{16}.

This algorithm only counts the least number of active S-boxes for each output truncated difference. To recover and to evaluate truncated differential paths which have the least number of active S-boxes, we have modified the algorithm additionally as follows.

- In Step 2, store all candidates of the output truncated difference of F-function.
- In Step 3, when the N round's output is searching, store all candidates of the $N-1$ round's output truncated difference which realize the least number of active S-boxes.

3.2 Connection Errors in Truncated Differential Paths

We show an example of a truncated differential path holding wrong properties. The search algorithm for the least number of active S-boxes found a Camellia's truncated differential path whose consecutive 3 rounds have truncated differential as in Fig. 2.

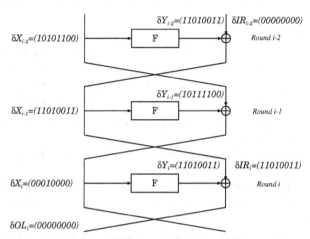

Fig. 2. A Example of Wrong Truncated Differential Path

The search algorithm exploits the following XOR rule of truncated difference based on a property of truncated difference.

XOR rule of truncated difference

$$\delta OL[j] = \begin{cases} 0 & if \quad \delta IR[j] = \delta Y[j] = 0 \\ * & if \quad \delta IR[j] = \delta Y[j] = 1 \\ 1 & else \end{cases} \quad (1)$$

for $(1 \leq j \leq m)$. Value $*$ means arbitrary selection of $0, 1$.

Obeying above rule, the XOR operation of $\delta IR_i = (11010011)$ and $\delta Y_i = (11010011)$ generates $\delta OL_i = (* * 0 * 00 * *)$. Thus $\delta OL_i = (00000000)$ can be obtained by the search algorithm.

But by considering a relationship of each output difference of F-function δY_i and δY_{i-2}, we can prove such an XOR operation cannot be realized.

Proof. From $\delta OL_i = (00000000)$ and $\delta IR_{i-2} = (00000000)$, we get an equation of differences $\Delta Y_i = \Delta Y_{i-2}$. And from the linear relation $\Delta Y = P(\Delta S)$, then $\Delta S_i[j] = \Delta S_{i-2}[j]$. But considering given input truncated differences of S4 in both round, we get $\Delta S_i[4] \neq \Delta S_{i-2}[4]$ because $\delta X_i[4] = 1$ and $\delta X_{i-2}[4] = 0$. This is contradiction. Also in S1, S3, S5, S6, there are contradictions. Thus such a path cannot exist in reality. □

From above observations, wrong truncated paths have some contradictions between input and output truncated differences of F-functions for every two round. In the next section, we propose an algorithm which determines if the target truncated differential path is wrong or not by checking solutions of the linear equations system constructed from the path.

4 Validity Checking Method for Truncated Differential Paths

We show a new algorithm to check the validity of truncated differential paths. This algorithm exploits a method of checking solutions of a linear equations system which is constructed from a truncated differential path. The basic steps of checking the validity of a truncated differential path is as follows:

- Represent all differences in Feistel network by linear forms using output difference of S-boxes
- Construct two linear equations systems and two sets of nonzero conditions using truncated differential values
- Check solutions of each linear equations system under corresponding nonzero conditions, and determine the validity of the truncated differential path.

4.1 Linear Forms Representation of Feistel Block Cipher

All differences in a Feistel cipher with SPN round function defined in Section 2 can be represented in linear form using differences $\Delta PL, \Delta PR$ and ΔS_i. To show this, we divide the Feistel block cipher in two parts **the left chain** and **the right chain** as shown in Fig.3.

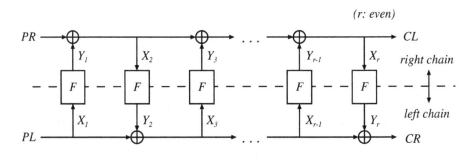

Fig. 3. The left chain and the right chain

Without loss of generality, we assume that the round number r is even.

The left chain is a data path which starts from left half of plain text PL, while the right chain starts from right half of plain text PR. In each chain, output of F-functions Y_i's are added to the ongoing data. The left chain includes the data $X_{2i-1}, Y_{2i}, CR(1 \le i \le \frac{r}{2})$, The right chain includes $PR, Y_{2i-1}, X_{2i}(1 \le i \le \frac{r}{2})$. From these definitions we can show the following lemma.

Lemma 1.
Let $\Delta X_1[j](= \Delta PL[j]), \Delta PR[j](1 \leq j \leq m)$ be variables which are set differences of plain text. And Let $\Delta S_i[j]$ be variables which are set differences of the output of j-th S-box Sj for i-th round $(1 \leq j \leq m, 1 \leq i \leq r)$, respectively. All differential data in the left chain are represented by linear form of elements in $\Delta X_i, \Delta S_{2i}(1 \leq i \leq \frac{r}{2})$. All differential data in the right chain are represented by linear form of elements in $\Delta PR, \Delta S_{2i-1}(1 \leq i \leq \frac{r}{2})$.

Proof. In the right chain, we show that $\Delta PR, \Delta Y_{2i-1}$ and ΔX_{2i} are represented by linear form. ΔPR are monomials. And using elements in linear transformation matrix P_{mat}, differences of output of F-functions $\Delta Y_{2i-1}[j](1 \leq i \leq r/2, 1 \leq j \leq m)$ can be described as following linear form:

$$\Delta Y_{2i-1}[j] = \sum_{k=1}^{m} a_{jk} \Delta S_{2i-1}[k] \qquad (2)$$

And the difference of the input data of the $2i$-th round $\Delta X_{2i}[j](1 \leq i \leq r/2, 1 \leq j \leq m)$ has the following equation:

$$\Delta X_{2i}[j] = \Delta PR[j] + \sum_{l=1}^{i} \Delta Y_{2l-1}[j] \qquad (3)$$

By combining (2) and (3), $\Delta X_{2i}[j]$ has the following linear form:

$$\Delta X_{2i}[j] = \Delta PR[j] + \sum_{l=1}^{i} \sum_{k=1}^{m} a_{jk} \Delta S_{2l-1}[k] \qquad (4)$$

All differential data in the left chain are also represented in same manner.   □

From the viewpoint of linear form representation, the sets of variables used in both chains are completely separated. The left chain uses a set of differential variables $(\Delta X_1, \Delta S_2, \Delta S_4, \ldots, \Delta S_r)$, and the right chain uses a set of differential variables $(\Delta PR, \Delta S_1, \Delta S_3, \ldots, \Delta S_{r-1})$. Thus, we can treat both chains separately. Then we can describe the relation between the variables and the differences in the right chain as follows.

$$
\begin{pmatrix}
{}^t\Delta PR \\
{}^t\Delta Y_1 \\
{}^t\Delta X_2 \\
{}^t\Delta Y_3 \\
{}^t\Delta X_4 \\
\vdots \\
{}^t\Delta Y_{r-3} \\
{}^t\Delta X_{r-2} \\
{}^t\Delta Y_{r-1} \\
{}^t\Delta X_r
\end{pmatrix}
=
\begin{pmatrix}
I & 0 & 0 & \cdots & 0 & 0 \\
0 & P_{mat} & 0 & \cdots & 0 & 0 \\
I & P_{mat} & 0 & \cdots & 0 & 0 \\
0 & 0 & P_{mat} & \cdots & 0 & 0 \\
I & P_{mat} & P_{mat} & \cdots & 0 & 0 \\
\vdots & \vdots & \vdots & \ddots & \vdots & \vdots \\
0 & 0 & 0 & \cdots & P_{mat} & 0 \\
I & P_{mat} & P_{mat} & \cdots & P_{mat} & 0 \\
0 & 0 & 0 & \cdots & 0 & P_{mat} \\
I & P_{mat} & P_{mat} & \cdots & P_{mat} & P_{mat}
\end{pmatrix}
\begin{pmatrix}
{}^t\Delta PR \\
{}^t\Delta S_1 \\
{}^t\Delta S_3 \\
\vdots \\
{}^t\Delta S_{r-3} \\
{}^t\Delta S_{r-1}
\end{pmatrix}
$$

I denotes a unit matrix of order m, 0 in the above matrix denotes zero matrix of order m. The total number of linear forms are $m(r+1)$, and the total number of variables are $m(r/2+1)$.

The data in left chain $(\Delta X_1, \Delta Y_2, \Delta X_3, \ldots, \Delta Y_r, \Delta CR)$ can be expressed by using $(\Delta X_1, \Delta S_2, \Delta S_4, \ldots, \Delta S_r)$ in the same way. Thus, all differences in the Feistel cipher with SPN round function defined in this paper have been represented by linear forms.

4.2   Evaluation Algorithm for Truncated Differential Paths

Let TP be a r round truncated differential path. To evaluate TP, we check whether there exist any differential characteristic which follow the truncated differential path TP.

For the right chain, truncated differences $\delta X_1, \delta X_3, \ldots, \delta X_{r-1}$ in TP determine whether S-boxes, whose outputs are added to the right chain, are active or not. If truncated difference $\delta X_i[j] = 0$ the output differences of j-th S-box in round i is always 0. Thus we can remove differential variables of such non active S-boxes from all linear forms. And if $\delta PR[i] = 0$, we can also remove the differential variable $\Delta PR[i]$ in linear forms. In this way, we obtain reduced linear forms.

Next we construct linear equations systems and sets of linear conditions. In the right chain, truncated differences $\delta Y_1, \delta Y_3, \ldots, \delta Y_{r-1}$ determine whether differential data of F-function's output have a differential value or not. Let $lf_{Y_i[j]}$ be a reduced linear form of differential data $\Delta Y_i[j]$. If $\delta Y_i[j] = 0$, we get a linear equation $lf_{Y_i[j]} = 0$. And if $\delta Y_i[j] = 1$, we get a nonzero condition $lf_{Y_i[j]} \neq 0$. The other reduced linear forms $\Delta PR, \Delta X_2, \Delta X_4, \ldots, \Delta X_r$ in the right chain can be sorted according to their truncated differential values in same way. Letting z be the total number of 0's in truncated differences $\delta PR, \delta Y_{2i-1}, \delta X_{2i}, (1 \leq i \leq r/2)$, we get a linear homogeneous equations system which has z equations and a set of nonzero conditions which has $m(r+1) - z$ conditions for the right chain.

Definition 5. *Let $\mathcal{F}$ be a r round Feistel cipher with SPN round function defined in Section 2, and let TP be a truncated differential path of $\mathcal{F}$. We define LES_{Right}, LES_{Left} as equation systems made of TP for the right chain and the left chain, respectively. And we define NC_{Right}, NC_{Left} as sets of nonzero conditions made of TP for the right chain and the left chain, respectively.*

Then we evaluate both chains by using $LES_{Right}, LES_{Left}, NC_{Right}, NC_{Left}$. Moreover, we use the following useful definition.

Definition 6. (reduced row-echelon matrix) *A matrix is a reduced row-echelon matrix if*

- *All rows of zero (if exists) are at the bottom of the matrix.*
- *The first nonzero number in a row is a 1 (leading 1).*
- *Each leading 1 is to the right of the leading 1's in the rows above it.*
- *Each column that contains a leading 1 has zeros everywhere else.*

Any matrix can be transformed into an unique reduced row-echelon matrix by performing a finite sequence of elementally row operations (sweep out method). Letting E_M be the set of all matrices which are obtained by performing elementally row operations on M, every matrix in E_M has the same reduced row-echelon matrix M'.

The following important properties related to a reduced row-echelon matrix and a homogeneous linear equations system are obtained.

Property 1. (partial solution) Let $Mx = 0$ be a matrix and vector representation of a linear homogeneous equations system with a vector of variables $x = {}^t(x_1, x_2, \ldots, x_k)$. And let M' be a reduced row-echelon matrix obtained from M. If there is any row containing only a leading 1 in M'. Letting i be the column index of the leading 1, the system has a solution $x_i = 0$.

Property 2. (property of kernel) Let $Mx = 0$ be a matrix and vector representation of a linear homogeneous equations system. And let $Z = \{z | Mz = 0\}$ be a set of solutions. And let M' be a reduced row-echelon matrix obtained from M. Let $v_1, \ldots, v_r$ be each nonzero row vector in M'. Let S be a subspace spanned by $(v_1, \ldots, v_r)$. Then $v_S \in S, z \in Z$ satisfies $v_S \cdot z = 0$. ($\cdot$ denotes inner products of vectors.)

Using the above properties we construct an evaluation algorithm as follows.

Algorithm
INPUT: $LES_{Right}, NC_{Right}, LES_{Left}, NC_{Left}$
OUTPUT: "wrong path" or "OK"

1. Represent linear equations system LES_{Right} in matrix and vector form as follows:

$$M_{Right} \, x_{Right} = 0$$

2. Perform elementary row operations on matrix M_{Right}, and get a reduced row-echelon matrix M'_{Right}
3. If there is any row containing only leading 1 in M'_{Right}, output "wrong path"; else go to the next step.
4. For each nonzero condition $nc \in NC_{Right}$ do the following:
   a) Represent nonzero condition nc by an inner product of vectors: $n \cdot x_{Right}$
   b) Let $v_1, \ldots, v_r$ be nonzero row vectors in M'_{Right}. If vector n can be represented by a linear combination of $v_1, \ldots, v_r$,

$$n = c_1 v_1 + \ldots + c_r v_r$$

   where $c_i \in GF(2^n), (1 \leq i \leq r)$,
   then output "wrong path"; else go to the next step.
5. For LES_{Left} and NC_{Left}, apply the same steps Step 1. $\sim$ Step 4.
6. Output "OK"

In Step 3. if such a row containing only a leading 1 exists, the corresponding variable in x_{Right} has to be always 0 (property 1). But all variables in x_{Right} have been selected to have a nonzero value, the solution cannot be realized.

In Step 4.(b) the check whether vector n can be represented as a linear combination can be achieved by easy row operations of the matrix because $v_1, \ldots, v_r$ are linearly independent and the vectors have the 4th condition of row echelon matrix. And if such a linear combination exists, nc must have difference 0 in spite of nonzero condition (property 2). This contradiction means the path is wrong.

4.3 Complexity

Let r be the round number of truncated differential path, and let m be a number of split data in F-function. Each chain has a $m(r+1) \times mr$ matrix of linear forms. If LES_{Right} has x rows, then NC_{Right} has $m(r+1) - x$ rows. Complexity to obtain reduced row-echelon matrix of LES_{Right} is proportional to xm^2r^2. And complexity to check NC_{Right} rows is proportional to $mrx(m(r+1) - x)$. Since we assume that $x \approx mr/c$ with some constant c, the total complexity of the proposed method is $O(m^3r^3)$.

4.4 Expansion to Truncated Linear Paths

The proposed method is also applicable to evaluation of truncated linear paths. Let ΓS be an input linear mask and let ΓY be an output linear mask of linear transformation layer P. Then there is a diffusion function P^* satisfying $\Gamma S = P^*(\Gamma Y)$ determined by P. If the diffusion function P^* can be expressed as a square matrix of order m on $GF(2^n)$, the proposed algorithm can be applied to truncated linear paths of the Feistel cipher exploiting the dual property between differential and linear mask[5].

In case of Camellia, matrix P^* can be expressed as ${}^tP_{Camellia}$ which is a square matrix of degree m on $GF(2^n)$. Thus, our method can be applied to evaluation for both truncated differential and linear paths for Camellia.

5 Evaluation for Camellia

We applied the proposed method to block cipher Camellia to reevaluate the upper bound of MDCP and MLCP. To obtain corrected values of the least numbers of active S-boxes, we also extracted truncated paths which have more than the least number of active S-boxes searched by the algorithm used in designer's (previous) evaluation. We used the following definitions and lemma to extract such truncated paths.

Definition 7. *Let $L_{act}(C, R)$ be the least number of active S-boxes of R round truncated path with output truncated difference C in the previous evaluation.*

The algorithm for counting the least number of active S-boxes outputs $L_{act}(C, R)$ for all $C \in \{0, 1\}^{16}$ with any round number R.

Definition 8. *Let* $Cipher(ACT, R) = \{c \mid L_{act}(c, R) = ACT\}$. *And let* $Paths(C, ACT, R)$ *be a set of all* R *round truncated differential paths which have* ACT *active S-boxes with output truncated difference* C *in the previous evaluation. And let* $Paths(ACT, R)$ *be a set of all* R *round truncated differential paths which have* ACT *active S-boxes in the previous evaluation.*

Lemma 2. *Letting* α *be the least number of active S-boxes for* r *round Camellia, truncated differential paths which have* $\beta(\geq \alpha)$ *active S-boxes can be written as follows:*

$$Paths(\beta, r) = \{p \mid p \in Paths(c, \beta, r), c \in \bigcup_{i=\alpha}^{\beta} Cipher(i, r)\} \tag{5}$$

Proof. Truncated differential paths which have β active S-boxes are also written as follows:

$$Paths(\beta, r) = \{p \mid p \in Paths(c, \beta, r), c \in \{0, 1\}^{16}\} \tag{6}$$

Let $D = \bigcup_{i=\beta+1}^{\infty} Cipher(i, r)$. For $d \in D$, we get following relations:

$$Paths(d, \beta, r) = \emptyset \tag{7}$$

Thus, we can reduce the candidates for c in (6) then obtain the equation (5). □

By using the above lemma, we obtain truncated paths which have variable number of active S-boxes. To obtain $Paths(c, \beta, r)$ of $c \in Cipher(\gamma, r)$ where $\beta > \gamma$, we extract paths by considering replacement of internal r' round paths $Paths(c', \gamma', r')$ in $Paths(c, \gamma, r)$ as subset of $Paths(\gamma' + \beta - \gamma, r')$, which are connectable to $r' + 1$ round's output, for $(1 \leq r' \leq r - 1)$.

5.1 Result

We show the result of our re-estimation of the upper bound of MDCP and MLCP in Table 1~3.

Tables show the evaluation results of truncated differential paths and truncated linear paths, respectively. The first row of each table indicates round number of reduced round Camellia without FL/FL^{-1}(we call it Camellia*). The rows below the first are divided into three parts. The upper part, the middle part and the lower part include information about the least number, the secondary least number, and the tertiary or the tertiary and the forth least number of active S-boxes, respectively. Each part has three rows. The first one indicates the number of active S-boxes, the second one indicates the number of truncated differential or linear paths which are evaluated to have the above number of active S-boxes by the previous(old) evaluation method. The last row of each part

Table 1. Check Result of Truncated Differential Paths (1~12 rounds)

Round	1	2	3	4	5	6	7	8	9	10	11	12
Active S-box	0	1	2	7	9	11	13	15	18	21	22	25
Path Number	255	16	8	1368	160	8	16	48	48	320	4	64
OK	255	16	8	1328	80	0	0	0	0	0	0	0
Active S-box	1	2	3	8	10	12	14	16	19	22	23	26
Path Number	19440	292	0	9817	5564	3136	1160	368	2428	14732	224	1736
OK	19440	292	0	7431	4408	1008	196	36	0	156	0	0
Active S-box	2	3	4	9	11	13	15	17	20	23	24	27
Path Number	182228	5000	112	114256	66624	41576	16748	4360	42456	302784	4784	38672
OK	182228	5000	112	72432	17772	4336	1568	328	812	2176	16	0

Table 2. Check Result of Truncated Linear Paths (1~12 rounds)

Round	1	2	3	4	5	6	7	8	9	10	11	12
Active S-box	0	1	2	6	9	11	13	14	18	20	22	25
Path Number	255	16	8	64	480	8	16	4	180	16	4	128
OK	255	16	8	16	240	0	0	0	0	0	0	0
Active S-box	1	2	3	7	10	12	14	15	19	21	23	26
Path Number	19440	292	0	1952	8732	4540	1208	24	4968	1040	448	3188
OK	19440	292	0	1656	4864	1948	192	0	144	0	0	0
Active S-box	2	3	4	8	11	13	15	16/17	20	22	24	27
Path Number	182228	4960	112	9733	76820	52072	21956	184/8142	63128	28640	10640	83800
OK	182228	4960	112	7327	26748	5680	5920	0/1560	4992	792	36	0

indicates the number of truncated differential or linear paths which are evaluated
as "OK" by proposed method.

In 6 ~ 12 rounds Camellia*, truncated differential and linear paths which are
evaluated to have the least number of active S-boxes are all evaluated as wrong.
At some rounds, all truncated paths with the secondary or the tertiary number
of active S-boxes are evaluated as wrong.

The left side of table 3 shows new upper bounds of MDCP and MLCP for
Camellia and Camellia* while the right side shows the old result in [2]. Letting
a be the number of active S-boxes, each probability is calculated by $(2^{-6})^a$. In
NEW table, $\star$ denotes the upper bounds which are updated. All upper bounds
of more than 6 round Camellia and Camellia* are as $2^{-6} \sim 2^{-18}$ times low
as previous upper bounds. The horizontal lines between upper bounds denotes
probability 2^{-128}. Thus, we obtain the new result that any characteristics of 10
rounds Camellia* cannot be distinguishable from random permutations.

Table 3. Revised Upper bounds MDCP and MLCP

	NEW				OLD			
	Differential		Linear		Differential		Linear	
	Camellia	Camellia*	Camellia	Camellia*	Camellia	Camellia*	Camellia	Camellia*
1	$1_{[0]}$	$1_{[0]}$	$1_{[0]}$	$1_{[0]}$	$1_{[0]}$	$1_{[0]}$	$1_{[0]}$	$1_{[0]}$
2	$2^{-6}_{[1]}$	$2^{-6}_{[1]}$	$2^{-6}_{[1]}$	$2^{-6}_{[1]}$	$2^{-6}_{[1]}$	$2^{-6}_{[1]}$	$2^{-6}_{[1]}$	$2^{-6}_{[1]}$
3	$2^{-12}_{[2]}$	$2^{-12}_{[2]}$	$2^{-12}_{[2]}$	$2^{-12}_{[2]}$	$2^{-12}_{[2]}$	$2^{-12}_{[2]}$	$2^{-12}_{[2]}$	$2^{-12}_{[2]}$
4	$2^{-42}_{[7]}$	$2^{-42}_{[7]}$	$2^{-36}_{[6]}$	$2^{-36}_{[6]}$	$2^{-42}_{[7]}$	$2^{-42}_{[7]}$	$2^{-36}_{[6]}$	$2^{-36}_{[6]}$
5	$2^{-54}_{[9]}$	$2^{-54}_{[9]}$	$2^{-54}_{[9]}$	$2^{-54}_{[9]}$	$2^{-54}_{[9]}$	$2^{-54}_{[9]}$	$2^{-54}_{[9]}$	$2^{-54}_{[9]}$
6	$\star 2^{-72}_{[12]}$	$\star 2^{-72}_{[12]}$	$\star 2^{-72}_{[12]}$	$\star 2^{-72}_{[12]}$ ⟸	$2^{-66}_{[11]}$	$2^{-66}_{[11]}$	$2^{-66}_{[11]}$	$2^{-66}_{[11]}$
7	$\star 2^{-78}_{[13]}$	$\star 2^{-84}_{[14]}$	$\star 2^{-78}_{[13]}$	$\star 2^{-84}_{[14]}$	$2^{-72}_{[12]}$	$2^{-78}_{[13]}$	$2^{-72}_{[12]}$	$2^{-78}_{[13]}$
8	$\star 2^{-78}_{[13]}$	$\star 2^{-96}_{[16]}$	$\star 2^{-78}_{[13]}$	$\star 2^{-102}_{[17]}$	$2^{-72}_{[12]}$	$2^{-90}_{[15]}$	$2^{-72}_{[12]}$	$2^{-84}_{[14]}$
9	$\star 2^{-84}_{[14]}$	$\star 2^{-120}_{[20]}$	$\star 2^{-84}_{[14]}$	$\star 2^{-114}_{[19]}$	$2^{-78}_{[13]}$	$2^{-108}_{[18]}$	$2^{-78}_{[13]}$	$2^{-108}_{[18]}$
10	$\star 2^{-114}_{[19]}$	$\star 2^{-132}_{[22]}$	$\star 2^{-108}_{[18]}$	$\star 2^{-132}_{[22]}$	$2^{-108}_{[18]}$	$2^{-126}_{[21]}$	$2^{-102}_{[17]}$	$2^{-120}_{[20]}$
11	$\star 2^{-126}_{[21]}$	$\star 2^{-144}_{[24]}$	$\star 2^{-126}_{[21]}$	$\star 2^{-144}_{[24]}$	$2^{-120}_{[20]}$	$2^{-132}_{[22]}$	$2^{-120}_{[20]}$	$2^{-132}_{[22]}$
12	$\star 2^{-144}_{[24]}$	—	$\star 2^{-144}_{[24]}$	—	$2^{-132}_{[22]}$	—	$2^{-132}_{[22]}$	—

Note: The numbers in brackets are the number of active S-boxes.

6   Conclusions and Future Research

In this paper we proposed a new algorithm to evaluate truncated differential/linear paths whether they are wrong or not. This algorithm is applicable to Feistel ciphers whose data are represented by linear forms of the output difference of S-boxes and the difference of plain text. By applying the proposed algorithm to evaluate the security of Camellia, we found tighter upper bounds of MDCP and MLCP because the least number of active S-boxes are updated. It is revealed that Camellia has a stronger immunity against differential attack and linear attack than before.

Note that even though a truncated differential path is not judged as a wrong path by proposed algorithm, it is not guaranteed that the path exists in reality. Because we have only excluded wrong paths in the context of truncated differential, we haven't taken into consideration of the variety of output differences of active S-boxes for fixed input difference. The direction of next research contains to find the MDCP, MLCP and paths which realize them by combining our proposed evaluation method and the properties of S-boxes.

References

1. K. Aoki, T. Ichikawa, M. Kanda, M. Matsui, S. Moriai, J. Nakajima and T. Tokita, "Specification of Camellia – a 128-bit Block Cipher," submitted to the First Open NESSIE Workshop, 13-14 November 2000, Leuven, Belgium – available at http://cryptonessie.org.
2. K. Aoki, T. Ichikawa, M. Kanda, M. Matsui, S. Moriai, J. Nakajima and T. Tokita, "Camellia: A 128-Bit Block Cipher Suitable for Multiple Platforms - Design and Analysis," Selected Area in Cryptography, SAC 2000, LNCS2012, pp.39-56, 2000.

3. E. Biham and A. Shamir, "Differential Cryptanalysis of DES-like Cryptosystems," CRYPTO '90, LNCS 537, pp.2-21, 1991.
4. Y. He and S. Qing, "Square Attack on Reduced Camellia Cipher," Information and Communications Security, ICICS 2001, LNCS 2229, pp.238-245, 2001.
5. M. Kanda and T. Matsumoto, "Security of Camellia against Truncated Differential Cryptanalysis," Fast Software Encryption, FSE2001, to appear
6. M. Kanda, S. Moriai, K. Aoki, H. Ueda, Y. Takashima, K. Ohta and T. Matsumoto, "E2 – A New 128-Bit Block Cipher," IEICE Transactions Fundamentals of Electronics, Communications and Computer Sciences, Vol. E83-A, No.1, pp.48-59, 2000.
7. T. Kawabata and T.Kaneko, "A study on higher order differential attack of Camellia," In Proceedings of the 2nd NESSIE workshop, 2001.
8. L.R.Knudsen, "Truncated and Higher Order Differentials," Fast Software Encryption – Second International Workshop, LNCS 1008, pp.196-211, 1995.
9. M.Matsui, "Differential Path Search of the Block Cipher E2," Technical Report ISEC99-19, IEICE, 1999.(written in Japanese)
10. M. Matsui and T. Tokita,"Cryptanalysis of Reduced Version of the Block Cipher E2," Fast Software Encryption, FSE'99, LNCS 1636, 1999.
11. M. Matsui, "Linear Cryptanalysis of the Data Encryption Standard," EUROCRYPT '93, LNCS 765, pp.386-397, 1994.
12. S. Moriai, M. Sugita, K. Aoki and M. Kanda, "Security of E2 against Truncated Differential Cryptanalysis," Selected Areas in Cryptography, SAC'99, LNCS 1758, pp.106-117, 2000.
13. M. Sugita, K. Kobara and H. Imai,"Security of Reduced Version of the Block Cipher Camellia against Truncated and Impossible Differential Cryptanalysis," ASISCRYPT 2001, LNCS 2248, pp.193-207.
14. New European Schemes for Signatures, Integrity, and Encryption, http://www.cryptonessie.org
15. CRYPTREC project, http://www.ipa.go.jp/security/enc/CRYPTREC/.

The Round Functions of RIJNDAEL Generate the Alternating Group

Ralph Wernsdorf

Rohde & Schwarz SIT GmbH, Agastraße 3,
D-12489 Berlin, Germany
Ralph.Wernsdorf@SIT.rohde-schwarz.com

Abstract. For the block cipher RIJNDAEL with a block length of 128 bits group theoretic properties of the round functions are derived. Especially it is shown that these round functions generate the alternating group.

Keywords. RIJNDAEL, permutation groups.

1 Introduction

The RIJNDAEL algorithm with a block length of 128 bits is a block cipher that was selected for a NIST standard [4] in the AES selection process. It was developed by J. Daemen and V. Rijmen [1].

In the following a proof is given that the round functions of RIJNDAEL with a block length of 128 bits generate the alternating group over the set $\{0,1\}^{128}$ of all 128-bit-vectors.

This result implies that from the algebraic point of view some thinkable weaknesses of RIJNDAEL can be excluded (if the generated group were smaller, then this would point to regularities in the algorithm, see for example [2], [5], [9]).

Similar properties are known for the DES ([6]) and for SAFER++ ([7]).

2 Definitions and Notations

The notation of the RIJNDAEL round function components will be similar to the RIJNDAEL definition given in [1]. One exception will be that the states are not given in a matrix form. They are given as 128-bit- or 16-byte-vectors, where the correspondence to the matrices in [1] is defined row by row from left to right. The round functions R_k are defined by

$$\forall k \in \{0,1\}^{128} \ \forall x \in \{0,1\}^{128} : R_k(x) := k \oplus mc(rs(s(x))),$$

J. Daemen and V. Rijmen (Eds.): FSE 2002, LNCS 2365, pp. 143–148, 2002.

where   "$\oplus$"   denotes the bitwise XOR-operation,

$k \in \{0,1\}^{128}$ denotes the corresponding round subkey,

$mc : \{0,1\}^{128} \to \{0,1\}^{128}$ denotes the *MixColumn*-transformation according to [1], p. 12,

$rs : \{0,1\}^{128} \to \{0,1\}^{128}$ denotes the *ShiftRow*-transformation according to [1], p. 11,

$s : \{0,1\}^{128} \to \{0,1\}^{128}$ denotes the application of the S-box S 16 times in parallel (the *ByteSub*-transformation, [1], p. 11).

The permutation group considered here is defined by:

$$G := \langle \{R_k : \{0,1\}^{128} \to \{0,1\}^{128} \mid k \in \{0,1\}^{128}\} \rangle,$$

where $\langle P \rangle$ denotes the closure of a permutation set P with respect to concatenation. The generating set of G given above contains 2^{128} permutations. Properties of the round subkeys caused by the key scheduling will be neglected here.

3   Some Properties of the Generated Group

Lemma 1. *The group G is transitive on the set $\{0,1\}^{128}$.*

Proof. By concatenations $R_k \circ R_{k'}^{-1}$ (where the transformation with k' is carried out at first) with suitable round subkeys k, k', each given element x of the set $\{0,1\}^{128}$ can obviously be transformed to each other arbitrarily given element x' of the set $\{0,1\}^{128}$. This can be achieved very simply by choosing $k \oplus k' = x \oplus x'$.
□

Lemma 2. *The group G contains only even permutations.*

Proof. In the following we will apply the well known fact that a permutation over a set with an even number of elements is even if and only if its cycle representation (including the fixed points) contains an even number of cycles.

The mappings $x \to k \oplus x$ are even permutations since for $k = (0,0,...,0)$ we obtain the identity permutation and for $k \neq (0,0,...,0)$ the cycle representation consists of 2^{127} cycles of length 2.

The linear transformations mc and rs are even permutations since binary one-to-one linear transformations over $\{0,1\}^n, n \geq 3$, are even permutations. This follows from the facts that each regular binary matrix can be obtained from the identity matrix by elementary row transformations (i.e., binary addition of one row to another row), and that these elementary row transformations (considered as linear mappings over $\{0,1\}^n$) are permutations with 2^{n-1} fixed points and 2^{n-2} cycles of length 2.

The permutation s is even because it can be represented as a concatenation of 16 permutations that carry out the S-box transformation for one byte and leave the other 15 bytes fixed. The cycle representation of each such permutation contains a number of cycles that is a multiple of 2^{120}.

Now the proof is complete since the concatenation of even permutations always yields an even permutation.   □

Lemma 3. *The S-box permutation $S : \{0,1\}^8 \to \{0,1\}^8$ is an odd permutation.*

Proof. The cycle representation of S consists of five cycles (with lengths 87, 81, 59, 27, and 2, respectively). Applying the fact mentioned at the beginning of the proof of Lemma 2, it follows that S is an odd permutation.   □

Lemma 4. *For all 16-tuples of even permutations $P_j : \{0,1\}^8 \to \{0,1\}^8$, $j = 0, 1, ..., 15$, the mapping*

$$M' : \{0,1\}^{128} \to \{0,1\}^{128}, (x_0, ..., x_{127}) \mapsto (y_0, ..., y_{127}),$$

defined by

$$(y_{8j}, y_{8j+1}, ..., y_{8j+7}) := P_j (x_{8j}, x_{8j+1}, ..., x_{8j+7})$$

for all $j = 0, 1, ..., 15$, is an element of G.

Proof. We consider products of the form $R_k^{-1} \circ R_{k'}$ and $R_k \circ R_{k'}^{-1}$ (where the transformations with k' are carried out first).
Let $j \in \{0, 1, ..., 15\}$ be arbitrarily fixed. Then we have for all $x \in \{0,1\}^{128}$:
    If $\forall\, i \in \{0, 1, ..., 127\} \setminus \{8j, 8j + 1, ..., 8j + 7\} :\ k_i = k'_i$,
    then $R_k \circ R_{k'}^{-1}$ changes no more than the j-th byte of x.
    If $\forall\, i \in \{0, 1, ..., 127\} \setminus \{8j, 8j + 1, ..., 8j + 7\} :$

$$(rs^{-1}(mc^{-1}(k)))_i = (rs^{-1}(mc^{-1}(k')))_i,$$

    then $R_k^{-1} \circ R_{k'}$ changes no more than the j-th byte of x.
This implies that we can define a subgroup of G, namely the group that is generated by all permutations $R_k^{-1} \circ R_{k'}$ and $R_k \circ R_{k'}^{-1}$ with the subkey pairs (k, k') as above. This subgroup acts only on the j-th byte and will therefore be considered as a permutation group on $\{0,1\}^8$. This group is generated by all byte-permutations of the form $x \to k \oplus x$ and all byte-permutations of the form $x \to S^{-1}(k \oplus S(x))$. Therefore it is transitive on this set and it contains only even permutations on this set.
After a random search for the following subkeys $(k^1, k^2, ..., k^8)$ and the following state x, a cycle of the permutation

$$R_{k^8} \circ R_{k^7}^{-1} \circ R_{k^6}^{-1} \circ R_{k^5} \circ R_{k^4} \circ R_{k^3}^{-1} \circ R_{k^2}^{-1} \circ R_{k^1}$$

with length 233 was found (bytes in hexadecimal notation):
    $k^1 = k^3 = k^5 = k^7 = (0, 0, ..., 0)$,
    the j-th byte of $rs^{-1}(mc^{-1}(k^2))$ is 0xb9, the other 15 bytes are zero,

the j-th byte of k^4 is 0x8b, the other 15 bytes are zero,

the j-th byte of $rs^{-1}(mc^{-1}(k^6))$ is 0xdd, the other 15 bytes are zero,

the j-th byte of k^8 is 0x1f, the other 15 bytes are zero,

the j-th byte of x is 0x9e, the other 15 bytes are randomly chosen.

Since the cycle length 233 is a prime greater than $256/2$ and less than 256, the subgroup is primitive on $\{0,1\}^8$ (such a cycle does not match to any structure of imprimitivity, see [7]).

Now Theorem 13.9 in [8], p. 39 can be applied: *"Let p be a prime and G be a group of degree $n = p + k$ with $k \geq 3$. If G contains an element of degree and order p, then G is either alternating or symmetric."* Here the *degree* of a permutation group over a finite set is defined as the number of elements that are changed by at least one permutation of the group. The *degree of a permutation* is defined as the degree of the cyclic group generated by this permutation.

It follows that the considered subgroup equals the alternating group on $\{0,1\}^8$. (Other cycles can be "cancelled" by considering suitable powers of the permutation.)

Since j and the remaining 15 bytes of x were chosen arbitrarily and the considered transformations have no influence on other bytes than the j-th byte, the proof of the Lemma is complete. □

Corollary 1. *The transformation $mc \circ rs \circ h$, where h denotes an arbitrarily fixed parallel application of 16 odd byte-permutations, is an element of G.*

Proof. We choose the round function R_k with the all zero subkey. From Lemma 4 we know that all parallel applications of 16 even byte-permutations are elements of G. Since the S-box is an odd byte-permutation and $mc \circ rs \circ s$ is an element of G, the proof can easily be completed by considering concatenations of group elements. □

4   Proof That the Round Functions Generate the Alternating Group

Lemma 5. *The group G is doubly transitive on the set $\{0,1\}^{128}$, i.e., each pair of different elements from $\{0,1\}^{128}$ can be mapped to each pair of different elements from $\{0,1\}^{128}$.*

Proof. Because the group G is transitive on the set $\{0,1\}^{128}$, it suffices to show that the subgroup G_0 of G containing all elements of G which let the all zero vector fixed, is transitive on $\{0,1\}^8 \setminus \{(0,0,...,0)\}$ (see for example [8], p. 19).

Let us start with an arbitrary non-zero vector $X \in \{0,1\}^{128}$. With the help of Lemma 4, it can be shown that it is possible to find an element of the subgroup G_0 that transforms X to a vector $X' \neq (0,0,...,0)$ with:

$$\forall j \in \{0,1,...,15\} \ : \ (X'_{8j}, X'_{8j+1}, ..., X'_{8j+7}) \in \{0\text{x}00, 0\text{x}01\}.$$

(Choose even permutations P_j that (*) let 0x00 fixed and that transform the non-all-zero-components $(X'_{8j}, X'_{8j+1}, ..., X'_{8j+7})$ to 0x01.)

By computations on a PC (there are only $2^{16} - 1$ such vectors X'), it was verified that it is possible to transform the mentioned X' to the vector $(0x01, 0x01, ..., 0x01)$ by repeated concatenations of $mc \circ rs \circ h'$ and permutations of the form (*) above. Here h' denotes the parallel application of 16 times the odd byte-permutation that changes 0x01 to 0x02, changes 0x02 to 0x01 and lets all other bytes fixed.

Because we have $mc \circ rs \circ h' \in G_0$ (see Corollary 1), it follows that G_0 is transitive on the set $\{0, 1\}^{128} \setminus \{(0, 0, ..., 0)\}$. Hence, G is doubly transitive on the set $\{0, 1\}^{128}$. $\qquad\square$

Theorem 1. *The group G is the alternating group on the set $\{0, 1\}^{128}$.*

Proof. From Lemma 4 it follows that the permutation $(P_0, P_1, P_1, ..., P_1)$, where P_1 is the identity permutation on $\{0, 1\}^8$ and where the cycle representation of P_0 contains a 3-cycle and 253 fixed points, is an element of G. This permutation lets exactly $253 \cdot 2^{15 \cdot 8}$ elements fixed. Hence, its degree is equal to $3 \cdot 2^{120}$.

The *minimal degree* of a permutation group is the smallest degree of the non-identity-permutations in the group. The minimal degree of G is not greater than $3 \cdot 2^{120}$.

Now, let us suppose that G is smaller than the alternating group. Then, (because of Lemma 5 G is doubly transitive) according to Theorem 15.1 in [8], p. 42 : *"Let G be a a k-ply transitive group, neither alternating nor symmetric. Let n be its degree, m its minimal degree. If $k \geq 2$, then $m \geq \frac{n}{3} - \frac{2\sqrt{n}}{3}$."*, we obtain the contradiction:

$$3 \cdot 2^{120} \geq \frac{2^{128}}{3} - \frac{2^{65}}{3}.$$

From this together with the result of Lemma 2, it follows that G is the alternating group on the set $\{0, 1\}^{128}$. $\qquad\square$

5   Conclusions and Remarks

By the result stated in the Theorem, several thinkable regularities like the existence of nontrivial factor groups or a too small diversity of occurring permutations in the RIJNDAEL algorithm can be excluded (the alternating group is a large, simple, primitive and $(2^{128} - 2)$-transitive permutation group).

With respect to the Markov approach to differential cryptanalysis, we obtain [2]: For all corresponding Markov ciphers the chain of differences is irreducible and aperiodic, i.e., after sufficiently many RIJNDAEL rounds all differences will be almost equally probable. If the hypothesis of stochastic equivalence holds for a part of the corresponding Markov ciphers, then for all of these Markov ciphers RIJNDAEL is secure against differential cryptanalysis attacks after a sufficient number of rounds.

The results give evidence that the S-box and the other transformations are well chosen from the algebraic point of view. Especially the results of [3] have no consequences with respect to the sizes of the considered permutation groups.

In the following remarks some generalizations (with proof sketches) are given:

Remark 1. Let us take a set of 129 pairwise different round subkeys that includes a basis of the binary vector space $\{0, 1\}^{128}$. Then the 129 corresponding round functions suffice to generate the alternating group. This follows from the fact that all round functions can be represented as concatenations of these 129 round functions and their inverses.

Remark 2. Let us take a byte permutation S' such that all byte-permutations of the form $x \to k \oplus x$ and all byte-permutations of the form $x \to S'^{-1}(k \oplus S'(x))$ generate the alternating group on $\{0, 1\}^8$. Then the round functions of (modified) RIJNDAEL with the S-box S' also generate the alternating group. For odd permutations S' this can be derived from the proofs given above. For even permutations S' Lemma 5 can be proved in a similar way and the other steps are the same (the result of Corollary 1 is not needed here).

References

1. Daemen, J. and Rijmen, V.: *AES-proposal RIJNDAEL, Version 2*, September 8, 1999. Available at http://www.esat.kuleuven.ac.be/~rijmen/rijndael/.
2. Hornauer, G., Stephan, W., and Wernsdorf, R.: Markov Ciphers and Alternating Groups. In *Advances in Cryptology – EUROCRYPT'93*, Lecture Notes in Computer Science, vol. 765, Springer 1994, 453–460.
3. Murphy, S. and Robshaw, M.: *New Observations on Rijndael, Preliminary Draft*, August 7, 2000.
4. National Institute of Standards and Technology (U.S.): *Advanced Encryption Standard (AES)*, FIPS Publication 197, November 26, 2001. Available at http://csrc.nist.gov/publications/fips/fips197/fips-197.pdf.
5. Paterson, K. G.: Imprimitive Permutation Groups and Trapdoors in Iterated Block Ciphers. In *Fast Software Encryption – FSE'99*, Lecture Notes in Computer Science, vol. 1636, Springer 1999, 201-214.
6. Wernsdorf, R.: The One-Round Functions of the DES Generate the Alternating Group. In *Advances in Cryptology - EUROCRYPT'92*, Lecture Notes in Computer Science, vol. 658, Springer 1993, 99–112.
7. Wernsdorf, R.: *IDEA, SAFER++ and Their Permutation Groups*, Second Open NESSIE Workshop, Royal Holloway University of London, Egham, September 2001.
8. Wielandt, H.: *Finite Permutation Groups*. Academic Press, New York and London, 1964.
9. Zieschang, T.: Combinatorial Properties of Basic Encryption Operations. In *Advances in Cryptology - EUROCRYPT'97*, Lecture Notes in Computer Science, vol. 1233, Springer 1997, 14–26.

Non-cryptographic Primitive for Pseudorandom Permutation

Tetsu Iwata[1], Tomonobu Yoshino[1], and Kaoru Kurosawa[2]

[1] Department of Communications and Integrated Systems,
Tokyo Institute of Technology
2–12–1 O-okayama, Meguro-ku, Tokyo 152-8552, Japan
tez@ss.titech.ac.jp
[2] Department of Computer and Information Sciences,
Ibaraki University
4–12–1 Nakanarusawa, Hitachi, Ibaraki 316-8511, Japan
kurosawa@cis.ibaraki.ac.jp

Abstract. Four round Feistel permutation (like DES) is super-pseudorandom if each round function is *random* or a *secret* universal hash function. A similar result is known for five round MISTY type permutation. It seems that each round function must be at least either *random* or *secret* in both cases.

In this paper, however, we show that the second round permutation g in five round MISTY type permutation need not be cryptographic at all, i.e., no randomness nor secrecy is required. g has only to satisfy that $g(x) \oplus x \neq g(x') \oplus x'$ for any $x \neq x'$. This is the first example such that a non-cryptographic primitive is substituted to construct the minimum round super-pseudorandom permutation. Further we show efficient constructions of super-pseudorandom permutations by using above mentioned g.

Keywords: Block cipher, pseudorandomness, MISTY type permutation.

1 Introduction

1.1 Super-Pseudorandomness

A secure block cipher should be indistinguishable from a truly random permutation. Consider an infinitely powerful distinguisher $\mathcal{D}$ which tries to distinguish a block cipher from a truly random permutation. It outputs 0 or 1 after making at most m queries to the given encryption and/or decryption oracles. We say that a distinguisher $\mathcal{D}$ is a pseudorandom distinguisher if it has oracle access to the encryption oracle. We also say that a distinguisher $\mathcal{D}$ is a super-pseudorandom distinguisher if it has oracle access to both the encryption oracle and the decryption oracle. Then a block cipher E is called pseudorandom if any pseudorandom distinguisher $\mathcal{D}$ cannot distinguish E from a truly random permutation. A block cipher E is called super-pseudorandom if any super-pseudorandom distinguisher $\mathcal{D}$ cannot distinguish E from a truly random permutation.

J. Daemen and V. Rijmen (Eds.): FSE 2002, LNCS 2365, pp. 149–163, 2002.

1.2   Previous Works

The super-pseudorandomness of Feistel permutation (like DES) has been studied extensively so far. Let $\phi(f_1, f_2, f_3)$ denote the three round Feistel permutation such that the i-th round function is f_i. Similarly, let $\phi(f_1, f_2, f_3, f_4)$ denote the four round Feistel permutation.

Suppose that each f_i is a random function. Then Luby and Rackoff proved that $\phi(f_1, f_2, f_3)$ is pseudorandom and $\phi(f_1, f_2, f_3, f_4)$ is super-pseudorandom [4]. Lucks showed that the $\phi(h_1, f_2, f_3)$ is pseudorandom even if h_1 is an ϵ-XOR universal hash function [5]. Suppose that h_1 and h_4 are uniform ϵ-XOR universal hash functions. Then Naor and Reingold proved that $h_4 \circ \phi(f_2, f_3) \circ h_1$ is super-pseudorandom [8], and Ramzan and Reyzin showed that $\phi(h_1, f_2, f_3, h_4)$ is super-pseudorandom even if the distinguisher has oracle access to f_2 and f_3 [9].

On the other hand, let $\psi(p_1, p_2, p_3, p_4, p_5)$ denote the five round MISTY type permutation such that the i-th round permutation is p_i. Suppose that each p_i is a random permutation. Then Iwata et al. [3] and Gilbert and Minier [2] independently showed that $\psi(p_1, p_2, p_3, p_4, p_5)$ is super-pseudorandom. More than that, let h_i be a uniform ϵ-XOR universal permutation. Iwata et al. proved that

1. $\psi(h_1, h_2, p_3, p_4, h_5^{-1})$ is super-pseudorandom even if the distinguisher has oracle access to p_3, p_3^{-1}, p_4 and p_4^{-1}.
2. $\psi(h_1, p_2, p_3, p_4, h_5^{-1})$ is super-pseudorandom even if the distinguisher has oracle access to p_2, p_2^{-1}, p_3, p_3^{-1}, p_4 and p_4^{-1}.

1.3   Our Contribution

Four round Feistel permutation (like DES) is super-pseudorandom if each round function is *random* or a *secret* universal hash function. A similar result is known for five round MISTY type permutation. It seems that each round function must be at least either *random* or *secret* in both cases.

In this paper, however, we show that the second round permutation g in five round MISTY type permutation need not be cryptographic at all, i.e., no randomness nor secrecy is required. g has only to satisfy that $g(x) \oplus x \neq g(x') \oplus x'$ for any $x \neq x'$. This is the first example such that a non-cryptographic primitive is substituted to construct the minimum round super-pseudorandom permutation. Further we show efficient constructions of super-pseudorandom permutations by using above mentioned g.

One might wonder if five rounds can be reduced to four rounds to obtain super-pseudorandomness of MISTY. However, it is not true because Sakurai and Zheng showed that the four round MISTY type permutation $\psi(p_1, p_2, p_3, p_4)$ is not super-pseudorandom [10].

More precisely, we prove that five round MISTY is super-pseudorandom if it is $\psi(h_1, g, p, p^{-1}, h_5^{-1})$, where g is the above mentioned permutation, h_1 is an ϵ-XOR universal permutation, h_5 is a uniform ϵ-XOR universal permutation, and p is a random permutation. Further, suppose that both h_1 and h_5 are *uniform*

ϵ-XOR universal permutations. Then we prove that it is super-pseudorandom even if the distinguisher has oracle access to p and p^{-1}.

More than that, we study the case such that the third and the fourth round permutations are both p. In this case, we show that it is not super-pseudorandom nor pseudorandom if a distinguisher has oracle access to p. More formally, we show that for any fixed and public g, $\psi(p_1, g, p, p, p_5)$ is not pseudorandom if a distinguisher has oracle access to p.

2 Preliminaries

2.1 Notation

For a bit string $x \in \{0,1\}^{2n}$, we denote the first (left) n bits of x by x_L and the last (right) n bits of x by x_R. If S is a probability space, then $s \overset{R}{\leftarrow} S$ denotes the process of picking an element from S according to the underlying probability distribution. The underlying distribution is assumed to be uniform (unless otherwise specified).

Denote by F_n the set of all functions from $\{0,1\}^n$ to $\{0,1\}^n$, which consists of $2^{n \cdot 2^n}$ functions in total. Similarly, denote by P_n the set of all permutations from $\{0,1\}^n$ to $\{0,1\}^n$, which consists of $(2^n)!$ permutations in total.

2.2 MISTY Type Permutation [6,7]

Definition 2.1 (The basic MISTY type permutation). *Let $x \in \{0,1\}^{2n}$. For any permutation $p \in P_n$, define the basic MISTY type permutation $\psi_p \in P_{2n}$ as $\psi_p(x) \overset{\text{def}}{=} (x_R, p(x_L) \oplus x_R)$. Note that it is a permutation since $\psi_p^{-1}(x) = (p^{-1}(x_L \oplus x_R), x_L)$.*

Definition 2.2 (The r round MISTY type permutation, ψ). *Let $r \geq 1$ be an integer, $p_1, \ldots, p_r \in P_n$ be permutations. Define the r round MISTY type permutation $\psi(p_1, \ldots, p_r) \in P_{2n}$ as $\psi(p_1, \ldots, p_r) \overset{\text{def}}{=} \rho \circ \psi_{p_r} \circ \cdots \circ \psi_{p_1}$, where $\rho(x_L, x_R) = (x_R, x_L)$ for $x \in \{0,1\}^{2n}$.*

See Fig. 1 (the five round MISTY type permutation) for an illustration. Note that p_i in Fig. 1 is a permutation. For simplicity, the left and right swaps are omitted.

2.3 Uniform ϵ-XOR Universal Permutation

Our definitions follow from those given in [1,3,9,11].

Definition 2.3. *Let H_n be a permutation family over $\{0,1\}^n$. Denote by $\#H_n$ the size of H_n.*

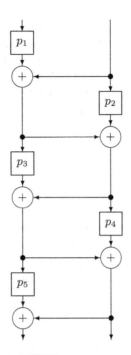

Fig. 1. MISTY type permutation

1. H_n is a uniform *permutation family if for any element* $x \in \{0,1\}^n$ *and any element* $y \in \{0,1\}^n$, *there exist exactly* $\frac{\#H_n}{2^n}$ *permutations* $h \in H_n$ *such that* $h(x) = y$.

2. H_n is an ϵ-XOR universal *permutation family if for any two distinct elements* $x, x' \in \{0,1\}^n$ *and any element* $y \in \{0,1\}^n$, *there exist at most* $\epsilon\#H_n$ *permutations* $h \in H_n$ *such that* $h(x) \oplus h(x') = y$.

Let $f_a(x) \stackrel{\text{def}}{=} a \cdot x$ over $\mathrm{GF}(2^n)$, where $a \neq 0$. Then $\{f_a(x)\}$ is a $\frac{1}{2^n-1}$-XOR universal permutation family.

Let $f_{a,b}(x) \stackrel{\text{def}}{=} a \cdot x + b$ over $\mathrm{GF}(2^n)$, where $a \neq 0$. Then $\{f_{a,b}(x)\}$ is a uniform $\frac{1}{2^n-1}$-XOR universal permutation family.

We will use the phrase "h is an ϵ-XOR universal permutation" to mean that "h is drawn uniformly from an ϵ-XOR universal permutation family". Similarly, we will use the phrase "h is a uniform ϵ-XOR universal permutation".

3   Improved Super-Pseudorandomness of MISTY Type Permutation

We say that a permutation g over $\{0,1\}^n$ is *XOR-distinct* if

$$g(x) \oplus x \neq g(x') \oplus x'$$

for any $x \neq x'$. Let $g(x) = a \cdot x$ over $GF(2^n)$, where $a \neq 0, 1$. Then this g is clearly XOR-distinct.

In this section, we prove that $\psi(h_1, g, p, p^{-1}, h_5^{-1})$ is super-pseudorandom even if the second round permutation g is fixed and publicly known. g has only to be XOR-distinct. This means that the five round MISTY type permutation is super-pseudorandom even if the second round permutation has no randomness nor secrecy.

Let H_n^0 be an ϵ-XOR universal permutation family over $\{0, 1\}^n$, and H_n^1 be a uniform ϵ-XOR universal permutation family over $\{0, 1\}^n$. Define

$$\begin{cases} \mathrm{MISTY}_{2n}^{01} \overset{\text{def}}{=} \{\psi(h_1, g, p, p^{-1}, h_5^{-1}) \mid p \in P_n, h_1 \in H_n^0, h_5 \in H_n^1\} \\ \mathrm{MISTY}_{2n}^{11} \overset{\text{def}}{=} \{\psi(h_1, g, p, p^{-1}, h_5^{-1}) \mid p \in P_n, h_1, h_5 \in H_n^1\} \end{cases}$$

3.1   Super-Pseudorandomness of MISTY_{2n}^{01}

Let $\mathcal{D}$ be a super-pseudorandom distinguisher for MISTY_{2n}^{01} which makes at most m queries in total. We consider two experiments, experiment 0 and experiment 1. In experiment 0, $\mathcal{D}$ has oracle access to ψ and ψ^{-1}, where ψ is randomly chosen from MISTY_{2n}^{01}. In experiment 1, $\mathcal{D}$ has oracle access to R and R^{-1}, where R is randomly chosen from P_{2n}.

Define the advantage of $\mathcal{D}$ as follows.

$$\mathrm{Adv}(\mathcal{D}) \overset{\text{def}}{=} |p_\psi - p_R|$$

where

$$\begin{cases} p_\psi \overset{\text{def}}{=} \Pr(\mathcal{D}^{\psi,\psi^{-1}}(1^{2n}) = 1 \mid \psi \overset{R}{\leftarrow} \mathrm{MISTY}_{2n}^{01}) \\ p_R \overset{\text{def}}{=} \Pr(\mathcal{D}^{R,R^{-1}}(1^{2n}) = 1 \mid R \overset{R}{\leftarrow} P_{2n}) \end{cases}$$

Lemma 3.1. *Fix* $x^{(i)} \in \{0, 1\}^{2n}$ *and* $y^{(i)} \in \{0, 1\}^{2n}$ *for* $1 \leq i \leq m$ *arbitrarily in such a way that* $\{x^{(i)}\}_{1 \leq i \leq m}$ *are all distinct and* $\{y^{(i)}\}_{1 \leq i \leq m}$ *are all distinct. Then the number of* $\psi \in \mathrm{MISTY}_{2n}^{01}$ *such that*

$$\psi(x^{(i)}) = y^{(i)} \text{ for } 1 \leq \forall i \leq m \tag{1}$$

is at least

$$(\#H_n^0)(\#H_n^1)(2^n - 2m)! \left(1 - 2\epsilon \cdot m(m-1) - \frac{2m^2}{2^n}\right) .$$

A proof is given in Appendix A.

Theorem 3.1. *For any super-pseudorandom distinguisher* $\mathcal{D}$ *that makes at most* m *queries in total,*

$$\mathrm{Adv}(\mathcal{D}) \leq 2\epsilon \cdot m(m-1) + \frac{2m^2}{2^n} .$$

Proof. Let $\mathcal{O} = R$ or ψ. The super-pseudorandom distinguisher $\mathcal{D}$ has oracle access to $\mathcal{O}$ and $\mathcal{O}^{-1}$.

There are two types of queries $\mathcal{D}$ can make: either $(+, x)$ which denotes the query "what is $\mathcal{O}(x)$?", or $(-, y)$ which denotes the query "what is $\mathcal{O}^{-1}(y)$?" For the i-th query $\mathcal{D}$ makes to $\mathcal{O}$ or $\mathcal{O}^{-1}$, define the query-answer pair $(x^{(i)}, y^{(i)}) \in \{0,1\}^{2n} \times \{0,1\}^{2n}$, where either $\mathcal{D}$'s query was $(+, x^{(i)})$ and the answer it got was $y^{(i)}$ or $\mathcal{D}$'s query was $(-, y^{(i)})$ and the answer it got was $x^{(i)}$. Define view v of $\mathcal{D}$ as $v = ((x^{(1)}, y^{(1)}), \ldots, (x^{(m)}, y^{(m)}))$.

Without loss of generality, we assume that $\{x^{(i)}\}_{1 \leq i \leq m}$ are all distinct, and $\{y^{(i)}\}_{1 \leq i \leq m}$ are all distinct.

Since $\mathcal{D}$ has unbounded computational power, $\mathcal{D}$ can be assumed to be deterministic. Therefore, the final output of $\mathcal{D}$ (0 or 1) depends only on v. Hence denote by $\mathcal{C}_\mathcal{D}(v)$ the final output of $\mathcal{D}$.

Let $\boldsymbol{v}_{one} \stackrel{\text{def}}{=} \{v \mid \mathcal{C}_\mathcal{D}(v) = 1\}$ and $N_{one} \stackrel{\text{def}}{=} \#\boldsymbol{v}_{one}$.

Evaluation of p_R. From the definition of p_R, we have

$$p_R = \Pr_R(\mathcal{D}^{R,R^{-1}}(1^{2n}) = 1)$$

$$= \frac{\#\{R \mid \mathcal{D}^{R,R^{-1}}(1^{2n}) = 1\}}{(2^{2n})!}.$$

For each $v \in \boldsymbol{v}_{one}$, the number of R such that

$$R(x^{(i)}) = y^{(i)} \text{ for } 1 \leq \forall i \leq m \tag{2}$$

is exactly $(2^{2n} - m)!$. Therefore, we have

$$p_R = \sum_{v \in \boldsymbol{v}_{one}} \frac{\#\{R \mid R \text{ satisfying } (2)\}}{(2^{2n})!}$$

$$= N_{one} \cdot \frac{(2^{2n} - m)!}{(2^{2n})!}.$$

Evaluation of p_ψ. From the definition of p_ψ, we have

$$p_\psi = \Pr_{h_1, p, h_5}(\mathcal{D}^{\psi,\psi^{-1}}(1^{2n}) = 1)$$

$$= \frac{\#\{(h_1, p, h_5) \mid \mathcal{D}^{\psi,\psi^{-1}}(1^{2n}) = 1\}}{(\#H_n^0)(2^n)!(\#H_n^1)}.$$

Similarly to p_R, we have

$$p_\psi = \sum_{v \in \boldsymbol{v}_{one}} \frac{\#\{(h_1, p, h_5) \mid (h_1, p, h_5) \text{ satisfying } (1)\}}{(\#H_n^0)(2^n)!(\#H_n^1)}.$$

Then from Lemma 3.1, we obtain that

$$p_\psi \geq \sum_{v \in \boldsymbol{v}_{one}} \frac{(2^n - 2m)! \left(1 - 2\epsilon \cdot m(m-1) - \frac{2m^2}{2^n}\right)}{(2^n)!}$$

$$= N_{one} \frac{(2^n - 2m)!}{(2^n)!} \left(1 - 2\epsilon \cdot m(m-1) - \frac{2m^2}{2^n}\right)$$

$$= p_R \frac{(2^{2n})!(2^n - 2m)!}{(2^{2n} - m)!(2^n)!} \left(1 - 2\epsilon \cdot m(m-1) - \frac{2m^2}{2^n}\right) .$$

Since $\frac{(2^{2n})!(2^n-2m)!}{(2^{2n}-m)!(2^n)!} \geq 1$ (This can be shown easily by an induction on m), we have

$$p_\psi \geq p_R \left(1 - 2\epsilon \cdot m(m-1) - \frac{2m^2}{2^n}\right)$$

$$\geq p_R - 2\epsilon \cdot m(m-1) - \frac{2m^2}{2^n} . \tag{3}$$

Applying the same argument to $1 - p_\psi$ and $1 - p_R$ yields that

$$1 - p_\psi \geq 1 - p_R - 2\epsilon \cdot m(m-1) - \frac{2m^2}{2^n} . \tag{4}$$

Finally, (3) and (4) give $|p_\psi - p_R| \leq 2\epsilon \cdot m(m-1) + \frac{2m^2}{2^n}$.

3.2   Super-Pseudorandomness of MISTY_{2n}^{11}

Let $\mathcal{D}$ be a super-pseudorandom distinguisher for MISTY_{2n}^{11}. $\mathcal{D}$ also has oracle access to p and p^{-1}, where p and p^{-1} are the third and fourth round permutations of MISTY_{2n}^{11} respectively. $\mathcal{D}$ makes at most m queries in total. We consider two experiments, experiment 0 and experiment 1. In experiment 0, $\mathcal{D}$ has oracle access to not only ψ and ψ^{-1}, but also p and p^{-1}, where ψ is randomly chosen from MISTY_{2n}^{11}. In experiment 1, $\mathcal{D}$ has oracle access to R, R^{-1}, p and p^{-1}, where R is randomly chosen from P_{2n} and p is randomly chosen from P_n.

Define the advantage of $\mathcal{D}$ as follows.

$$\text{Adv}(\mathcal{D}) \overset{\text{def}}{=} |p_\psi - p_R|$$

where

$$\begin{cases} p_\psi \overset{\text{def}}{=} \Pr(\mathcal{D}^{\psi,\psi^{-1},p,p^{-1}}(1^{2n}) = 1 \mid \psi \overset{R}{\leftarrow} \text{MISTY}_{2n}^{11}) \\ p_R \overset{\text{def}}{=} \Pr(\mathcal{D}^{R,R^{-1},p,p^{-1}}(1^{2n}) = 1 \mid R \overset{R}{\leftarrow} P_{2n}, p \overset{R}{\leftarrow} P_n) \end{cases}$$

Lemma 3.2. *Let m_0 and m_1 be integers. Fix $x^{(i)} \in \{0,1\}^{2n}$ and $y^{(i)} \in \{0,1\}^{2n}$ for $1 \leq i \leq m_0$ arbitrarily in such a way that $\{x^{(i)}\}_{1 \leq i \leq m_0}$ are all distinct and $\{y^{(i)}\}_{1 \leq i \leq m_0}$ are all distinct. Similarly, fix $X^{(i)} \in \{0,1\}^n$ and $Y^{(i)} \in \{0,1\}^n$ for $1 \leq i \leq m_1$ arbitrarily in such a way that $\{X^{(i)}\}_{1 \leq i \leq m_1}$ are all distinct and $\{Y^{(i)}\}_{1 \leq i \leq m_1}$ are all distinct.*

Then the number of $\psi \in \text{MISTY}_{2n}^{11}$ such that

$$\psi(x^{(i)}) = y^{(i)} \text{ for } 1 \leq \forall i \leq m_0 \text{ and } p(X^{(i)}) = Y^{(i)} \text{ for } 1 \leq \forall i \leq m_1 \tag{5}$$

is at least

$$(\#H_n^1)^2 (2^n - 2m_0 - m_1)! \left(1 - 2\epsilon \cdot m_0(m_0 - 1) - \frac{4m_0 m_1}{2^n} - \frac{2m_0^2}{2^n}\right) .$$

A proof is given in Appendix B.

Theorem 3.2. *For any super-pseudorandom distinguisher $\mathcal{D}$ that also has oracle access to p and p^{-1} and makes at most m queries in total,*

$$\mathrm{Adv}(\mathcal{D}) \leq 2\epsilon \cdot m(m-1) + \frac{6m^2}{2^n} \ .$$

Proof. Let $\mathcal{O} = R$ or ψ. The super-pseudorandom distinguisher $\mathcal{D}$ has oracle access to $\mathcal{O}$, $\mathcal{O}^{-1}$, p and p^{-1}. Assume that $\mathcal{D}$ makes m_0 queries to $\mathcal{O}$ or $\mathcal{O}^{-1}$, and m_1 queries to p or p^{-1}, where $m = m_0 + m_1$.

There are four types of queries $\mathcal{D}$ can make: either $(+, x)$ which denotes the query "what is $\mathcal{O}(x)$?", $(-, y)$ which denotes the query "what is $\mathcal{O}^{-1}(y)$?", $(+, X)$ which denotes the query "what is $p(X)$?", or $(-, Y)$ which denotes the query "what is $p^{-1}(Y)$?" For the i-th query $\mathcal{D}$ makes to $\mathcal{O}$ or $\mathcal{O}^{-1}$, define the query-answer pair $(x^{(i)}, y^{(i)}) \in \{0,1\}^{2n} \times \{0,1\}^{2n}$, where either $\mathcal{D}$'s query was $(+, x^{(i)})$ and the answer it got was $y^{(i)}$ or $\mathcal{D}$'s query was $(-, y^{(i)})$ and the answer it got was $x^{(i)}$. Similarly for the i-th query $\mathcal{D}$ makes to p or p^{-1}, define the query-answer pair $(X^{(i)}, Y^{(i)}) \in \{0,1\}^n \times \{0,1\}^n$, where either $\mathcal{D}$'s query was $(+, X^{(i)})$ and the answer it got was $Y^{(i)}$ or $\mathcal{D}$'s query was $(-, Y^{(i)})$ and the answer it got was $X^{(i)}$. Define view v and V of $\mathcal{D}$ as $v = ((x^{(1)}, y^{(1)}), \ldots, (x^{(m_0)}, y^{(m_0)}))$ and $V = ((X^{(1)}, Y^{(1)}), \ldots, (X^{(m_1)}, Y^{(m_1)}))$. Without loss of generality, we assume that $\{x^{(i)}\}_{1 \leq i \leq m_0}$ are all distinct, $\{y^{(i)}\}_{1 \leq i \leq m_0}$ are all distinct, $\{X^{(i)}\}_{1 \leq i \leq m_1}$ are all distinct and $\{Y^{(i)}\}_{1 \leq i \leq m_1}$ are all distinct.

Then similarly to the proof of Theorem 3.1, denote by $\mathcal{C}_\mathcal{D}(v, V)$ the final output of $\mathcal{D}$.

Let $(v, V)_{one} \overset{\mathrm{def}}{=} \{(v, V) \mid \mathcal{C}_\mathcal{D}(v, V) = 1\}$ and $N_{one} \overset{\mathrm{def}}{=} \#(v, V)_{one}$.

Evaluation of p_R. From the definition of p_R, we have

$$p_R = \Pr_{R,p}(\mathcal{D}^{R,R^{-1},p,p^{-1}}(1^{2n}) = 1)$$

$$= \frac{\#\{(R,p) \mid \mathcal{D}^{R,R^{-1},p,p^{-1}}(1^{2n}) = 1\}}{(2^{2n})!(2^n)!} \ .$$

For each $(v, V) \in (v, V)_{one}$, the number of (R, p) such that

$$R(x^{(i)}) = y^{(i)} \text{ for } 1 \leq \forall i \leq m_0 \text{ and } p(X^{(i)}) = Y^{(i)} \text{ for } 1 \leq \forall i \leq m_1 \qquad (6)$$

is exactly $(2^{2n} - m_0)!(2^n - m_1)!$. Therefore, we have

$$p_R = \sum_{(v,V) \in (v,V)_{one}} \frac{\#\{(R,p) \mid (R,p) \text{ satisfying } (6)\}}{(2^{2n})!(2^n)!}$$

$$= N_{one} \cdot \frac{(2^{2n} - m_0)!}{(2^{2n})!} \cdot \frac{(2^n - m_1)!}{(2^{2n})!} \ .$$

Evaluation of p_ψ. From the definition of p_ψ, we have

$$p_\psi = \Pr_{h_1,p,h_5}(\mathcal{D}^{\psi,\psi^{-1},p,p^{-1}}(1^{2n}) = 1)$$

$$= \frac{\#\{(h_1,p,h_5) \mid \mathcal{D}^{\psi,\psi^{-1},p,p^{-1}}(1^{2n}) = 1\}}{(\#H_n^1)^2(2^n)!} \ .$$

Similarly to p_R, we have

$$p_\psi = \sum_{(v,V)\in(\boldsymbol{v},\boldsymbol{V})_{one}} \frac{\#\left\{(h_1,p,h_5) \mid (h_1,p,h_5) \text{ satisfying } (5)\right\}}{(\#H_n^1)^2(2^n)!} .$$

Then from Lemma 3.2, we obtain that

$$p_\psi \geq \sum_{(v,V)\in(\boldsymbol{v},\boldsymbol{V})_{one}} \frac{(2^n - 2m_0 - m_1)!\left(1 - 2\epsilon \cdot m_0(m_0 - 1) - \frac{4m_0m_1}{2^n} - \frac{2m_0^2}{2^n}\right)}{(2^n)!}$$

$$= N_{one}\frac{(2^n - 2m_0 - m_1)!}{(2^n)!}\left(1 - 2\epsilon \cdot m_0(m_0 - 1) - \frac{4m_0m_1}{2^n} - \frac{2m_0^2}{2^n}\right)$$

$$= p_R\frac{(2^{2n})!(2^n - 2m_0 - m_1)!}{(2^{2n} - m_0)!(2^n - m_1)!}\left(1 - 2\epsilon \cdot m_0(m_0 - 1) - \frac{4m_0m_1}{2^n} - \frac{2m_0^2}{2^n}\right) .$$

Since $\frac{(2^{2n})!(2^n - 2m_0 - m_1)!}{(2^{2n} - m_0)!(2^n - m_1)!} \geq 1$ (This can be shown easily by an induction on m_0), we have

$$p_\psi \geq p_R\left(1 - 2\epsilon \cdot m_0(m_0 - 1) - \frac{4m_0m_1}{2^n} - \frac{2m_0^2}{2^n}\right)$$

$$\geq p_R - 2\epsilon \cdot m_0(m_0 - 1) - \frac{4m_0m_1}{2^n} - \frac{2m_0^2}{2^n}$$

$$\geq p_R - 2\epsilon \cdot m(m - 1) - \frac{6m^2}{2^n} . \tag{7}$$

Applying the same argument to $1 - p_\psi$ and $1 - p_R$ yields that

$$1 - p_\psi \geq 1 - p_R - 2\epsilon \cdot m(m - 1) - \frac{6m^2}{2^n} . \tag{8}$$

Finally, (7) and (8) give $|p_\psi - p_R| \leq 2\epsilon \cdot m(m - 1) + \frac{6m^2}{2^n}$.

4   Negative Result

Let g be a fixed and publicly known XOR-distinct permutation. In Theorem 3.2, we showed that $\psi(h_1, g, p, p^{-1}, h_5^{-1})$ is super-pseudorandom even if the distinguisher has oracle access to p and p^{-1}, where h_1 and h_5 are uniform ϵ-XOR universal permutations, and p is a random permutation.

One might think that $\psi(h_1, g, p, p, h_5^{-1})$ is super-pseudorandom even if the distinguisher has oracle access to p and p^{-1}. In this section, however, we show that this is not true. We can distinguish $\psi(h_1, g, p, p, h_5^{-1})$ from a random permutation with advantage very close to 1.

More generally, let $p_1, p_2, p, p_5 \in P_n$ be random permutations and $\psi = \psi(p_1, p_2, p, p, p_5)$. We prove that ψ is not pseudorandom if the distinguisher has oracle access to p_2, p_2^{-1} and p. This proof implies that for any fixed and public g, $\psi(p_1, g, p, p, p_5)$ is not super-pseudorandom nor pseudorandom if the distinguisher has oracle access to p.

Define the advantage of $\mathcal{D}$ as follows.

$$\texttt{Adv}(\mathcal{D}) \stackrel{\text{def}}{=} |p_\psi - p_R|$$

where

$$\begin{cases} p_\psi \stackrel{\text{def}}{=} \Pr(\mathcal{D}^{\psi, p_2, p_2^{-1}, p}(1^{2n}) = 1 \mid p_1, p_2, p, p_5 \stackrel{R}{\leftarrow} P_n, \psi = \psi(p_1, p_2, p, p, p_5)) \\ p_R \stackrel{\text{def}}{=} \Pr(\mathcal{D}^{R, p_2, p_2^{-1}, p}(1^{2n}) = 1 \mid R \stackrel{R}{\leftarrow} P_{2n}, p_2, p \stackrel{R}{\leftarrow} P_n) \end{cases}$$

Theorem 4.1. *There exists a pseudorandom distinguisher $\mathcal{D}$ that has oracle access to p_2, p_2^{-1} and p and makes 6 queries in total,*

$$\texttt{Adv}(\mathcal{D}) \geq 1 - \frac{2}{2^n} \ .$$

Proof. Let $\mathcal{O} = R$ or ψ. Our distinguisher $\mathcal{D}$ has oracle access to $\mathcal{O}, p_2, p_2^{-1}$ and p. Consider the following $\mathcal{D}$:

1. Ask $(0, \ldots, 0) \in \{0,1\}^n$ to p_2^{-1} and obtain A.
2. Pick $X, A' \in \{0,1\}^n$ such that $A \neq A'$ arbitrarily.
3. Ask (X, A) to $\mathcal{O}$ and obtain (Y, B).
4. Ask $A \oplus A'$ to p_2 and obtain C.
5. Ask $A' \oplus B$ to p and obtain D.
6. Ask $A' \oplus B \oplus C$ to p and obtain E.
7. Ask $(X, A \oplus A')$ to $\mathcal{O}$ and obtain (Z, F).
8. Output "1" if and only if $F = A' \oplus B \oplus C \oplus D \oplus E$.

If $\mathcal{O} = \psi$, then B is the input to p in both third round and fourth round at step 3 since $p_2(A) = (0, \ldots, 0)$. Therefore we have $p_1(X) \oplus A = B$. Now the input to p in the third round at step 7 is $p_1(X) \oplus A \oplus A'$ which is equivalent to $A' \oplus B$. Next the input to p in the fourth round at step 7 is $A' \oplus B \oplus C$ since $p_2(A \oplus A') = C$. Then we always have $F = A' \oplus B \oplus C \oplus D \oplus E$ at step 8. Hence we have $p_\psi = 1$.

If $\mathcal{O} = R$, we have $p_R = \frac{2^n}{2^{2n}-1} \leq \frac{2}{2^n}$.

Corollary 4.1. *For any fixed and public g, $\psi(p_1, g, p, p, p_5)$ is not super-pseudorandom if the distinguisher has oracle access to p.*

Proof. From the proof of Theorem 4.1.

5   Conclusion

In this paper, we proposed more efficient constructions of super-pseudorandom permutations based on the five round MISTY type permutation than those given in [3] .

In particular, we showed that the second round permutation g need not be cryptographic at all, i.e., no randomness nor secrecy is required.

More precisely, let p and p_i be random permutations, then we proved that

1. $\psi(h_1, g, p, p^{-1}, h_5^{-1})$ is super-pseudorandom, where h_1 is an ϵ-XOR universal permutation, g is a (publicly known and fixed) XOR-distinct permutation, and h_5 is a uniform ϵ-XOR universal permutation (Theorem 3.1),

2. $\psi(h_1, g, p, p^{-1}, h_5^{-1})$ is super-pseudorandom, even if the adversary has oracle access to p and p^{-1}, where h_1 and h_5 are uniform ϵ-XOR universal permutations, and g is a (publicly known and fixed) XOR-distinct permutation (Theorem 3.2),

3. but $\psi(p_1, p_2, p, p, p_5)$ is *not* pseudorandom *nor* super-pseudorandom, if the adversary has oracle access to p_2, p_2^{-1} and p (Theorem 4.1).

References

1. J. L. Carter and M. N. Wegman. Universal classes of hash functions. *JCSS*, vol. 18, no. 2, pp. 143–154, 1979.

2. H. Gilbert and M. Minier. New results on the pseudorandomness of some block cipher constructions. Pre-proceedings of *Fast Software Encryption, FSE 2001*, pp. 260–277 (to appear in LNCS, Springer-Verlag).

3. T. Iwata, T. Yoshino, T. Yuasa and K. Kurosawa. Round security and super-pseudorandomness of MISTY type structure. Pre-proceedings of *Fast Software Encryption, FSE 2001*, pp. 245–259 (to appear in LNCS, Springer-Verlag).

4. M. Luby and C. Rackoff. How to construct pseudorandom permutations from pseudorandom functions. *SIAM J. Comput.*, vol. 17, no. 2, pp. 373–386, April 1988.

5. S. Lucks. Faster Luby-Rackoff ciphers. *Fast Software Encryption, FSE '96, LNCS 1039*, pp. 189–203, Springer-Verlag.

6. M. Matsui. New structure of block ciphers with provable security against differential and linear cryptanalysis. *Fast Software Encryption, FSE '96, LNCS 1039*, pp. 206–218, Springer-Verlag.

7. M. Matsui. New block encryption algorithm MISTY. *Fast Software Encryption, FSE '97, LNCS 1267*, pp. 54–68, Springer-Verlag.

8. M. Naor and O. Reingold. On the construction of pseudorandom permutations: Luby-Rackoff revised. *J. Cryptology*, vol. 12, no. 1, pp. 29–66, Springer-Verlag, 1999.

9. Z. Ramzan and L. Reyzin. On the round security of symmetric-key cryptographic primitives. *Advances in Cryptology — CRYPTO 2000, LNCS 1880*, pp. 376–393, Springer-Verlag, 2000.

10. K. Sakurai and Y. Zheng. On non-pseudorandomness from block ciphers with provable immunity against linear cryptanalysis. *IEICE Trans. Fundamentals*, vol. E80-A, no. 1, pp. 19–24, April 1997.

11. M. N. Wegman and J. L. Carter. New hash functions and their use in authentication and set equality. *JCSS*, vol. 22, no. 3, pp. 265–279, 1981.

Appendix A. Proof of Lemma 3.1

In ψ, we denote by $I_3^{(i)} \in \{0,1\}^n$ the input to p in the third round, and denote by $O_3^{(i)} \in \{0,1\}^n$ the output of it. Similarly, $I_4^{(i)}, O_4^{(i)} \in \{0,1\}^n$ are the input and output of p in the fourth round, respectively. That is, $p(I_3^{(i)}) = O_3^{(i)}$ and $p(I_4^{(i)}) = O_4^{(i)}$.

Number of h_1. First, for any fixed i and j such that $1 \leq i < j \leq m$:

- if $x_L^{(i)} = x_L^{(j)}$, then there exists no h_1 such that

$$h_1(x_L^{(i)}) \oplus x_R^{(i)} = h_1(x_L^{(j)}) \oplus x_R^{(j)} \tag{9}$$

  since $x_L^{(i)} = x_L^{(j)}$ implies $x_R^{(i)} \neq x_R^{(j)}$;
- if $x_L^{(i)} \neq x_L^{(j)}$, then the number of h_1 which satisfies (9) is at most $\epsilon \# H_n^0$ since h_1 is an ϵ-XOR universal permutation.

Therefore, the number of h_1 such that

$$h_1(x_L^{(i)}) \oplus x_R^{(i)} = h_1(x_L^{(j)}) \oplus x_R^{(j)} \text{ for } 1 \leq \exists i < \exists j \leq m \tag{10}$$

is at most $\epsilon \binom{m}{2} \# H_n^0$.

Next, for any fixed i and j such that $1 \leq i < j \leq m$:

- if $x_L^{(i)} = x_L^{(j)}$, then there exists no h_1 such that

$$h_1(x_L^{(i)}) \oplus g(x_R^{(i)}) \oplus x_R^{(i)} = h_1(x_L^{(j)}) \oplus g(x_R^{(j)}) \oplus x_R^{(j)} \tag{11}$$

  since $x_L^{(i)} = x_L^{(j)}$ implies $x_R^{(i)} \neq x_R^{(j)}$, and our XOR-distinct g guarantees $g(x_R^{(i)}) \oplus x_R^{(i)} \neq g(x_R^{(j)}) \oplus x_R^{(j)}$;
- if $x_L^{(i)} \neq x_L^{(j)}$, then the number of h_1 which satisfies (11) is at most $\epsilon \# H_n^0$ since h_1 is an ϵ-XOR universal permutation.

Therefore, the number of h_1 such that

$$h_1(x_L^{(i)}) \oplus g(x_R^{(i)}) \oplus x_R^{(i)} = h_1(x_L^{(j)}) \oplus g(x_R^{(j)}) \oplus x_R^{(j)} \text{ for } 1 \leq \exists i < \exists j \leq m \tag{12}$$

is at most $\epsilon \binom{m}{2} \# H_n^0$.

Then, from (10) and (12), the number of h_1 such that

$$\left. \begin{array}{l} h_1(x_L^{(i)}) \oplus x_R^{(i)} \neq h_1(x_L^{(j)}) \oplus x_R^{(j)} \text{ for } 1 \leq \forall i < \forall j \leq m, \text{ and} \\ h_1(x_L^{(i)}) \oplus g(x_R^{(i)}) \oplus x_R^{(i)} \neq h_1(x_L^{(j)}) \oplus g(x_R^{(j)}) \oplus x_R^{(j)} \text{ for } 1 \leq \forall i < \forall j \leq m \end{array} \right\} \tag{13}$$

is at least $\# H_n^0 - 2\epsilon \binom{m}{2} \# H_n^0$. Fix h_1 which satisfies (13) arbitrarily. This implies that $I_3^{(1)}, \ldots, I_3^{(m)}$ and $O_4^{(1)}, \ldots, O_4^{(m)}$ are fixed in such a way that:

- $I_3^{(i)} \neq I_3^{(j)}$ for $1 \leq \forall i < \forall j \leq m$, and
- $O_4^{(i)} \neq O_4^{(j)}$ for $1 \leq \forall i < \forall j \leq m$.

Number of h_5. Similarly, the number of h_5 such that

$$\left.\begin{aligned}
h_5(y_L^{(i)} \oplus y_R^{(i)}) \oplus y_R^{(i)} &\neq h_5(y_L^{(j)} \oplus y_R^{(j)}) \oplus y_R^{(j)} \text{ for } 1 \leq \forall i < \forall j \leq m, \\
h_5(y_L^{(i)} \oplus y_R^{(i)}) \oplus O_4^{(i)} &\neq h_5(y_L^{(j)} \oplus y_R^{(j)}) \oplus O_4^{(j)} \text{ for } 1 \leq \forall i < \forall j \leq m, \\
h_5(y_L^{(i)} \oplus y_R^{(i)}) \oplus O_4^{(i)} &\neq O_4^{(j)} \text{ for } 1 \leq \forall i, \forall j \leq m, \text{ and} \\
h_5(y_L^{(i)} \oplus y_R^{(i)}) \oplus y_R^{(i)} &\neq I_3^{(j)} \text{ for } 1 \leq \forall i, \forall j \leq m,
\end{aligned}\right\} \quad (14)$$

is at least $\#H_n^1 - 2\epsilon\binom{m}{2}\#H_n^1 - \frac{2m^2\#H_n^1}{2^n}$. Fix h_5 which satisfies (14) arbitrarily. This implies that $O_3^{(1)}, \ldots, O_3^{(m)}$ and $I_4^{(1)}, \ldots, I_4^{(m)}$ are fixed in such a way that:

- $I_4^{(i)} \neq I_4^{(j)}$ for $1 \leq \forall i < \forall j \leq m$,
- $O_3^{(i)} \neq O_3^{(j)}$ for $1 \leq \forall i < \forall j \leq m$,
- $O_3^{(i)} \neq O_4^{(j)}$ for $1 \leq \forall i, \forall j \leq m$, and
- $I_4^{(i)} \neq I_3^{(j)}$ for $1 \leq \forall i, \forall j \leq m$.

Number of p. Now h_1 and h_5 are fixed in such a way that

$$I_3^{(1)}, \ldots, I_3^{(m)}, I_4^{(1)}, \ldots, I_4^{(m)}$$

(which are inputs to p) are all distinct and

$$O_3^{(1)}, \ldots, O_3^{(m)}, O_4^{(1)}, \ldots, O_4^{(m)}$$

(which are corresponding outputs of p) are all distinct. In other words, for p, the above $2m$ input-output pairs are determined. The other $2^n - 2m$ input-output pairs are undetermined. Therefore we have $(2^n - 2m)!$ possible choice of p for any such fixed h_1 and h_5.

To summarize, we have:

- at least $\#H_n^0 - 2\epsilon\binom{m}{2}\#H_n^0$ choice of h_1,
- at least $\#H_n^1 - 2\epsilon\binom{m}{2}\#H_n^1 - \frac{2m^2\#H_n^1}{2^n}$ choice of h_5 when h_1 is fixed, and
- $(2^n - 2m)!$ choice of p when h_1 and h_5 are fixed.

Then the number of $\psi \in \text{MISTY}_{2n}^{01}$ which satisfy (1) is at least

$$(\#H_n^0)(\#H_n^1)(2^n - 2m)! \left(1 - 2\epsilon\binom{m}{2}\right)\left(1 - 2\epsilon\binom{m}{2} - \frac{2m^2}{2^n}\right)$$

$$\geq (\#H_n^0)(\#H_n^1)(2^n - 2m)! \left(1 - 2\epsilon \cdot m(m-1) - \frac{2m^2}{2^n}\right)$$

This concludes the proof of the lemma.

Appendix B. Proof of Lemma 3.2

We use the same definition of $I_3^{(i)}$, $O_3^{(i)}$, $I_4^{(i)}$ and $O_4^{(i)}$ as in the proof of Lemma 3.1.

Number of h_1. First, similarly to the proof of Lemma 3.1, the number of h_1 such that

$$
\left.
\begin{aligned}
&h_1(x_L^{(i)}) \oplus x_R^{(i)} \neq h_1(x_L^{(j)}) \oplus x_R^{(j)} \text{ for } 1 \leq \forall i < \forall j \leq m_0, \\
&h_1(x_L^{(i)}) \oplus x_R^{(i)} \neq X^{(j)} \text{ for } 1 \leq \forall i \leq m_0 \text{ and } 1 \leq \forall j \leq m_1, \\
&h_1(x_L^{(i)}) \oplus g(x_R^{(i)}) \oplus x_R^{(i)} \neq h_1(x_L^{(j)}) \oplus g(x_R^{(j)}) \oplus x_R^{(j)} \text{ for } 1 \leq \forall i < \forall j \leq m_0, \\
&h_1(x_L^{(i)}) \oplus g(x_R^{(i)}) \oplus x_R^{(i)} \neq Y^{(j)} \text{ for } 1 \leq \forall i \leq m_0 \text{ and } 1 \leq \forall j \leq m_1
\end{aligned}
\right\}
\tag{15}
$$

is at least $\#H_n^1 - 2\epsilon\binom{m_0}{2}\#H_n^1 - \frac{2m_0 m_1 \#H_n^1}{2^n}$. Fix h_1 which satisfies (15) arbitrarily. This implies that $I_3^{(1)}, \ldots, I_3^{(m_0)}$ and $O_4^{(1)}, \ldots, O_4^{(m_0)}$ are fixed in such a way that:

- $I_3^{(i)} \neq I_3^{(j)}$ for $1 \leq \forall i < \forall j \leq m_0$,
- $I_3^{(i)} \neq X^{(j)}$ for $1 \leq \forall i \leq m_0$ and $1 \leq \forall j \leq m_1$,
- $O_4^{(i)} \neq O_4^{(j)}$ for $1 \leq \forall i < \forall j \leq m_0$, and
- $O_4^{(i)} \neq Y^{(j)}$ for $1 \leq \forall i \leq m_0$ and $1 \leq \forall j \leq m_1$.

Number of h_5. Similarly, the number of h_5 such that

$$
\left.
\begin{aligned}
&h_5(y_L^{(i)} \oplus y_R^{(i)}) \oplus y_R^{(i)} \neq h_5(y_L^{(j)} \oplus y_R^{(j)}) \oplus y_R^{(j)} \text{ for } 1 \leq \forall i < \forall j \leq m_0, \\
&h_5(y_L^{(i)} \oplus y_R^{(i)}) \oplus y_R^{(i)} \neq X^{(j)} \text{ for } 1 \leq \forall i \leq m_0 \text{ and } 1 \leq \forall j \leq m_0, \\
&h_5(y_L^{(i)} \oplus y_R^{(i)}) \oplus O_4^{(i)} \neq h_5(y_L^{(j)} \oplus y_R^{(j)}) \oplus O_4^{(j)} \text{ for } 1 \leq \forall i < \forall j \leq m_0, \\
&h_5(y_L^{(i)} \oplus y_R^{(i)}) \oplus O_4^{(i)} \neq Y^{(j)} \text{ for } 1 \leq \forall i \leq m_0 \text{ and } 1 \leq \forall j \leq m_0, \\
&h_5(y_L^{(i)} \oplus y_R^{(i)}) \oplus O_4^{(i)} \neq O_4^{(j)} \text{ for } 1 \leq \forall i, \forall j \leq m_0, \text{ and} \\
&h_5(y_L^{(i)} \oplus y_R^{(i)}) \oplus y_R^{(i)} \neq I_3^{(j)} \text{ for } 1 \leq \forall i, \forall j \leq m_0,
\end{aligned}
\right\}
\tag{16}
$$

is at least $\#H_n^1 - 2\epsilon\binom{m_0}{2}\#H_n^1 - \frac{2m_0 m_1 \#H_n^1}{2^n} - \frac{2m_0^2 \#H_n^1}{2^n}$. Fix h_5 which satisfies (16) arbitrarily. This implies that $O_3^{(1)}, \ldots, O_3^{(m_0)}$ and $I_4^{(1)}, \ldots, I_4^{(m_0)}$ are fixed in such a way that:

- $I_4^{(i)} \neq I_4^{(j)}$ for $1 \leq \forall i < \forall j \leq m_0$,
- $I_4^{(i)} \neq X^{(j)}$ for $1 \leq \forall i \leq m_0$ and $1 \leq \forall j \leq m_1$,
- $O_3^{(i)} \neq O_3^{(j)}$ for $1 \leq \forall i < \forall j \leq m_0$,
- $O_3^{(i)} \neq Y^{(j)}$ for $1 \leq \forall i \leq m_0$ and $1 \leq \forall j \leq m_1$,
- $O_3^{(i)} \neq O_4^{(j)}$ for $1 \leq \forall i, \forall j \leq m_0$, and
- $I_4^{(i)} \neq I_3^{(j)}$ for $1 \leq \forall i, \forall j \leq m_0$.

Number of p. Now h_1 and h_5 are fixed in such a way that

$$
I_3^{(1)}, \ldots, I_3^{(m_0)}, I_4^{(1)}, \ldots, I_4^{(m_0)}, X^{(1)}, \ldots, X^{(m_1)}
$$

(which are inputs to p) are all distinct and

$$O_3^{(1)}, \ldots, O_3^{(m_0)}, O_4^{(1)}, \ldots, O_4^{(m_0)}, Y^{(1)}, \ldots, Y^{(m_1)}$$

(which are corresponding outputs of p) are all distinct. Then we have $(2^n - 2m_0 - m_1)!$ possible choice of p for any such fixed h_1 and h_5.

To summarize, we have:

- at least $\#H_n^1 - 2\epsilon\binom{m_0}{2}\#H_n^1 - \frac{2m_0 m_1 \#H_n^1}{2^n}$ choice of h_1,
- at least $\#H_n^1 - 2\epsilon\binom{m_0}{2}\#H_n^1 - \frac{2m_0 m_1 \#H_n^1}{2^n} - \frac{2m_0^2 \#H_n^1}{2^n}$ choice of h_5 when h_1 is fixed, and
- $(2^n - 2m_0 - m_1)!$ choice of p when h_1 and h_5 are fixed.

Then the number of $\psi \in \text{MISTY}_{2n}^{11}$ which satisfy (5) is at least

$$(\#H_n^1)^2(2^n - 2m_0 - m_1)!$$
$$\times \left(1 - 2\epsilon\binom{m_0}{2} - \frac{2m_0 m_1}{2^n}\right)\left(1 - 2\epsilon\binom{m_0}{2} - \frac{2m_0 m_1}{2^n} - \frac{2m_0^2}{2^n}\right)$$
$$\geq (\#H_n^1)^2(2^n - 2m_0 - m_1)!\left(1 - 2\epsilon \cdot m_0(m_0 - 1) - \frac{4m_0 m_1}{2^n} - \frac{2m_0^2}{2^n}\right)$$

This concludes the proof of the lemma.

BeepBeep: Embedded Real-Time Encryption

Kevin Driscoll

Honeywell Laboratories, 3660 Technology Drive, Minneapolis, MN 55418, USA
kevin.driscoll@Honeywell.com

Abstract. The BeepBeep algorithm is designed to supply secrecy and integrity for embedded real-time systems. These systems must achieve their required timing performance under all conditions, while operating in a multi-tasking environment with tightly constrained CPU, memory, and bandwidth resources. BeepBeep was designed to be implemented as software on the processors most commonly used for embedded controllers. It uses little program memory, no data memory (its state fits into most processors' register sets), and no inherent message padding (ciphertext is a 1:1 replacement for plaintext). It is significantly faster than existing algorithms (e.g. AES) in this environment and includes mechanisms to support integrity as part of its basic secrecy operation.

1   Motivation and Requirements

Examples of embedded real-time applications requiring security include wireless communications (cell phones, pagers, aircraft, Bluetooth, IEEE 802.11), remote management of control systems for chemical and power plants, distributed management of distribution networks (pipelines and electrical grids, remote meter readers), access and control of remote sites (physical security management, electrical load shedding, medical equipment). Typical real-time cryptography requirements differ significantly from conventional cryptography in a number of ways.

- 1:1 message size: Varying-length byte streams must be encrypted with minimal message expansion, particularly in retrofit applications.
- varying integrity requirements: Detection of message corruption is essential, particularly for actions with serious consequences.
- key agility: Each message can have a different key for each unicast/multicast address, requiring rapid key change after a message header is processed.
- low latency: Input to output delay is more important than throughput.
- low jitter: Processing time for each message packet should be the same. (There is little or no time for per message key scheduling.)
- small memory footprint: Many high-volume cost-sensitive applications use single-chip microcomputers with a total RAM of 128 or 256 bytes.
- security time horizon: "tactical" rather than a "strategic"
- compatibility: Embedded systems tend to be closed communities, with little need to be compatible with the rest of the world

J. Daemen and V. Rijmen (Eds.): FSE 2002, LNCS 2365, pp. 164–178, 2002.
© Springer-Verlag Berlin Heidelberg 2002

A search of existing cryptographic algorithms failed to find any that met these requirements. The algorithms' problems include large latency and slow speed (particularly for the small messages typical of real-time systems), large data structures, expansion of messages, and significant cost to switch keys in RAM-constrained implementations.

1.1 Deadlines

Real-time cryptography must live within deadlines, typically that repeat with fixed periodicity. Overstepping the deadline frequently is not possible. Only worst-case execution times factor are important; average performance better than worst-case has limited utility. Missed deadlines can cause catastrophic failures in safety critical systems. Most real-time systems are heavily multi-tasked and the time slices allocated to a cryptographic task may be very small.

1.2 Context Switching and State Size

The heavily multi-tasked and interrupt-driven nature of real-time systems, coupled with their tight latency requirements, means that such systems do a lot of context switching. Frequent switching reduces the utility of modern processor cache technology, and can even lead to counter-productive cache thrashing. Indeed, partly for this reason, microprocessors used in real-time systems often have no data cache.

A real-time cryptographic algorithm should be designed to minimize memory access penalties. Ideally, the crypto state of the algorithm should fit within the register set of the target CPUs. But, small 8-bit and 16-bit microcomputers often do not have enough register space to hold the minimum crypto state needed to be secure.

Given the general trend of real-time controllers increasing their word size to 32 bits and with most 32-bit controllers' register sets having sufficient size, it makes sense to size an encryption algorithm's state to fit into as many of the 32-bit microcomputers' register sets as possible and be resigned to the fact that smaller processors will have store part of its state in RAM.

A survey of 32-bit CPUs found that most have at least seven 32-bit registers available to hold crypto state data, when leaving enough other registers to hold the rest of algorithm's state and temporary values. The non-ignorable exception is the Intel x86 family. For the x86 family, use of MMX registers, or of a single on-chip cache line, can provide the same storage. Overall, then, an algorithm's crypto state should not exceed the magical number seven[1] 32-bit words.

1.3 Message Size

Most real-time communication products are designed to minimize energy use, size, weight and cost, while providing an acceptable bit error or message loss

---

[1] With apologies to Miller[8].

rate. When cryptographically-based communications security is included in these products, either as a new design or a retrofit, there is seldom an available budget for cryptographic overhead. Instead, every increase in message size leads to a decrease in the functionality for which the product was purchased.

In some retrofit applications, where correctness of system communications behavior has previously been certified, changes in system timing due to cryptographic expansion of messages is not tolerable. In other networks, where users pay by the byte (e.g., with aircraft or LEO satellites), cryptographic expansion impacts profitability.

All of these situations create a need to minimize message growth due to cryptography. This need is amplified by the very short message sizes of many real-time systems, frequently on the order of 10 bytes, where even small per-message overhead due to cryptography can be a large burden.

For all of the just-stated reasons, a major goal of a real-time cryptographic algorithm design is to minimize or eliminate message expansion. This requirement eliminates the use of block cipher modes such as ECB, which round messages up to the next block size; the chosen cipher must either be a stream cipher, or a block cipher used in a stream cipher mode. The former is typically more efficient.

Stream ciphers also have communications overhead, in the form of an initialization vector (IV). Luckily, most real time communication messages are individually identified through extrinsic or intrinsic means which can be used as the IV without creating any additional overhead.

1.4   Security

Secrecy. Most real-time communications have a need for short-term secrecy, to deny an attacker knowledge of current control system state. The need for long-term secrecy in such systems is infrequent, but it does exist. Information of major economic value, such as trade secret process parameters or inventory levels of arbitrage-able commodities, requires long-term secrecy. Such secrecy can always be provided by super-encryption of the information at risk, though this is undesirable. Ideally, a single cipher should meet both short-term and long-term secrecy needs.

Authentication and Integrity. Real-time systems usually need to prevent message forgeries and unauthorized message modification. Corrupt control messages can cause disasters directly. Corrupt reports of current state can lead to disasters indirectly. Authentication and integrity can be supported by including predictable values in the (extended) plaintext message. The classical way of doing this is by appending a cryptographic hash of the plaintext to the message. A less computationally costly alternative is possible when the cipher provides suitable feedback of the plaintext (or a derivative text) into subsequent ciphertext, eventually affecting an expected value at the end of the message. In many real-time systems, particularly those involving retrofit or rollover, existing frame check data can be included in the encryption as the predictable postfix integrity value. This can reduce or eliminate message size expansion.

Existing real-time systems often can add cryptography only as new "lump in the cable" hardware. Coupled with latency restrictions, this requires on-the-fly cryptography – secrecy and integrity have to be done in one pass.

Where on-the-fly cryptography isn't needed, a second encryption pass over the message with a different starting point and/or direction, can distribute an integrity check over the entire message. This permits all predictable values within the plaintext to be used for integrity without regard to their location, including data that can do double duty as an IV as well integrity checking.

Real-time systems typically are autonomous and do not accept for encryption and transmission messages from untrusted sources. This precludes many "oracle" and related types of attack.

1.5   Asset Exploitation

Embedded real-time cryptography is a struggle of economics, in which the goal is to make an adversary incur more cost than the effort is worth while not imposing prohibitive cost on authorized users. A design should attempt to include assets which are already available to authorized users in a way that prevents an adversary from exploiting alternate technologies to gain an advantage. The greatest perceived threat is the conversion of a weakness in an algorithm into a workable break by using custom integrated circuit (ICs) or field programmable gate arrays (FPGAs) to greatly speed up trial decodes.[4][1] This suggests a design goal for the algorithm to include elements which are cheap or free when implemented in software on real-time controllers but are expensive to implement in ICs or FPGAs. The latter expense is primarily "silicon area" in terms of gates and routing. This goal can be accomplished by incorporating in the algorithm the use of hardware elements which exist in most real-time control CPUs and require a large number of gates and/or routing. The most obvious big consumer of "silicon" in a CPU is memory (including caches). For reasons given above, memory assets for data storage cannot be cheaply exploited. The next largest silicon element is a fast multiplier. CPUs in real-time controllers have 32-bit arithmetic units, including a multiplier. The multiply speed has been getting faster in each generation of CPUs. This trend is expected to continue.

2   BeepBeep Description

BeepBeep is a new algorithm designed specifically to meet the above requirements. Its main elements (described in detail below) are a 127-bit primitive Linear Feedback Shift Register (LFSR), clock control, non-linear filter, and two stage combiner. The LFSR provides a stream of pseudo-random values. The clock control and non-linear filter protect against known LFSR attacks. The two stage combiner mixes the algorithm's "text" input with the nonlinear filter output and a word from the LFSR to produce the algorithm's final output. A block diagram of BeepBeep's main loop is shown in Fig. 1 and pseudo C-code for BeepBeep is shown in Fig. 2. For simplicity, this pseudo C-code assumes the output buffer is

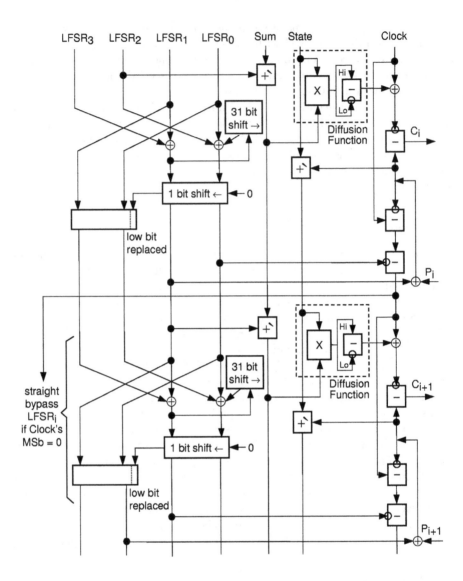

Fig. 1. Encryption Block Diagram

a multiple of 8 bytes and BeepBeep's execution will change bytes in this buffer following the last byte of the message for messages sizes not a multiple of 8.

The symbol $+'$ in the block diagram and the $+'$ routine in the pseudo-C code means unsigned ones complement addition. Most CPUs today use twos complement addition. Conversion to ones complement is done by wrapping the carry out of the most significant bit of a twos complement sum back into its least significant bit. The carry wrap is done using a "with carry" variant of add (which

```
#define +'(a, b)    // unsigned ones complement sum
  ( (a + b) < a ? a + b + 1 : a + b )

#define step0_32(ctl)  // if ctl = 1, advance LFSR by 32 bits
  i        = lfsr[3] ^ lfsr[1];  // 32 new LFSR bits
  lfsr[3] = lfsr[3 - ctl];  // shift LFSR by 0 or 32 bits
  lfsr[2] = lfsr[2 - ctl];                  // "
  lfsr[1] = ctl ? lfsr[0] | i >> 31 : lfsr[1];  // "
  lfsr[0] = ctl ?            i << 1 : lfsr[0];  // "

#define crypt(src, ctl)  // process one word, update variables
  sum = +'(sum, lfsr[src]);
  step0_32(ctl); step0_32(ctl);  // if ctl, advance LFSR by 64 bits
  m.f = sum * state;  // entwine and nonlinearize LFSR output bits
  i = (m.h.u - m.h.l) ^ clock;  // " and whiten
  if (encrypt) { *c++ = (j = (*p++        ^ lfsr[3 - src])) - i; }
  else         { *p++ = (j = (*c++ + i)) ^ lfsr[3 - src]; }
  state  = +'(j, state);
  clock -= j + (src == 2 ? lfsr[0] : lfsr[2]);

void BeepBeep(lfsr, clock, sum, state, bytes, p, c, encrypt)
  uint32 lfsr[4]; // left-justified 127-bit LFSR; 128 key bits
  uint32 clock;    // LFSR clocking control; 5th word of key
  uint32 sum;       // running sum of LFSR ouput; 6th word of key
  uint32 state;    // filter state feedback; 7th word of key
  sint32 bytes;    // number of bytes in message
  uint32 *p, *c;  // plaintext and ciphertext word pointers
  char    encrypt;
{
  uint32 i, j;   // short term temporaries
  union { struct {uint32 u, l;} h;  // (u)pper and (l)ower (h)alves
                 uint64 f; }    m;  // of (f)ull 64-bit (m)ultiply

  // Optional initialization goes here

  if (big_endian)
     (encrypt ? *p : *c)[(bytes+3)/4] >>>= (4 - bytes % 4) * 8;
  for (; bytes > 4; bytes -= 8) { crypt(2, 1);  // 8 bytes per loop
                                  crypt(1, clock >> 31); }
  if (bytes > 0)                 crypt(2, 1);  // 1 to 4 bytes left
  if (big_endian)
     (encrypt ? *p : *c)[-1] <<<= (4 - bytes % 4) * 8;
}
```

Fig. 2. BeepBeep Pseudo-C Code

exists on all twos complement CPUs) on the immediately following add to the sum. If the next operation isn't an add to the sum, a instruction must be inserted which adds a (twos complement) zero to the sum. Thus, the +' subroutine uses only one or two machine instructions. Ones complement format has two zeros (+0 and -0). As used here, +0 is not possible if either input is non-zero. A ones complement -0 is a -1 when used in twos complement operations.

2.1   Crypto State

BeepBeep uses seven 32-bit words as its crypto state, which is held in three 32-bit variables (*clock*, *sum*, and *state*) and one array of four 32-bit elements (*lfsr*). BeepBeep's key is simply its entire initial crypto state. The size of the key is be larger than the security of the algorithm. The "excess" key size is used to speed up the initialization of the key generator when key changes are made.

IV methods are discussed in the Initialization section following the description of BeepBeep's main processing loop. In the following description, the term "processed" means that all operations required to produce an output (ciphertext for encryption and plaintext for description) have been performed.

2.2   Clock Controlled Linear Feedback Shift Register

The polynomial for BeepBeep's LFSR is $x^{127} + x^{63} + 1$. The LFSR is left-justified in four 32-bit words (*lfsr[3]* down to *lfsr[0]*) and is left shifted. It operates in Tausworthe (full words) fashion, producing three 32-bit output words (the right-most three words). The right-most bit of the right-most word contains one bit of key before the first LFSR shift and is zero thereafter. The LFSR advances 64 bits iff the most significant bit (MSb) of the *clock* variable is a one or an even number of text words have been processed.

LFSR outputs are used for inputs to three functions: non-linear filter, clocking, and two stage combiner. Which outputs are used for which function were chosen to avoid the reuse of any output value for the same function when clock control does not advance the LFSR, to avoid the use of LFSR[0] where knowledge of it LSb being zero would be a weakness, and to allow efficient implementation on 64 bit CPUs.

After each text word is processed, the autokey feedback (ciphertext + *i*) is subtracted from *clock*. If an even number of words have been processed, *lfsr[0]* is subtracted from *clock*, else *lfsr[2]* is subtracted from *clock*. This provides an autokey influence on LFSR clocking.

2.3   Nonlinear Filter with State and Diffusion Function

Before each text word is processed, one of the LFSR outputs is added to *sum* using ones complement addition. If an even number of words have been processed, *lfsr[2]* is added, else *lfsr[1]* is added. Then *sum* and *state* are multiplied together. After the multiplication, the lower half of the product is subtracted from the upper half. This multiply and subtract function provides most of BeepBeep's diffusion. The diffusion function output is then XORed with *clock* to create *i*.

After each text word is processed, the autokey feedback (ciphertext + *i*) is added to *state* using ones complement addition.

2.4 Two Stage Combiner

The output of the filter (i) goes to the final section of this algorithm, which is a two stage combiner.

For encryption, the plaintext is first combined with an LFSR word using XOR and then the result is combined with the nonlinear filter's output (i) using twos complement subtraction. The result of this operation is the ciphertext. If an even number of words have been processed, *lfsr[1]* is used, else *lfsr[2]* is used for the XOR.

For decryption, the order of the combiner's two operations is reversed and addition is used instead of subtraction.

2.5 Initialization

BeepBeep currently has five options for performing the IV function, depending on on system requirements and available resources. The first option is the traditional explicit IV value prepended to each message. The second option uses an implicit value available in most real-time systems, such as frame sequence number or system time. For these two options, the *state* variable is initialized with the IV value instead of key. (Such systems use only 192 bits of key.) Fig. 3 shows the method for incorporating this IV into the key.

```
if (bytes < 1) {           // used only for new key or IV
  bytes    = - bytes;
  m.f      = (state + lfsr[3]) * (2 * (state + lfsr[3]) + 1);
  clock    += m.h.l;   // mix in IV changes
  lfsr[2]  -= m.h.l;   // mix in IV changes
  m.h.l <<<= 16;       // circularly left shift by 16 bits
  state    = +'(m.h.l, bytes);   // +' ensures state is not +0
  lfsr[1]  = +'(lfsr[1], state); // ensure LFSR is not +0, vary with IV
  sum     |= state; }            // ensure sum  is not +0, vary with IV
```

Fig. 3. Initialization example

The IV (held in *state*) is added with *sum* to form a value x, which is transformed via the bijective polynomial $2x^2 + x$. The lower 32 bits of the resulting value ($m.f$) is then added into *clock* and subtracted from *lfsr[2]* both using twos complement arithmetic. The $m.f$ value is then circularly rotated left by 16 bits. The result is added with the message size (in *bytes*) to create a new *state* value. This *state* value is added to *lfsr[1]* using ones complement addition and ORed into sum.

Initialization or key distribution must ensure a non-zero starting value for at least one word of the LFSR and preferably for *sum* and *state*. At the beginning of the algorithm, the only variable that is known to not be zero is *bytes* (the count of bytes remaining to be processed). This fact is exploited in the last three lines of this initialization option to create the desired non-zero values. The OR function is used for *sum* because it is a single instruction where the ones complement addition requires two instructions. The downside of the OR function is that its result is too rich in one bits. This bias is not a problem because the next operation on the variable *sum* will always be a ones complement addition with one of the LFSR words. The IV is added into *lfsr[1]* because its change diffuses the fastest into the other LFSR words.

The third IV option is to just prepend truly random data to the plaintext before encryption and discarded it after decryption. Because BeepBeep uses an autokey, this data performs the role of an IV (without revealing its actual value). This option creates only a small variation in the LFSR's state, which could lead to related key attacks. Thus, this option should only be used where code space is extremely tight on a device that does decryption only and the random data can be made large relative the actual message data.

The fourth IV option is to derive an IV from the message itself. For applications that can't use either of the above IV options, a "hash IV" can be used. To use this, a one-way hash of the message's tail is done. The resulting hash is added to the message's head just for encryption (discarded before the message is transmitted). The combination of hashing and BeepBeep's forward diffusion causes all bits of the message's tail to be diffused throughout the whole message. The "tail" could be the whole message, if the operation starts at the middle of the message and wraps around, ending back at the middle. Because BeepBeep is faster than known hash algorithms, it can be used to create the hash and pre-encrypt the message at the same time. This effectively converts BeepBeep into a variable sized block cipher (no padding). This is similar to the old idea of using an autokeyed cipher to make two passes over a message (one pass in each direction), which also could be used. Of course, using an IV derived from the message body means that a repetition of a message is detectable by an adversary.

The fifth IV option is to use crypto-state carry over between messages. This can be used when BeepBeep is implemented on a reliable message delivery service, which guarantees in-order and error-free reception (as is the case in many real-time systems). This option can be viewed as all messages being just packets of one large virtual message. The initial key is never reused; so, no IV is needed.

3   Design Rational for Performance

3.1   Linear Feedback Shift Register

An LFSR may seem like an odd choice because LFSRs are notoriously slow in software. This problem is solved by using a trinomial with the taps spaced exactly a multiple of a word-width apart. Using word-wide operations creates a variant of a Tausworthe generator that can produce a word of new bits with fewer instructions than is used to produce one bit in typical software LFSRs. This trick increases the LFSR's speed by over 60 times.

BeepBeep's loop encrypts two text words per iteration, each word using different LFSR outputs. The selection can be virtual, with no software execution cost, by the loop being a 2x unroll of basic 32-bit encryption.

3.2 Clock Control

The minimum possible clock control scheme is self-decimation using one bit from the LFSR as the clock control. But, this scheme has been successfully attacked for even bit-wise clocking. BeepBeep's 64-bit step size would be even weaker. By adding just one instruction, a stronger clocking mechanism can be built. BeepBeep subtracts one of the LFSR's words from a running difference (*clock*). On top of this, BeepBeep includes an autokey feedback into the clock control using just one more instruction.

The most significant bit of *clock* is used as the clocking control to allow efficient implementation on most CPU types. Implementing clock control using branching instructions is very slow on most modern high performance CPUs. This is because the direction taken at each branch will be unpredictable. Unpredicted branches usually cause pipeline flushes and refills. Many CPU types (such as ARM, MIPS, and Pentium) have conditional instructions. For these CPUs, the new value of an LFSR word can be conditionally moved into the register that holds the old value, based on the sign of *clock*. Of all the bits in a register, the sign bit is the one that is universally the easiest to test. For CPUs without conditional instructions, the following trick can be used. First, convert the sign bit to a full boolean word by doing an arithmetic right shift by 31 bits. Then, replace the conditional expression "ctl ? new : old" with the logic expression "(ctl & (new XOR old)) XOR old" for each LFSR word.

To further reduce the performance cost of clock control and to increase the "decimation rate" (if that concept is even applicable to a Tausworthe generator where three of the four LFSR words are used for something in the remainder of the algorithm), the clock control is applied only to the second text word of each loop; the LFSR is always advanced for the first word.

The LFSR words used for the *clock* subtrahend are *lfsr[0]* and *lfsr[2]*. Because the least significant bit (LSb) of *lfsr[0]* is always 0 (required by the fast LFSR trick), it cannot be used as the LFSR value used for the ciphertext XOR. To balance LFSR word usage, *lfsr[0]* is used here where knowledge that the LSb is 0 does not create a possible weakness. The other subtrahend variant is *lfsr[2]* because it is easy to access on a hybrid 32/64 bit CPU.

4 Design Rationale for Security

The best known attacks against BeepBeep currently take on the order of 2^{96} work. However, this is the result of limited analysis. The remainder of this section describes some of BeepBeep's security considerations.

4.1   Clock Controlled Linear Feedback Shift Register

This LFSR was chosen to get a keystream sequence which is long enough to effectively never repeat and has known good statistics while using a minimum of storage resources. But, the Tausworthe speed-up trick's use of a trinomial exacerbates the LFSR's vulnerability to well known attacks. BeepBeep's defense against these attacks include clock control and a nonlinear filter with state. Other defenses (such as those using multiple LFSRs) were found to be too expensive to implement in real-time software. The LFSR's clock control is anemic. It adds only 1/2 bit of uncertainty for each 32 bits of text. This could lead to an attack by exhaustive enumeration. But, this mechanism is designed only to cover any weaknesses that may be found in the nonlinear filter which require large amounts of text to be successful. If an attack against the nonlinear filter gains less than 1/2 bit of information per word encrypted, this clocking mechanism may defeat that attack.

The clock control is of the stop-and-go type, which has known attacks. The LFSR output selection covers this by not reusing the same value for the same function whenever the LFSR is stopped.

Both *clock* and the *sum* variables act as integrators of the LFSR outputs. This stops the majority of attacks against clock controlled LFSRs, which assume the current crypto state is a function just of the number of deletions that have occurred. With the integrators, the current crypto state is dependent not only on the number of deletions, but on their specific history as well.

The author is unaware of any attacks against an LFSR of this size and has both clock control and a nonlinear filter with state.

4.2   Nonlinear Filter with State

One of desired characteristics of a nonlinear filter is a high nonlinear order. The nonlinearity (in GF_2) of this filter begins with the carry chain in the ones complement additions. Each bit of each sum is 33rd order (as opposed to a twos complement sum which would be 32nd order in its most significant bit and decreasing one order for each lesser significant bit). Each bit of each partial product in the multiply is 66th order. The carry chain for the multiply partial product additions and the borrow chain in the subtraction of the full product halves, complete the 128th order of the filter. This high nonlinear order makes higher order differential analysis infeasible (viewing the filter with feedback as a round in a 32-bit block cipher, ignoring the fact that the "key" changes for every block).[7]

The nonlinear filter with state has three design goals beyond those normally seen in a filter generator. The first is to exploit the use of the 32-bit multiplier that exists on almost all real-time controller CPUs. The second is to maintain real-time performance by working on full 32-bit words. The third is to provide forward text diffusion as part of the autokey mechanism.

The autokey feedback to the filter is first added into *state*. Because of the wrap-around carry, any bit change in the feedback can affect any bit in the result, although with decreasing probably along the carry propagation chains. Each bit of *state* is paired with each bit of *sum* in the multiply's partial products. Given that *sum* is nearly uniform (except that it can't be zero), each bit you flip in the ciphertext has about a 50% chance of affecting any bit in the filter output, and thus also any succeeding bit in the remainder of the message. This is one of BeepBeep's mechanisms for providing integrity.

Because the multiply filter is nonlinear and one-way, including its output makes recovery of the previous states difficult.

The use of full words means each LFSR input to the filter is decimated by a factor of 32, even if there weren't the additional decimation from clock controlled LFSR stepping. That is, without clock control, each bit position of an LFSR output word would "see" an LFSR sequence decimated by 32. This means that current nonlinear filter analyses (such as [3]) don't hold here.

As with any LFSR filter, the output should be uniformly distributed. Without the ones complement addition on the inputs to the multiply, a zero output would be much more frequent than the mean output frequency. To solve this problem, the ones complement running sums are used. These sums are never zero if initialized to a non-zero value. Even with this "correction", there is still some bias. This "multiplication followed by subtracting the product halves" has a distribution which is closely related to the factorization of $2^{32} + 1$ (641 and 6700417). That is, values which are multiples of these factors will occur more often than the average. The XOR of *clock* into the feedback path around the multiplier "whitens" the output and prevents any possible short cycles in the multiply's feedback path.

DIEHARD (http://stat.fsu.edu/~geo/diehard.html) showed no problems, even when parts of BeepBeep disabled (e.g. keying the LFSR to zero, forcing *state* to zero).

Ignoring carry and borrow, each of the filter's 32 output bits is a perfect nonlinear function of all its 64 input bits. The effect of carry and borrow on nonlinearity and correlation has not been analyzed.

4.3   Two Stage Combiner

This two stage combiner is used for the following reasons: (1) Addition and subtraction provide some lateral plaintext diffusion. (2) Using non-associative operations provides some integrity protection. With a simple XOR combiner (or any linear combiner), an adversary knowing a plaintext can manipulate the ciphertext bits to make the plaintext resulting from decryption be anything the adversary wants. But, not with a two stage combiner having non-associative functions; the most significant bit of each word is the only bit vulnerable. This is the lessor of BeepBeep's mechanisms for providing integrity protection.

4.4   Keying and IV

BeepBeep uses seven words (224 bits) for keying, or 192 bits if *state* is used as an IV. Some of these keys are so weak as to be illegal to use. These are the keys which make the LFSR all zero. The IV initialization described above prevents this. Therefore, any bit pattern can be used for keying BeepBeep with this initialization.

An autokey function is included to compensate for the very small amount of crypto state that can be held and to provide forward diffusion as part of the integrity protection. As with any autokey, adaptive chosen plaintext attacks are a concern. Such attacks are not possible for most applications in the intended domain. For those applications where such attacks are possible, the implementation should not allow messages to be sent such that they are associated with an IV that has known properties.

While BeepBeep has several IV options, ignoring the requirement for an IV is not one of them. Without an IV, BeepBeep can be the subject of "walking one" chosen plaintext attack. With just 32 one-word messages, the lower 31 bits of i and $lfsr[1]$ and be found. The attack can proceed to find all of the LFSR and some bits of $clock$. An adaptive chosen attack needs only 32 messages minus the Hamming weight of the lower 30 bits of i.

4.5 Integrity

During both encryption and decryption, the middle value of the two-stage combiner (ciphertext + i) is fed back into $state$ and $clock$. This autokey mechanism propagates errors forward to provide integrity. The $state$ and $clock$ variables provide only 64 bits of change propagation through the message. But, the most significant bit of $clock$ controls the state of the LFSR and sum, which adds another $127 + 32$ bits to the error propagation state. Information is accumulated into these latter 159 bits at a rate of 0.5 bits per word encrypted.

Integrity loss is detected by checking a known value (check data) at the end of a message after decryption. For most real-time communication, check data is already used in messages to detect naturally occurring errors. For messages without existing check data or if the size of the check data is too small for integrity checking, additional check data has to be appended to a message prior to encryption. Given the high diffusion in the feedback loop, any change in the ciphertext will have a 50% change of affecting each bit of the check data.

The two-stage combiner is another integrity mechanism. Because the operations are not associative, this is not equivalent to a simple additive combiner and the typical integrity attacks do not work. The "lateral diffusion" of the 32-bit twos complement addition is hidden by the 32-bit XOR super-encipherment. This leaves only the most significant bit vulnerable to attack. Given complete knowledge of a message's plaintext and ciphertext, an attacker still cannot manipulate the other decrypted plaintext bits in a word to be a value of his choosing, even if the autokey mechanism weren't used. Thus, BeepBeep has double integrity coverage for most of a message's bits.

Via the autokey feedbacks to $state$ and $clock$, any change in a message will eventually affects the entire crypto state. Because most modern communications systems and virtually all communications in real-time control have error detection or correction schemes, which accept messages only if they are error free, the historic concern of plain-fed autokey propagating errors is rarely applicable today.

The autokey eventually has the enciphered text affecting all bits of the crypto state, which includes the LFSR's 127 bits, the filter's state of 64 bits and the clock control's 32 bits. All of these are interconnected with multiple paths to prevent divide and conquer attacks.

5   Applications and Performance

BeepBeep is being included in several product developments. One is to encrypt radio communications with commercial aircraft. Another is the remote control of buildings' safety, security and other automation functions, including meter reading, load shedding, and other gas and electric network management functions.

BeepBeep was first implemented on a Pentium II in assembly. Encryption took about 6.5 clocks per byte. A hand analysis of the assembly code showed it should have taken about 4.4 clocks. The reason for the difference is not known, but the most likely suspect is costly cache misses while reading in the plaintext on Windows NT. Start up time was under 100 clock cycles. The code space was 460 bytes each for encryption and decryption (they were coded separately), with the main loop being 184 bytes. The algorithm's entire data state was held in the Pentium's registers (including MMX).

One of the remote control application is interesting because its CPU is only an 8/16-bit hybrid. But, it is a typical heavily multi-tasked embedded control system. The requirements were that in the residual approximately 50 bytes of RAM and 1,638 bytes of ROM, the "security layer" of the protocol stack had to provide secrecy, integrity, authentication, and key management while consuming minimal communication bandwidth (including idle/turnaround time). Rijndael[2] exceeded the memory limits just trying to do secrecy;[6] XTEA (tean)[10] exceeded the limits when simple integrity was added; Skipjack[9] was better than XTEA for RAM but not ROM. Surprisingly, a BeepBeep based solution fit into 28 bytes of RAM and 1,628 bytes of ROM (954 of it for BeepBeep) even though BeepBeep's 32 bit operations had to be synthesized out of 8 and 16 bit instructions. This application also had the constraints of minimizing bandwidth and execution time because the communication rate can be slow (2400 baud) and some customers are charged for communication time.

6   Conclusion

The need for an encryption algorithm designed specifically embedded real-time systems has been identified. An algorithm to meet the unique requirements for these systems has been designed and is being fielded. This algorithm exceeds the performance of other algorithms in several areas including memory, time (particularly latency and jitter), and message size.

References

1. Bond M., Clayton R.: Extracting a 3DES key from an IBM 4758
   http://www.cl.cam.ac.uk/~rnc1/descrack/
2. Daemen, J., Rijmen, V.: AES Proposal: Rijndael. AES Submission. (June 1998)
3. Dichtl, M.: On Non-linear Filter Generators. FSE '97, Lecture Notes in Computer
   Science, Vol. 1267. Springer-Verlag, Berlin Heidelberg New York (1997) 103–106
4. Electronic Frontier Foundation: Cracking DES: Secrets of Encryption Research,
   Wiretap Politics & Chip Design. 1st Edition O'Reilly & Associates, Sebastopol
   CA (July 1998)
5. Gollmann, D., Chambers W.: Clock-Controlled Shift Registers: A Review. IEEE
   Journal on Selected Areas in Communications. (1989) 7: 525–533.
6. Keating, G.: Performance analysis of AES candidates on the 6805 CPU core. Proceedings of The Second AES Candidate Conference. (1999) 109-114.
   http://www.ozemail.com.au/~geoffk/aes-6805/paper.pdf
7. Knudsen, L. R.: The interpolation attack on block ciphers. FSE '95, Lecture Notes
   in Computer Science, Vol. 1008. Springer-Verlag, Berlin Heidelberg New York
   (1995) 196–211
8. Miller, G. A.: The Magical Number Seven, Plus or Minus Two: Some Limits on Our
   Capacity for Processing Information. Psychology Review, American Psychological
   Association Inc. **63** No 2. (1956)
9. United States National Security Agency: Skipjack and KEA algorithm specifications, Version 2.0. (29 May 1998)
   http://csrc.nist.gov/encryption/skipjack/skipjack.pdf
10. Needham, R., Wheeler, D.: Tea Extensions. Draft technical report, Computer Lab-
    oratory, University of Cambridge. (October 1997)
    http://www.ftp.cl.cam.ac.uk/ftp/users/djw3/xtea.ps

A New Keystream Generator MUGI

Dai Watanabe[1], Soichi Furuya[1], Hirotaka Yoshida[1],
Kazuo Takaragi[1], and Bart Preneel[2]

[1] Systems Development Laboratory, Hitachi, Ltd.,
292 Yoshida-cho, Totsuka-ku, Yokohama, 244-0817, Japan
{daidai, soichi, takara}@sdl.hitachi.co.jp
[2] Katholieke Universiteit Leuven, Dept. Electrical Engineering-ESAT,
Kasteelpark Arenberg 10, B-3001 Heverlee, Belgium
Bart.Preneel@esat.kuleuven.ac.be

Abstract. We present a new keystream generator (KSG) MUGI, which is a variant of PANAMA proposed at FSE '98. MUGI has a 128-bit secret key and a 128-bit initial vector as parameters and generates a 64-bit string per round. The design is particularly suited for efficient hardware implementations, but the software performance of MUGI is excellent as well. A speed optimized implementation in hardware achieves about 3 Gbps with 26 Kgates, which is several times faster than AES. On the other hand the security was evaluated according to re-synchronization attack, related-key attack, and linear correlation of an output sequence. Our analysis confirms that MUGI is a secure KSG.

Keywords. Keystream generator, Block cipher, PANAMA, Re-synchronization attack, Related-key attack.

1 Introduction

This paper presents a new keystream generator MUGI that is designed for use as a stream cipher. MUGI has a 256-bit input (consisting of a 128-bit secret key and a 128-bit public parameter IV) and outputs a 64-bit random data block for each round.

Several approaches are known in the literature to the design of KSGs. One particularly popular approach is based on Linear Feedback Shift Registers (LFSRs). They are suitable for very compact hardware implementations and provide good randomness. However, due to their linearity and predictability, they cannot be used in their pure forms. Several techniques have been developed to improve their security, such as the combination generator, non-linear filtering, and clock control. A substantial amount of research has been spent on the security of these schemes. But LFSRs are not suited for efficient software implementations.

On the other hand software-oriented stream ciphers seem to be designed in an *ad hoc* way, and we do not seem to have the appropriate tools to evaluate them. The most important criterion is to verify deviations from randomness. Examples in this class include (Alleged) RC4 [Sc96] (security analysis in [FS01]), SEAL 3.0

J. Daemen and V. Rijmen (Eds.): FSE 2002, LNCS 2365, pp. 179–194, 2002.

[RC98] (security analysis in [Fl01]), LEVIATHAN [McF00] (security analysis in [CL01]), and LILI-128 [CGMPS00] (security analysis in [MFI01]).

In this paper we focus on PANAMA, designed by J. Daemen and C. Clapp [DC98]. PANAMA is based on generic design principles, comparable to those of block ciphers. PANAMA can be used both as a KSG and as a hash function. However, recently Rijmen *et al.* [RRPV01] have exposed security weaknesses in the security of PANAMA as a hash function. But these weaknesses have no impact at all on the security of PANAMA as a KSG.

MUGI is a variant of PANAMA which is only suitable as a KSG. The design goal is to make MUGI suitable for many platforms. As a result, MUGI achieves a performance that is equal to or even better than AES [DR99], especially the hardware performance is excellent. MUGI can be implemented in hardware with 18 Kgates. In terms of security, we evaluate the security against re-synchronization attacks [DGV94] and related-key attacks. Furthermore we calculate the linear correlation of the output sequence. As the result, we conclude that MUGI is a reliable and efficient cryptographic primitive that can be used to provide encryption and message authentication.

This article is organized as follows: in Sect. 2 we describe the generalization of a PANAMA-like structure and discuss the security of this type of KSG. The specification of MUGI is given in Sect. 3. In Sect. 4 we present some results about the security of MUGI. In Sect. 5 we discuss the implementation of MUGI both in software and hardware. In Appendix, we give test vectors of MUGI. You can find the perfect version of this paper at http://www.sdl.hitachi.co.jp/crypto/mugi/index-e.html.

2   Design Policy

2.1   PANAMA-**Like Keystream Generator**

The principal part of a KSG is a set $(\mathcal{S}, \Upsilon, f)$ which consists of an internal state $\mathcal{S}$, its update function Υ, and the output filter f which abstracts the output sequence from the internal state $\mathcal{S}$. We call the set $(\mathcal{S}, \Upsilon)$ the *internal-state machine*. In addition we call a single application of the state update function a *round*. $\mathcal{S}^{(t)}$ refers to the internal state at round t.

For PANAMA, the internal state is divided into two parts, state a and buffer b. The update function of PANAMA depends in a different way on different parts of the internal state. Note that each update function uses another internal state as a parameter. We denote the update functions of state a and buffer b with ρ and λ, respectively.

The noteworthy characteristic of PANAMA's ρ-function is its use of an SPN structure. Such a KSG design must be motivated by the following simple question: how can a secure cryptographic function be constructed from insecure cryptographic components? For block ciphers (or pseudorandom permutations) there is a *de facto* standard construction, which uses a Feistel network or an SPN as a component (called a round function) and iterates it for mixing. PANAMA is an answer for a KSG. It uses a core mixing function ρ similar to the round function of a block cipher and a large buffer instead of fixed extended keys and iterations of a round function.

On the other hand the function λ is a simple linear transformation. The output filter f drops about half of the bits of state a for each round. We call a KSG which satisfies such characteristics PANAMA-like keystream generator (PKSG). This can be formalized in the definition of a PKSG as follows.

Definition 1 *Consider an internal-state machine consisting of two internal states, namely the state a, the buffer b, and their update functions ρ, λ. The keystream generator which consists of an internal-state machine $((a, b), (\rho, \lambda))$ and an output filter f is called a* PANAMA-***like keystream generator*** *if it satisfies the following conditions:*

(1) ρ includes an SPN transformation that uses parts of buffer b as a parameter.

$$a^{(t+1)} = \rho(a^{(t)}, b^{(t)}).$$

(2) λ is a linear transformation that uses a part of state a as a parameter.

$$b^{(t+1)} = \lambda(b^{(t)}, a^{(t)}).$$

(3) f outputs a part of state a, which is typically no more than $1/2$ of the bits of a.

The first condition characterizes a PKSG, but the other conditions are also necessary. For example, not updating the buffer or outputting all of the state significantly decrease the security [FWT00].

2.2   Selection of Components

MUGI is a KSG and has a PKSG structure. In order to select the components for MUGI, we want to build on other strong cryptographic primitives in the literature. As a result we use some components of AES [DR99], which are well evaluated. For example the substitution table S-box and the linear transformation are the same as for AES. Although currently the design of a PKSG is not as straight forward as that of block ciphers, this selection should make MUGI more secure.

2.3   The Difference between PANAMA and MUGI

The MUGI design aims to achieve the following two points:

1. Efficiency in hardware implementations. Particularly a gate-efficient implementation must be possible.
2. To make evaluation easier than PANAMA.

To achieve these properties, the basic data size is decreased from 256-bit to 64-bit. And an 8-bit substitution table is adopted to improve the security of ρ. In addition, an extended Feistel network is adopted in ρ instead of a simple SPN-structure, in order to simplify the evaluation.

3 Specification of MUGI

In this section we give a description of MUGI. MUGI is a KSG with a 128-bit secret key K (a secret parameter) and a 128-bit initial vector I (a public parameter). It generates a 64-bit length random bit string $Out[t]$ for each round.

As we mention in Sect. 2.1 any KSG can be described as the combination of an internal-state machine and an output filter.

First we describe the internal state of MUGI in Sect. 3.2 and the update function in Sect. 3.3. Then we discuss the initialization in Sect. 3.4 and the random number generation in Sect. 3.5.

3.1 Input

The basic data size of MUGI is 64 bits, called a *unit* in this paper. MUGI has two inputs as a parameter. One is a 128-bit secret key K and the other one is a 128-bit initial vector I. The left and right units of K are denoted by K_0 and K_1, respectively. I_0 and I_1 are used in a similar way.

3.2 Internal State

MUGI has two internal states, state a and buffer b. The state a consists of 3 units denoted by a_0, a_1, a_2 from left to right. On the other hand, the buffer b consists of 16 units. Each of them is denoted by $b_0, \ldots, b_{15}$ in the same manner as state a.

3.3 Update Function

The update function of PKSG consists of ρ and λ, the update functions of state a and buffer b, each of which uses the other internal state as a parameter. In other words the update function Υ of the complete internal state is described as follows:

$$(a^{(t+1)}, b^{(t+1)}) = \Upsilon(a^{(t)}, b^{(t)}) = (\rho(a^{(t)}, b^{(t)}), \lambda(a^{(t)}, b^{(t)})).$$

In the following we explain ρ and λ of MUGI.

Core mixing function ρ. ρ is the update function of state a. It is a kind of target-heavy Feistel structure [SK96] with two identical F-functions (Fig. 1), it uses buffer b as a parameter. The function ρ can be described as follows:

$$a_0^{(t+1)} = a_1^{(t)}$$
$$a_1^{(t+1)} = a_2^{(t)} \oplus F(a_1^{(t)}, b_4^{(t)}) \oplus C_1$$
$$a_2^{(t+1)} = a_0^{(t)} \oplus F(a_1^{(t)}, b_{10}^{(t)} <<< 17) \oplus C_2$$

C_1, C_2 in the equations above are constants.

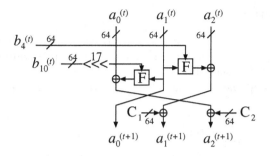

Fig. 1. ρ-function of MUGI

F-function. The F-function consists of a key addition (the data addition from the buffer), a non-linear transformation using the S-box, a linear transformation using the MDS matrix M and a byte shuffling (Fig. 2). The S-box and the MDS matrix are the same as for AES.

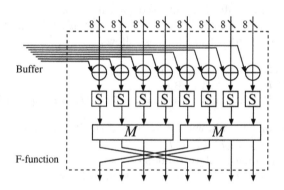

Fig. 2. F-function of MUGI

Buffer update function λ. The function λ is the update function of buffer b, it uses a part of state a as a parameter. λ is the linear transformation of b and can be described as follows:

$$b_j^{(t+1)} = b_{j-1}^{(t)} \quad (j \neq 0, 4, 10)$$
$$b_0^{(t+1)} = b_{15}^{(t)} \oplus a_0^{(t)}$$
$$b_4^{(t+1)} = b_3^{(t)} \oplus b_7^{(t)}$$
$$b_{10}^{(t+1)} = b_9^{(t)} \oplus (b_{13}^{(t)} <<< 32)$$

3.4 Initialization

The initialization of MUGI is divided into three steps. The first step initializes the buffer b with a secret key K. The second initializes state a with an initial vector I. Finally the whole internal state is mixed.

In the first step the secret key K is extended to 192 bits and it is put into state a as follows:

$$
\begin{aligned}
a_0^{(t_0)} &= K_0, \\
a_1^{(t_0)} &= K_1, \\
a_2^{(t_0)} &= (K_0 <<< 7) \oplus (K_1 >>> 7) \oplus C_0,
\end{aligned}
$$

Here time t_0 denotes the start of the initialization. The value C_0 in the above equation is a constant (see Sect. 3.6). Then follow, a mixing step with only a ρ iteration and the left side unit of each $a^{(t)}$, $a_0^{(t)}$ is put into the buffer b as follows:

$$
b_{15-i} = (\rho^{i+1}(a^{(t_0)}, 0))_0
$$

In the above equations ρ^i denotes the i-th iteration of ρ and $\rho(a, 0)$ means that the data stored into buffer b is not used for this step.

In the second step the mixed state $a(K) := \rho^{16}(a^{(t_0)}, 0)$ and the initial vector I are required. I is added to state a as follows:

$$
\begin{aligned}
a(K, I)_0 &= a(K)_0 \oplus I_0, \\
a(K, I)_1 &= a(K)_1 \oplus I_1, \\
a(K, I)_2 &= a(K)_2 \oplus (I_0 <<< 7) \oplus (I_1 >>> 7) \oplus C_0,
\end{aligned}
$$

Then state a is mixed again with 16 rounds of the iteration ρ. So the mixed state a can be represented as $\rho^{16}(a(K, I), 0)$.

The last step consists of 16 rounds of the whole update function Υ, so $a^{(1)}$, the initialized state with K, I, can be written as follows:

$$
a^{(1)} = \Upsilon^{16}(\rho^{16}(a(K, I), 0), b(K)),
$$

where $b(K)$ denotes the buffer b initialized by the secret key K.

3.5 Random Number Generation

After the initialization, MUGI generates 64-bit random numbers and transforms the internal state in every iteration. Denote the output of round t as $Out[t]$, then the output is given as follows:

$$
Out[t] = a_2^{(t)}
$$

In other words MUGI outputs 64 bits of the right side of state a at the beginning of the round process.

The processes from the initialization to the random number generation are summarized in Table 1.

Table 1. Schedule of MUGI

	Round t	Process	Input	Output
	-49	Inputting Key	K	$-$
	$-48, \ldots, -33$	Mixing (by ρ)	$-$	$-$
Initialization	-32	Inputting IV	I	$-$
	$-31, \ldots, -16$	Mixing (by ρ)	$-$	$-$
	$-15, \ldots, 0$	Mixing (by Υ)	$-$	$-$
Generating bit strings	$1, \ldots$	Mixing and Outputting	$-$	$Out[t]$

3.6 Constants

The MUGI algorithm uses three constants: C_0 in the initialization, and C_1, C_2 in ρ. They have the following values:

$$C_0 = \text{0x6A09E667F3BCC908},$$
$$C_1 = \text{0xBB67AE8584CAA73B},$$
$$C_2 = \text{0x3C6EF372FE94F82B}.$$

These are hexadecimal values of $\sqrt{2}$, $\sqrt{3}$, and $\sqrt{5}$ multiplied by 2^{64}. These constants aim to prevent the invariance of byte-wise equality, and are chosen to ensure that there is no trap-door.

4 Security

The security of KSG is reduced to the relationship between input and output bits (or relationship between output bits). All attacks to KSG that improve over exhaustive key search and over exhaustive search over the internal state use some of these relationships and guess the internal state. We consider the possibility that the attacker can observe any kind of relationship, i.e. the condition that the attacker can observe some deviation between input and output bits (or between only output bits) is identified with the success of the attack, even if the attacker cannot get any information about the internal state. This identification comes from the philosophy that the output sequence of the secure KSG should be *unpredictable*. The relationship mentioned above is divided into three cases as follows:

Randomness. An attacker fixes a secret key and an initial vector, and then he observes the relation in the output sequence.

Re-synchronization attack. An attacker fixes a secret key, and then he observes the relation between initial vectors and output sequences.

Related-key. An attacker fixes an initial vector, and then he observes the relation between keys and output sequences. The related-key attack includes observing the relation between keys and initial vectors.

On the other hand exhaustive key searching require 2^{127} computations on average to find correct key. So we say an attack is efficient when it costs less than 2^{127} encryptions on average to find the correct key.

4.1   Randomness of an Output Sequence

The linearity should be one of the most important characteristics in the known evaluation methods. Here 'linearity' does not imply linear complexity, but maximum probability of linear combinations of output bits. We note that searching this linear combination is analogous to search for the best approximation for a block cipher, and apply the evaluation method used in linear cryptanalysis [Ma94]. More specifically, this corresponds to counting active S-boxes in linear approximations for evaluating the linearity of a MUGI output sequence. At the same time applying this technique to PKSGs is more difficult than applying it to block ciphers because the buffer is updated dinamically. Therefore, we give up constructing actual linear approximations and calculate the lower bound of the number of active S-boxes required for any linear approximation.

We denoted the number of active S-boxes of a linear approximation with $\mathcal{AS}$. The maximum linear probability of the MUGI S-box is 2^{-6}, so it can be assumed that the linear characteristic of the output sequence of MUGI is sufficiently small if there is no linear approximation with $\mathcal{AS} < 22$. Applying this method to MUGI results in the following theorem:

Theorem 1 *For all linear approximations of MUGI, $\mathcal{AS} \geq 22$.*

Here, we present the proof of this theorem. Constructing a linear approximation which consists of output units can be separated into two steps as follows:

1. Construct a linear approximation of ρ.
2. Search a path including the buffer.

We illustrate each step below.

The linear approximation of ρ. Before starting the evaluation, we give an equivalent transformation of ρ for easier analysis. Figure 3 shows the digest of the transformation. In Fig. 3 the left side F-function is denoted by G; we use this notation just for convinience. First, F can be moved to the left side in the next round. Next, the mask corresponding to an output unit can assume all values, so we separate this part into two masks, an output mask and an input mask. This transformation is not 'equivalent' in the common sense, but it is equivalent in the sense that mask patterns are not changed by the transformation. After that, we remove unnecessary branches. The right side of Fig. 3 shows the transformed ρ. Hereafter, "ρ" represents the transformed ρ. Note that the number of branches drops off to two, and the output masks of the F- and G-functions come directly from the 'input' and 'output' masks, which the attacker can choose.

Figure 4 shows some important paths of ρ. Only the five paths shown there assure that the number of active S-boxes is greater than five. The branch number of the matrix M is defined by $\min_{x \neq 0}(w_H(x) + w_H(Mx))$, where $w_H(x)$ denotes the byte-wise Hamming weight of x [Da95]. The branch number of the linear transformation is an important characteristic for the diffusion properties of a block cipher. But for PKSGs, the branch number of the matrix M does

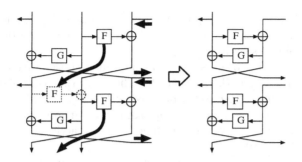

Fig. 3. Equivalent transformation of ρ

not guarantee a lower bound on the number of active S-boxes for a linear approximation, even if it includes several active F-functions. This property is quite different from that of block ciphers.

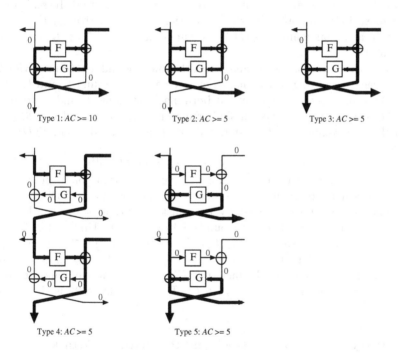

Fig. 4. Linear approximation of ρ

Linear path trail of MUGI. Next, we search for a path including the buffer that gives a linear approximation consisting of only output bits. For PKSGs, the attacker can observe any number of rounds. So it is possible to construct the linear approximation with the outputs of any rounds. Furthermore, some linear approximations may skip intermediate ρ-functions, it implies that there is a possibility that to observe more rounds increases the deviation. This feature makes it difficult to search all paths.

Before the starting discussion we define some notation. We denote the first and last round of the path as t_s and t_e. The mask that is applied to the data XOR-ed from state a to buffer b is denoted as $\Gamma(D)^{(t)}$. In addition we denote an active F-function as 1, and a zero approximated F-function as 0. For example, when an F-function is active and a G-function is not active in round t, we denote this as $\Gamma(a)^{(t)} = (1, 0)$.

First, we pay special attention to the first round and the last round of the path. The value of the input mask for all units of the buffer and their state is zero in the first round, and only the mask for an output unit $Out[t_s]$ is active. Only two paths, Type 1 and 3 in Figure 4, satisfy this condition. The last round is the same as the first round, so the possible paths in round t_e are only those shown as Type 1 and 2.

Next we consider the influence of the buffer to ρ. The $\Gamma(D)^{(t)}$ is 0 from round t_s to $t_s + 4$ because all input masks for the first round are 0. In addition, the input mask from the buffer to the G-function must be active, so $\Gamma(D)^{(t_s+5)}$ is active. In a similar manner, $\Gamma(D)^{(t)}$ is 0 at round $t_e - 5 \le t \le t_e$ and is active at round $t_e - 6$.

The path search (or the calculation of the lower bound for $\mathcal{AS}$) is divided into several cases according to the mask before or after the round t_s and t_e. But we show the proof only for the case that both masks at round t_s and t_e are of Type 1. Other cases can be proved in similar manner. In this case $\mathcal{AS} \ge 20$ because both the first round and the last round are Type 1. In addition, $\Gamma(D)^{(t_e-6)}$ is active.

If there is a round i $(1 \le i \le 4)$ such that $(\Gamma(a)^{(t_s+i)}, \Gamma(a)^{(t_s+i+1)}) = ((0,0), (1,1))$, the path includes more active F-functions of Type 1 or 3, so $\mathcal{AS} \ge 25$ is derived. Hence we consider only the case that $\Gamma(a)^{(t_s+i)} = 0$ for all rounds i from 1 to 4. Similarly the mask condition before the last round must be $\Gamma(a)^{(t_e-i)} = 0$ for all rounds i from 1 to 6. Under this condition, $\Gamma(a)^{(t_s+5)} \ne (0,0)$. Additionally, $\Gamma(D)^{(t_s+6)}$ and $\Gamma(D)^{(t_e-6)}$ are active and $\Gamma(D)^{(t_e-7)}$ is equal to 0. So the number of rounds $t_e - t_s$ must be greater than 14. These results and the fact that $\Gamma(D)^{(t_e-6)}$ is active demonstrate that $\Gamma(a)^{(t_e-6)} \ne (0,0)$ or $\Gamma(a)^{(t_e-7)} \ne (0,0)$. Therefore $\mathcal{AS} \ge 22$ is shown in this case.

4.2   Re-synchronization Attack and Related-Key Attack

The re-synchronization attack [DGV94] should be the most effective attack against PKSGs, hence we try to apply it to MUGI. Before starting the discussion we give a brief explanation of this type of attack. A re-synchronization attack can be used against keystream generators, which have not only a secret key, but also a public parameter. It is an effective attack if the initialization of the algorithm is too simple. Under the assumption that the secret key is fixed, the attacker first searches for some relationship between the public parameters and the corresponding outputs. If some relationship has a high probability, one can guess information about the secret key from it. For example, linear cryptanalysis on the counter mode of a block cipher is a kind of re-synchronization attack. Estimating the security against related-key attack is the same as re-synchronization, by interchanging initial vectors with secret keys.

We chose differential and linear characteristics, and SQUARE attack [DKR97] variants for evaluating the relationship between inputs and outputs of MUGI. The attacks against block ciphers using these characteristics are well known as differential cryptanalysis [BS93] and linear cryptanalysis [Ma94]. The design of a PKSG, especially its ρ function, is quite similar to a block cipher design. This suggests that the above two statistical properties are well suited for evaluating the relationship between the initial vector I and a corresponding internal state.

Maximum differential and linear characteristics of iterations of ρ. Now we ignore the XOR to the buffer and output generation, i.e., we consider only the iteration of ρ and evaluate its differential and linear characteristics. We can apply these evaluation methods in the same way as they are applied to block ciphers.

Table 2 shows the minimum number of active F-functions in all units of state a for each attack.

Table 2. Number of active F-functions in the differential and linear paths of ρ

Number of rounds	$\cdots$	11	12	13	14	15	16	17	18	19	20	21	22	23
Differential	$\cdots$	10	12	12	12	14	16	16	16	18	20	20	20	22
Linear	$\cdots$	10	12	12	13	14	16	16	17	18	20	20	21	22

Resistance against a re-synchronization attack: Table 2 shows the relationship between the initial vector I and corresponding state $a^{(t)}$ transformed by t iterations of ρ. It implies that more than 23 iterations of ρ have no differential and linear characteristics with a probability higher than 2^{-128}.

In the initialization of MUGI, 16 rounds transformed only by ρ are applied after setting the initial vector I. Afterwards, 16 rounds transformed by Υ are applied. However, the buffer b influences the differential and linear characteristics of state a only after round -9, i.e., 22 rounds after setting I. Therefore, we conclude that to observe the deviation due to these characteristics is difficult after round $t > 0$.

On the other hand Table 2 suggests that there are some correlation between initial vector I and some units of corresponding buffer b at round 0. However, the differential characteristic consists of an output sequence and the buffer has more than two buffer-units. The correlation between one of them and I is too small to observe. Therefore, no attacker can exploit that correlation. The conditions for linear cryptanalysis are similar.

Related-key attacks: To observe the correlation between keys and the corresponding outputs is more difficult than the correlation between initial vectors and the corresponding outputs because of the first mixing step. So no security flaw can be found by using differential and linear cryptanalysis.

Square-attack variants. Because of the highly byte-oriented structure, some of the SQUARE attack [DKR97] variants can be considered. The SQUARE attack is currently the most successful attack against block ciphers with an SPN-structure, e.g., Rijndael, the AES. We examine the applicability of the attack and investigate the possible relations. Consequently we conclude that no variants of the SQUARE attack can reduce the security of MUGI PKSG.

The SQUARE attack against a block cipher is a chosen-plaintext attack where an attacker chooses a number of related plaintext blocks each of which is typically different only in a byte or a word. We call these chosen plaintext set Λ-set if the word has all values, and we say that the word is *saturated*. Because of the saturation at the input of a non-linear function, the attacker can expect to control the intermediate values to some extent. From the ciphertext side, the attacker partially decrypts the intermediate value which is still controlled because of the saturated plaintext blocks. If the attacker guesses the key for the partial decryption, then the attacker can distinguish the correct and incorrect round keys.

In a stream cipher, an attacker must try to select different value of either key or initial vector values to mount this attack. Therefore the possible applications of the SQUARE attack must be either a related-key cryptanalysis or a chosen initial vector attack.

Related-key attacks: At first, we define the model of the attack. We assume that the attacker does not know the key value. To obtain the saturation property, the attacker can *run* a number of key initializations, the keys of which differ only in a part of the key value; in this discussion we will concentrate on key-dependent runs where the keys differ in one word. The attacker cannot observe anything until the pseudorandom number sequence comes out. We check if the attacker may find any properties at the output sequences amongst a number of runs.

The saturated key set will inject the saturation property during the buffer initialization. At first, we investigate how buffers are initialized with the properties. For simplicity, we ignore the key padding rule so that we give the attacker the maximum flexibility for setting the initial state values. Let Λ denote the property of an intermediate word such that in each run the concerning word has a different value, i.e., the word is saturated. Let O denote the property that for all runs the value is constant. Also we introduce the weakest property "balanced" denoted by Φ that means that the XOR-summation over all runs is zero. If the word is neither of them, namely uncontrollable, then we use the notation $*$. If the word triple (A, B, C) has the properties of Λ, O, and Φ for the word A, B, C, then we write $(A, B, C) \xrightarrow{p} (\Lambda, O, \Phi)$, or $A \xrightarrow{p} \Lambda, B \xrightarrow{p} O$, and $C \xrightarrow{p} \Phi$.

Obviously the most effective word in which to inject the saturation is the word that affects other words the last. We analyze the case of $(a_0, a_1, a_2) \xrightarrow{p} (\Lambda, O, O)$. Remember the output of the t-th round is denoted by $(a_0^{(t)}, a_1^{(t)}, a_2^{(t)})$. We simply trace the property and show the results in Table 3.

Table 3. The word properties in each intermediate values

Intermediate value	Word property
$(a_0^{(0)}, a_1^{(0)}, a_2^{(0)})$	(Λ, O, O)
$(a_0^{(1)}, a_1^{(1)}, a_2^{(1)})$	(O, O, Λ)
$(a_0^{(2)}, a_1^{(2)}, a_2^{(2)})$	(O, Λ, O)
$(a_0^{(3)}, a_1^{(3)}, a_2^{(3)})$	$(\Lambda, \Lambda, \Lambda)$
$(a_0^{(4)}, a_1^{(4)}, a_2^{(4)})$	(Λ, Φ, Φ)
$(a_0^{(5)}, a_1^{(5)}, a_2^{(5)})$	$(\Phi, *, *)$
$(a_0^{(6+)}, a_1^{(6+)}, a_2^{(6+)})$	$(*, *, *)$

Hence, the initial values of the buffer b_i have the following properties depending on the index i:

$$b_i \xrightarrow{p} \begin{cases} O : i = 15, 14, \\ \Lambda : i = 13, 12, \\ \Phi : i = 11, \\ * : i = 10, 9, ..., 0 \end{cases} \tag{1}$$

Note that this does not mean that the attacker is able to control the intermediate value up to b_{11}. In fact, b_{11} can be expressed by other buffer values and a single F-function evaluation (see the discussion above concerning non-linear buffer relation). However, thanks to the subsequent randomization after initial vector injection, this property must be destroyed before the output sequence is generated. Therefore we believe the related-key attack based on the SQUARE attack does not pose any threat.

Re-synchronization attacks: This attack may be more practical than the above related-key cryptanalysis. However, the initial vector does not inject any value to the buffer until the 16-round mixing completes. Taking the number of controllable rounds shown above into account, 16-round mixing is sufficient to destroy the saturation property due to initial vector.

5   Implementation

MUGI is designed to be suitable both in software and hardware implementations. In both cases, the implementation achieves a high performance and a low implementation cost. Table 4 and 5 summarize the software and hardware performance respectively. Table 4 shows that the performance in C is a little bit faster than AES.

Table 4. Software performance

Processor	Frequency	OS	Compiler	Performance (cycle/byte)
Alpha 21164	600MHz	Digital UNIX V4.0B	DEC cc	9.8
Intel Pentium III	500MHz	Windows NT 4.0	Visual C++ 6.0	17.7

Table 5. Hardware performance (Hitachi 0.35 μm CMOS ASIC library)

Optimization	Gate size (K gate)	Clock cycle (MHz)	Throughput (Mbps)	Initialization (ns)
speed opt.	26.1	45.7	2922	1095
gate cnt. opt.	18.0	42.3	676	4590
(3 layers pipelining)	($\geq$ 19.0)	(126.6)	(2025)	(1531)

The hardware implementation of MUGI achieves excellent performance, several time faster than AES.

6   Conclusion

We have proposed a new keystream generator MUGI built on the idea of PANAMA. MUGI is efficient in both hardware and software. Our security analysis indicates that MUGI is resistant against related-key attacks and resynchronization attacks. But the security of MUGI should be evaluated more. We invite the reader to explore the security of MUGI.

References

[BS93] E. Biham, A. Shamir, "Differential Cryptanalysis of the Data Encryption Standard," Springer-Verlag, 1993

[CGMPS00] A. Clark, J. Golic, W. Millan, L. Penna, L. Simpson, "The LILI-128 Keystream Generator," *NESSIE project submission*, 2000, available at http://www.cryptonessie.org.

[CL01] P. Crowley, S. Lucks, "Bias in the LEVIATHAN Stream Cipher," *Fast Software Encryption, FSE 2001*, Proceedings, pp. 223–230, 2001.

[Da95] J. Daemen, "Cipher and hash function design strategies based on linear and differential cryptanalysis," Doctoral Dissertation, March 1995, K. U. Leuven.

[DC98] J. Daemen, C. Clapp, "Fast Hashing and Stream Encryption with PANAMA," *Fast Software Encryption, FSE'98*, Springer-Verlag, LNCS 1372, pp.60–74, 1998.

[DGV94] J. Daemen, R. Govaerts, J. Vandewalle, "Resynchronization weaknesses in synchronous stream ciphers," *Advances in Cryptology, Proceedings Eurocrypt'93*, Springer-Verlag, LNCS 765, pp. 159-169, 1994.

[DKR97] J. Daemen, L. Knudsen, V. Rijmen, "The Block Cipher SQUARE," *Fast Software Encryption*, Springer-Verlag, LNCS 1267, pp. 149–165, 1997.

[DR99] J. Daemen, V. Rijmen, "AES Proposal: Rijndael," AES algorithm submission, September 3, 1999, available at `http://www.nist.gov/aes/`.

[Fl01] S. Fluhrer, "Cryptanalysis of the SEAL 3.0 Pseudorandom Function Family," *Fast Software Encryption, FSE 2001*, Proceedings, pp. 142–151, 2001.

[FS01] S. Fluhrer, M. Shamir, "Weaknesses in the Key Scheduling Algorithm of RC4," *Selected in Areas in Cryptography, SAC 2001*, Springer-Verlag, LNCS 2259, pp. 1–24, 2001.

[FWT00] S. Furuya, D. Watanabe, K. Takaragi, "Self-Evaluation Report MULTI-S01," 2000, available at `http://www.sdl.hitachi.co.jp/crypto/s01/index.html`

[JK97] T. Jacobsen and L. R. Knudsen, "The Interpolation Attack on Block Ciphers," *Fast Software Encryption, FSE'97*, Springer-Verlag, LNCS 1267, pp. 28–40, 1997.

[Ku94] L. R. Knudsen, "Truncated and Higher Order Differentials," *Fast Software Encryption, FSE'94*, Springer-Verlag, LNCS 1008, pp. 196–211, 1995.

[Ma94] M. Matsui, "Linear cryptanalysis method for DES cipher," *Advances in Cryptology, Eurocrypt'93*, Springer-Verlag, LNCS 765, pp. 159–169, 1994.

[McF00] D. McGrew, S. Fluhrer, "The stream cipher LEVIATHAN," *NESSIE project submission*, 2000, available at `http://www.cryptonessie.org/`.

[MFI01] M. Mihaljevic, M. Fossorier, H. Imai, "Fast Correlation Attack Algorithm with List Decoding and an Application," *Fast Software Encryption, FSE 2001*, Proceedings, pp. 208–222, 2001.

[RC94] P. Rogaway, D. Coppersmith, "A Software-Optimized Encryption Algorithm," *Fast Software Encryption, FSE'94*, Springer-Verlag, LNCS 809, pp. 56–63, 1994.

[RC98] P. Rogaway, D. Coppersmith, "A Software-Optimized Encryption Algorithm," *Journal fo Cryptography*, Vol. 11, No. 4, pp. 273–287, 1998.

[RRPV01] V. Rijmen, B. Van Rompay, B. Preneel, J. Vandewalle, "Producing Collisions for PANAMA," *Fast Software Encryption, FSE 2001*, proceedings, pp. 39–53, 2001.

[Sc96] B. Schneier, *Applied Cryptography*, Second Edition, John Wiley & Sons, pp. 397-398, 1996.

[SK96] B. Schneier, J. Kelsey, "Unbalanced Feistel Networks and Block Cipher Design," *Fast Software Encryption, FSE'96*, Springer-Verlag, LNCS 1039, pp. 121–144, 1996.

[Spec] D. Watanabe, S. Furuya, H. Yoshida, K. Takaragi, *MUGI Pseudorandom number generator, Specification*, 2001, available at `http://www.sdl.hitachi.co.jp/crypto/mugi/index-e.html`.

[Eval] D. Watanabe, S. Furuya, H. Yoshida, K. Takaragi, *MUGI Pseudorandom number generator, Self Evaluation*, 2001, available at `http://www.sdl.hitachi.co.jp/crypto/mugi/index-e.html`.

A Test Vector

```
key[16] =
{0x00 0x01 0x02 0x03 0x04 0x05 0x06 0x07 0x08 0x09 0x0a 0x0b 0x0c 0x0d 0x0e 0x0f}
iv[16]  =
{0xf0 0xe0 0xd0 0xc0 0xb0 0xa0 0x90 0x80 0x70 0x60 0x50 0x40 0x30 0x20 0x10 0x00}

after key input:
state  a = 0001020304050607  08090a0b0c0d0e0f  7498f5f1e727d094
buffer b =
0000000000000000  0000000000000000  0000000000000000  0000000000000000
0000000000000000  0000000000000000  0000000000000000  0000000000000000
0000000000000000  0000000000000000  0000000000000000  0000000000000000
```

```
0000000000000000   0000000000000000   0000000000000000   0000000000000000

after the first 16 rounds mixing:
state  a = 7dea261cb61d4fea  eafb528479bb687d  eb8189612089ff0b
buffer b =
7dea261cb61d4fea  bfe2485ac2696cc7  c905d08f50fa71db  fd5755df9cc0ceb9
5cc4835080bc5321  dfbbb88c02c9c80a  591a6857e3112cee  20ead0479e63cdc3
2d13c00221057d8d  b36b4d944f5d04cb  738177859f3210f6  c08ee4dcb2d08591
9c0c2097edb20067  09671cfbcfaa95fb  9724d9144c5d8926  08090a0b0c0d0e0f

after iv input:
state  a = 8d0af6dc06bddf6a  9a9b02c4499b787d  f100cffe031d365b
buffer b =
7dea261cb61d4fea  bfe2485ac2696cc7  c905d08f50fa71db  fd5755df9cc0ceb9
5cc4835080bc5321  dfbbb88c02c9c80a  591a6857e3112cee  20ead0479e63cdc3
2d13c00221057d8d  b36b4d944f5d04cb  738177859f3210f6  c08ee4dcb2d08591
9c0c2097edb20067  09671cfbcfaa95fb  9724d9144c5d8926  08090a0b0c0d0e0f

after the second 16 rounds mixing:
state  a = 4e466dffcb92db48  f5eb67b928359d8b  5d3c31a0af9cd78f
buffer b =
7dea261cb61d4fea  bfe2485ac2696cc7  c905d08f50fa71db  fd5755df9cc0ceb9
5cc4835080bc5321  dfbbb88c02c9c80a  591a6857e3112cee  20ead0479e63cdc3
2d13c00221057d8d  b36b4d944f5d04cb  738177859f3210f6  c08ee4dcb2d08591
9c0c2097edb20067  09671cfbcfaa95fb  9724d9144c5d8926  08090a0b0c0d0e0f

after the whole initialization:
state  a = 0ce5a4d1a0cbc0f7  316993816117e50f  bc62430614b79b71
buffer b =
d25c6643a9dabd67  e893c5b5a5b2ff2b  ce840df556562dc6  4210def4ccf1b145
5eda7c5b0dbf1554  d3e8a809b214218a  d42bcb0bb4811480  76d9c281df20192d
3dc6c6bc876beb72  39d84df58f8840e2  cd7fe2794367de6c  680920245819a4f5
f5e9e609dd8e3cc3  9cf94157cf512603  871323e1d70caa2b  0b6bb4c0466c7aba

output =
bc62430614b79b71  71a66681c35542de  7aba5b4fb80e82d7  0b96982890b6e143
4930b5d033157f46  b96ed8499a282645  dbeb1ef16d329b15  34a9192c4ddcf34e
...
```

Scream: A Software-Efficient Stream Cipher

Shai Halevi, Don Coppersmith, and Charanjit Jutla

IBM T.J. Watson Research Center, Yorktown Heights, NY 10598, USA,
{shaih,copper,csjutla}@watson.ibm.com

Abstract. We report on the design of Scream, a new software-efficient stream cipher, which was designed to be a "more secure SEAL". Following SEAL, the design of Scream resembles in many ways a block-cipher design. The new cipher is roughly as fast as SEAL, but we believe that it offers a significantly higher security level. In the process of designing this cipher, we re-visit the SEAL design paradigm, exhibiting some tradeoffs and limitations.

Keywords: Stream ciphers, Block ciphers, Round functions, SEAL.

1 Introduction

A stream cipher (or pseudorandom generator) is an algorithm that takes a short random string, and expands it into a much longer string, that still "looks random" to adversaries with limited resources. The short input string is called the seed (or key) of the cipher, and the long output string is called the output stream (or key-stream). Stream ciphers can be used for shared-key encryption, by using the output stream as a one-time-pad. In this work we aim to design a secure stream cipher that has very fast implementations in software.

1.1 A More Secure SEAL

The starting point of our work was the SEAL cipher. SEAL was designed in 1992 by Rogaway and Coppersmith [5], specifically for the purpose of obtaining a software efficient stream cipher. Nearly ten years after it was designed, SEAL is still the fastest steam cipher for software implementations on contemporary PC's, with "C" implementations running at 5 cycle/byte on common PC's (and 3.5 cycle/byte on some RISC workstations).

The design of SEAL shares many similarities with the design of common block ciphers. It is built around a repeating *round function*, which provides the "cryptographic strength" of the cipher. Roughly speaking, the main body of SEAL keeps a state which is made of three parts: an *evolving state*, some *round keys*, and a *mask table*. The output stream is generated in steps (or rounds). In each step, the round function is applied to the evolving state, using the round keys. The new evolving state is then masked by some of the entries in the mask

J. Daemen and V. Rijmen (Eds.): FSE 2002, LNCS 2365, pp. 195–209, 2002.

table and this value is output as a part of the stream. The mask table is fixed, and some of the round keys are be changed every so often (but not every step).[1]

In terms of security, SEAL is somewhat of a mixed story. SEAL is designed to generate up to 2^{48} bytes of output per seed. In 1997, Handschuh and Gilbert showed, however, that the output stream can be distinguished from random after seeing roughly 2^{34} bytes of output [4]. SEAL was slightly modified after that attack, and the resulting algorithm is known as SEAL 3.0. Recently, Fluhrer described an attack on SEAL 3.0, that can distinguish the output stream from random after about 2^{44} output bytes [3]. Hence, it seems prudent to avoid using the same seed for more than about 2^{40} bytes of output.

The goal of the current work was to come up with a "more secure SEAL". As part of that, we studied the advantages, drawbacks, and tradeoffs of this style of design. More specifically, we tried to understand what makes a "good round function" for a stream cipher, and to what extent a "good round function" for a block cipher is also good as the basis for a stream cipher. We also studied the interaction between the properties of the round function and other parts of the cipher. Our design goals for the cipher were as follows:

- Higher security than SEAL: It should be possible to use the same seed for 2^{64} bytes of output. More precisely, an attacker that sees a total of 2^{64} bytes of output (possibly, using several IV's of its choice), would be forced to spend an infeasible amount of time (or space) in order to distinguish the cipher from a truly random function. A reasonable measure of "infeasibility" is, say, 2^{80} space and 2^{96} time, so we tried to get the security of the cipher comfortably above these values.[2]
- Comparable speed to SEAL, i.e., about 5 cycles per byte on common PC's.
- We want to allow a full 128-bit input nonces (vs. 32-bit nonce in SEAL).
- Other, secondary, goals were to use smaller tables (SEAL uses 4KB of secret tables), get faster initialization (SEAL needs about 200 applications of SHA to initialize the tables), and maybe make the cipher more amenable to implementation in other environments (e.g., hardware, smartcard, etc.) We also tried to make the cipher fast on both 32-bit and 64-bit architectures.

1.2   The End Result(s)

In this report we describe three variants of our cipher. The first variant, which we call Scream-0, should perhaps be viewed as a "toy cipher". Although it may be secure enough for some applications, it does not live up to our security goals. In the full version of this report we describe a "low-diffusion attack" that works in time 2^{79} and space 2^{50}, and distinguishes Scream-0 from random after seeing about 2^{44} bytes of the output stream.

---

[1] In SEAL, the evolving state is the words A, B, C, D, the round keys consists of the table T and the n_i's, and the mask table is S.

[2] This security level is arguably lower than, say, AES. This seems to be the price that one has to pay for the increased speed. We note that the "obvious solution" of using Rijndael with less rounds, fails to achieve the desired security/speed tradeoff.

We then describe Scream, which is the same as Scream-0, except that it replaces the fixed S-boxes of Scream-0 by key-dependent S-boxes. Scream has very fast software implementations, but to get this speed one has to use secret tables roughly as large as those of SEAL (mainly, in order to store the S-boxes). On our Pentium-III machine, an optimized "C" implementation of Scream runs at 4.9 cycle/byte, slightly faster than SEAL. On a 32-bit PowerPC, the same implementation runs at 3.4 cycle/byte, again slightly faster than SEAL. This optimized implementation of Scream uses about 2.5 KB of secret tables. Scream also offers some space/time tradeoffs. (In principle, one could implement Scream with less than 400 bytes of memory, but using so little space would imply a slowdown of at least two orders of magnitude, compared to the speed-optimized implementation.) In terms of security, if the attacker is limited to only 2^{64} bytes of text, we do not know of any attack that is faster than exhaustively searching for the 128-bit key. On the other hand, we believe that it it possible to devise a linear attack to distinguish Scream from random, with maybe 2^{80} bytes of text.

At the end of this report we describe another variant, called Scream-F (for Fixed S-box), that does not use secret S-boxes, but is slower than Scream (and also somewhat "less elegant"). An optimized "C" implementation of Scream-F runs at 5.6 cycle/byte on our Pentium-III, which is 12% slower than SEAL. On our PowerPC, this implementation runs at 3.8 cycle/byte, 10% slower than SEAL. This implementation of Scream-F uses 560 bytes of secret state. We believe that the security of Scream-F is roughly equivalent to that of Scream.

1.3 Organization

In Section 2 below we first describe Scream-0 and then Scream. In Section 3 we discuss implementation issues and provide some performance measurements. In Section 4 we briefly discuss the cryptanalysis of Scream-0. (A more detailed analysis can be found in the full version.) Finally, in Section 5, we describe the cipher Scream-F. In the appendix we give the constants that are used in Scream, and also provide some "test vectors".

2   The Design of Scream

We begin with the description of Scream-0. As with SEAL, this cipher too is built around a "round function" that provides the cryptographic strength. Early in our design, we tried to use an "off the shelf" round function as the basis for the new cipher. Specifically, we considered using the Rijndael round function [2], which forms the basis of the new AES. However, as we discuss in the full paper, the "wide trail strategy" that underlies the design of the Rijndael round function is not a very good match for this type of design. We therefore designed our own round function.

At the heart of our round function is a scaled-down version of the Rijndael function, that operates on 64-bit blocks. The input block is viewed as a 2×4 matrix of bytes. First, each byte is sent through an S-box, $S[\cdot]$, then the second row in the matrix is shifted cyclically by one byte to the right, and finally each column is multiplied by a fixed 2×2 invertible matrix M. Below we call this function the "half round function", and denote it by $G_{S,M}(x)$. A pictorial description of $G_{S,M}$ can be found in Figure 1.

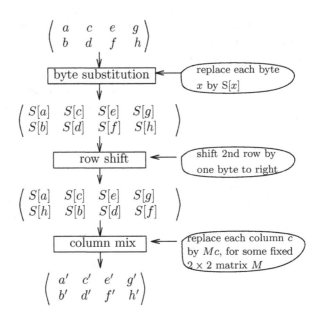

Fig. 1. The "half round" function $G_{S,M}$

Our round function, denoted $F(x)$, uses two different instances of the "half-round" function, G_{S_1,M_1} and G_{S_2,M_2}, where S_1, S_2 are two different S-boxes, and M_1, M_2 are two different matrices. The S-boxes S_1, S_2 in Scream-0 are derived from the Rijndael S-box, by setting $S_1[x] = S[x]$, and $S_2[x] = S[x \oplus 00010101]$, where $S[\cdot]$ is the Rijndael S-box. The constant 00010101 (decimal 21) was chosen so that S_2 will not have a fixed-point or an inverse fixed-point.[3] The matrices M_1, M_2 were chosen so that they are invertible, and so that neither of M_1, M_2 and $M_2^{-1} M_1$ contains any zeros. Specifically, we use

$$M_1 = \begin{pmatrix} 1 & x \\ x & 1 \end{pmatrix} \qquad M_2 = \begin{pmatrix} 1 & x+1 \\ x+1 & 1 \end{pmatrix}$$

where $1, x, x+1$ are elements of the field $GF(2^8)$, which is represented as $\mathbb{Z}_2[x]/(x^8 + x^7 + x^6 + x + 1)$.

The function F is a mix of a Feistel ladder and an SP-network. A pseudocode of F is provided below, and a pictorial description can be found in Figure 2.

---

[3] An inverse fixed-point is some x such that $S[x] = \bar{x}$.

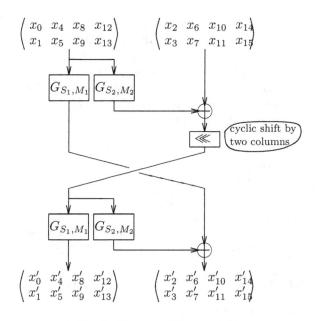

Fig. 2. The round function, F

<u>Function $F(x)$:</u>

1. Partition x into two 2×4 matrices

$$A := \begin{pmatrix} x_0 & x_4 & x_8 & x_{12} \\ x_1 & x_5 & x_9 & x_{13} \end{pmatrix} \qquad B := \begin{pmatrix} x_2 & x_6 & x_{10} & x_{14} \\ x_3 & x_7 & x_{11} & x_{15} \end{pmatrix}$$

2. $B := B \oplus G_{S_2,M_2}(A)$ // use A to modify A, B
3. $A := G_{S_1,M_1}(A)$

4. $B := \begin{pmatrix} B_{0,2} & B_{0,3} & B_{0,0} & B_{0,1} \\ B_{1,2} & B_{1,3} & B_{1,0} & B_{1,1} \end{pmatrix}$ // rotate B by two columns

5. Swap $A \leftrightarrow B$
6. $B := B \oplus G_{S_2,M_2}(A)$ // use A to modify A, B
7. $A := G_{S_1,M_1}(A)$

8. Collect the 16 bytes in A, B back into x
$$x' := (A_{0,0}\ A_{1,0}\ B_{0,0}\ B_{1,0}\ A_{0,1}\ A_{1,1}\ B_{0,1}\ B_{1,1}\ A_{0,2}\ A_{1,2}\ B_{0,2}\ B_{1,2}\ A_{0,3}\ A_{1,3}\ B_{0,3}\ B_{1,3})$$

The main loop of Scream-0. As with SEAL, the cipher Scream-0 maintains a state that consists of the "evolving state" x, some round keys y, z, and a "mask table" W. In Scream-0, x, y and z are 16-byte blocks, and the table W consists of 16 blocks, each of 16 bytes. In step i of Scream-0, the evolving state is modified by setting $x := F(x \oplus y) \oplus z$, and we then output $x \oplus W[i \bmod 16]$.

    In Scream-0, both the mask table and the round keys are modified, albeit slowly, throughout the computation. Specifically, after every pass through the mask table (i.e., every 16 steps), we modify y, z and one entry in W, by passing them through the F function. The entries of W are modified in order: after the j'th pass through the table we modify the entry $W[j \bmod 16]$. Moreover, instead of keeping both y, z completely fixed for 16 rounds, we rotate y by a few bytes after each use. The rotation amounts were chosen so that the rotation would be "almost for free" on 32-bit and 64-bit machines. This simple measure provides some protection against "low-diffusion attacks" and linear analysis. A pseudocode of the body of Scream-0 is described in Figure 3.

The main loop of Scream:
State: x, y, z – three 16-byte blocks
   W – a table of 16 16-byte blocks
   i_w – an index into W (initially $i_w = 0$)
1. repeat (until you get enough output bytes)
2. for $i = 0$ to 15   // generate the next 16 output blocks
3. $x := F(x \oplus y)$ // modify the "evolving state" x
4. $x := x \oplus z$
5.   output $x \oplus W[i \bmod 16]$
6.   if $i = 0$ or $2 \bmod 4$ // rotate y
7.   rotate y by 8 bytes, $y := y_{8..15,0..7}$
8.   else if $i = 1 \bmod 4$
9.   rotate each half of y by 4 bytes, $y := y_{4..7,0..3,12..15,8..11}$
10.   else if $i < 15$ // no point in rotating when $i = 15$
11.   rotate each half of y by three bytes to the right, $y := y_{5..7,0..4,13..15,8..12}$
12.   end-if
13. end-for
14. $y := F(y \oplus z)$ // modify y, z, and $W[i_w]$
15. $z := F(z \oplus y)$
16. $W[i_w] := F(W[i_w])$
17. $i_w := i_w + 1 \bmod 16$
18. end-repeat

Fig. 3. The main body of Scream and Scream-0

Key- and nonce-setup. The key- and nonce-setup procedures of Scream-0 are quite straightforward: We just use the round function F to derive all the quantities that we need. The key-setup routine fills the table W with some initial values. These values are later modified during the nonce-setup routine, and they also double as the equivalent of a "key schedule" for the nonce-setup routine. A pseudocode for these two routines is provided in Figures 4 and 5.

2.1   The Ciphers Scream

The cipher Scream is the same as Scream-0, except that we derive the S-boxes $S_1[\cdot], S_2[\cdot]$ from the Rijndael S-box $S[\cdot]$ in a key-dependent fashion. We replace line 0a in Figure 4 by the following

  0a. set $S_1[x] := S[\ldots S[S[x + \text{seed}_0] + \text{seed}_1] \ldots + \text{seed}_{15}]$ for all x

Key-setup:

Input: seed   – a 16-byte block
State: a, b – temporary variables, each a 16-byte block
Output: $W0$ – a table of sixteen 16-byte blocks

0a. set $S_1[x] := S[x]$ for all x // $S[\cdot]$ is the Rijndael S-box
0b. set $S_2[x] := S_1[x \oplus 00010101]$ for all x

1. $a := $ seed
2. $b := F(a \oplus pi)$ // pi is a constants: the first 16 bytes in the binary expansion of π
3. for $i = 0$ to 15
4. $a := F^4(a) \oplus b$ // four applications of the function F
5. $W0[i] := a$
6. end-for

Fig. 4. The key-setup of Scream-0

Nonce-setup:

Input: nonce   – a 16-byte block
State: $W0$ – a table of sixteen 16-byte blocks
          a, b – temporary variables, each a 16-byte block
Output: x, y, z – three 16-byte blocks
          W – a table of sixteen 16-byte blocks

1. $z := F^2(\text{nonce} \oplus W0[1])$ // two applications of the function F
2. $y := F^2(z \oplus W0[3])$
3. $a := F^2(y \oplus W0[5])$
4. $x := F(a \oplus W0[7])$ // only one application of F
5. $b := x$
6. for $i = 0$ to 7   // set W as a modification of $W0$
7. $b := F(b \oplus W0[2i])$
8. $W[\ 2i\ ]\ \ \ := W0[\ 2i\ ] \oplus a$
9. $W[2i + 1] := W0[2i + 1] \oplus b$
10. end-for

Fig. 5. The nonce-setup of Scream and Scream-0

(Notice that + denotes integer addition mod 256, rather then exclusive-or.) In terms of speed (in software), Scream-S is just as fast as Scream-0, except for the key-setup. However, it has a much larger secret state (a speed-optimized software implementation of Scream-S uses additional 2Kbyte of secret tables). We note that we still have $S_2[x] = S_1[x \oplus 00010101]$, so a space-efficient implementation need only store S_1.

3   Implementation and Performance

Software implementation of the F function. A fast software implementation of the F function uses tricks similar to Rijndael: Namely, we can implement the two "half round" functions $G_{S_1, M_1}, G_{S_2, M_2}$ together, using just eight lookup operations into two tables, each consisting of 256 four-byte words. Let the eight-byte input to G_{S_1,M_1}, G_{S_2,M_2} be denoted $(x_0, x_1, x_4, x_5, x_8, x_9, x_{12}, x_{13})$, the output of G_{S_1,M_1} be denoted $(u_0, u_1, u_4, u_5, u_8, u_9, u_{12}, u_{13})$, and the output of G_{S_2,M_2} be denoted $(u_2, u_3, u_6, u_7, u_{10}, u_{11}, u_{14}, u_{15})$. Then we can write:

$$u_0 = M_1(0,0) \cdot S1[x_0] \oplus M_1(0,1) \cdot S1[x_{13}]$$
$$u_1 = M_1(1,0) \cdot S1[x_0] \oplus M_1(1,1) \cdot S1[x_{13}]$$
$$u_2 = M_2(0,0) \cdot S2[x_0] \oplus M_2(0,1) \cdot S2[x_{13}]$$
$$u_3 = M_2(1,0) \cdot S2[x_0] \oplus M_2(1,1) \cdot S2[x_{13}]$$

(where $M(i, j)$ is the entry in row i, column j of matrix M, indexing starts from zero). Similar expressions can be written for the other bytes of u. Therefore, if we set the tables T_0, T_1 as

$$T_0(x) = \left\langle\; M_1(0,0) \cdot S1[x] \;\middle|\; M_1(1,0) \cdot S1[x] \;\middle|\; M_2(0,0) \cdot S2[x] \;\middle|\; M_2(1,0) \cdot S2[x] \;\right\rangle$$

$$T_1(x) = \left\langle\; M_1(0,1) \cdot S1[x] \;\middle|\; M_1(1,1) \cdot S1[x] \;\middle|\; M_2(0,1) \cdot S2[x] \;\middle|\; M_2(1,1) \cdot S2[x] \;\right\rangle$$

Then we can compute $u_{0..3} := T_0[x_0] \oplus T_1[x_{13}]$, $u_{4..7} := T_0[x_4] \oplus T_1[x_1]$, $u_{8..11} := T_0[x_8] \oplus T_1[x_5]$, and $u_{12..15} := T_0[x_{12}] \oplus T_1[x_9]$. A "reasonably optimized" implementation of the round function F (on a 32-bit machine) may work as follows:

Function $F(x_0, x_1, x_2, x_3)$:   // each x_i is a four-byte word

Temporary storage: u_0, u_1, u_2, u_3, each a four-byte word
1. $u_0 := T_0[\text{byte0}(x_0)] \oplus T_1[\text{byte1}(x_3)]$ // first "half round"
2. $u_1 := T_0[\text{byte0}(x_1)] \oplus T_1[\text{byte1}(x_0)]$
3. $u_2 := T_0[\text{byte0}(x_2)] \oplus T_1[\text{byte1}(x_1)]$
4. $u_3 := T_0[\text{byte0}(x_3)] \oplus T_1[\text{byte1}(x_2)]$
5. $[\text{byte2}(u_0) \mid \text{byte3}(u_0)] := [\text{byte2}(u_0) \mid \text{byte3}(u_0)] \;\oplus\; [\text{byte2}(x_0) \mid \text{byte3}(x_0)]$
6. $[\text{byte2}(u_1) \mid \text{byte3}(u_1)] := [\text{byte2}(u_1) \mid \text{byte3}(u_1)] \;\oplus\; [\text{byte2}(x_1) \mid \text{byte3}(x_1)]$
7. $[\text{byte2}(u_2) \mid \text{byte3}(u_2)] := [\text{byte2}(u_2) \mid \text{byte3}(u_2)] \;\oplus\; [\text{byte2}(x_2) \mid \text{byte3}(x_2)]$
8. $[\text{byte2}(u_3) \mid \text{byte3}(u_3)] := [\text{byte2}(u_3) \mid \text{byte3}(u_3)] \;\oplus\; [\text{byte2}(x_3) \mid \text{byte3}(x_3)]$

9. $u_0 := u_0 \lll 2 \text{ bytes}$ // swap the two halves
10. $u_1 := u_1 \lll 2 \text{ bytes}$
11. $u_2 := u_2 \lll 2 \text{ bytes}$
12. $u_3 := u_3 \lll 2 \text{ bytes}$

13. $x_0 := T_0[\text{byte0}(u_2)] \oplus T_1[\text{byte1}(u_2)]$ // second "half round"
14. $x_1 := T_0[\text{byte0}(u_3)] \oplus T_1[\text{byte1}(u_3)]$
15. $x_2 := T_0[\text{byte0}(u_0)] \oplus T_1[\text{byte1}(u_0)]$
16. $x_3 := T_0[\text{byte0}(u_1)] \oplus T_1[\text{byte1}(u_1)]$
17. $[\text{byte2}(x_0) \mid \text{byte3}(x_0)] := [\text{byte2}(x_0) \mid \text{byte3}(x_0)] \;\oplus\; [\text{byte2}(u_0) \mid \text{byte3}(u_0)]$
18. $[\text{byte2}(x_1) \mid \text{byte3}(x_1)] := [\text{byte2}(x_1) \mid \text{byte3}(x_1)] \;\oplus\; [\text{byte2}(u_1) \mid \text{byte3}(u_1)]$
19. $[\text{byte2}(x_2) \mid \text{byte3}(x_2)] := [\text{byte2}(x_2) \mid \text{byte3}(x_2)] \;\oplus\; [\text{byte2}(u_2) \mid \text{byte3}(u_2)]$

20. $[\text{byte2}(x_3) \mid \text{byte3}(x_3)] := [\text{byte2}(x_3) \mid \text{byte3}(x_3)] \oplus [\text{byte2}(u_3) \mid \text{byte3}(u_3)]$

21. output (x_0, x_1, x_2, x_3)

We note the need for explicit swapping of the two halves above (lines 9-12). The reason for that is that the tables T_0, T_1 are arranged so that the part corresponding to G_{S_1,M_1} is in the first two bytes of each entry, and the part of G_{S_2,M_2} is in the last two bytes. The code above can be optimized further, combining the rotation in these lines with the masking, which is implicit in lines 5-8, 17-20. Hence, the rotation becomes essentially "for free".

This structure provides a space/time tradeoff similar to Rijndael: Since the matrices M_1, M_2 are symmetric, one can obtain $T_2(x)$ from $T_1(x)$ using a few shift operations. Hence, it is possible to store only one table, at the expense of some slowdown in performance. This tradeoff is particularly important for Scream, where the tables T_0, T_1 are key-dependent.

The nonce-setup routine. The nonce-setup routine was designed so that the first output block can be computed as soon as possible. Although all the entries of the table W have to be modified during the nonce-setup, an application that does not use all of them can modify only as many as it needs. Hence an application that only outputs a few blocks per input nonce, does not have to complete the entire nonce-setup. Alternatively, an application can execute the nonce-setup together with the first "chunk" of 16 steps, modifying each mask of W just before this mask is needed.

Performance in software. We tested the software performance of Scream and Scream-F on two platforms, both with word-length of 32 bits: One platform is an IBM PC 300PL, with a 550MHz Pentium-III processor, running Linux and using the gcc compiler, version 3.0.3. The other platform is an RS/6000 43P-150 workstation, with a 375MHz 304e PowerPC processor, running AIX 4.3.3 and using the IBM C compiler (xlc) version 3.6.6. On both platforms, we measured peak throughput, and also timed the key-setup and nonce-setup routines. To measure peak throughput, we timed a procedure that produces 256MB of output (all with the same key and nonce). Specifically, the procedure makes one million calls to a function that outputs the next 256 bytes of the cipher. To eliminate the effect of cache misses, we used the same output buffer in all the calls. We list our test results in the table below.

Platform	Operation	Scream-F	Scream	SEAL
Pentium-III	throughput	5.6 cycle/byte	4.9 cycle/byte	5.0 cycle/byte
550 MHz	key-setup	3190 cycles	27500 cycles	
Linux, gcc	nonce-setup	1276 cycles	1276 cycles	
604e PowerPC	throughput	3.8 cycle/byte	3.4 cycle/byte	3.45 cycle/byte
375 MHz	key-setup	1950 cycles	16875 cycles	
AIX, xlc	nonce-setup	670 cycles	670 cycles	

Implementation in different environments. Being based on a Rijndael-like round function, Scream is amenable for implementations in many different environments. In particular, it should be quite easy to implement it in hardware, and the area/speed tradeoff in such implementation may be similar to Rijndael (except that Scream needs more memory for the mask table). Also, it should be quite straightforward to implement it for 8- and 16-bit processors (again, as long as the architecture has enough memory to store the internal state). Scream is clearly not suited for environments with extremely small memory, but it can be implemented with less than 400 bytes of memory (although such implementation would be quite slow).

4   Security Analysis

Below we examine some possible attacks on Scream-0 and Scream. The discussion below deals mostly with Scream-0. At the end we briefly discuss the effect of Scream's key-dependent S-boxes on these attacks. We examine two types of attacks, one based on linear approximations of the F function, and the other exploits the low diffusion provided by a single application of F. In both attacks, the goal of the attacker is to distinguish the output of the cipher from a truly random stream.[4]

4.1   Linear Attacks

It is not hard to see that the F function has linear approximations that approximate only three of the 8-by-8 S-boxes. Since the S-boxes in Scream-0 are based on the Rijndael S-box, the best approximation of them has bias 2^{-3}, so we can probably get a linear approximation of the F function with bias 2^{-9}. Namely, there exists a linear function L such that $\Pr_x[L(x, F(x)) = 0] = 1/2 \pm 2^{-9}$. (In the full version of this report we show that bias of 2^{-9} is indeed the best possible.)

To use this approximation, we need to eliminate the linear masking, introduced by the y, z and the $W[i]$'s. Here we use the fact that each one of these masks is used sixteen times before it is modified. For each step of the cipher, the attacker sees a pair $(x \oplus y \oplus W[i], \ F(x) \oplus z \oplus W[i+1])$, where x is random. Applying the function L to this pair, we get the bit

$$\sigma = L(x, F(x)) \oplus L(y, z) \oplus L(W[i], W[i+1])$$

For simplicity, we ignore for the moment the rotation of the y block after each step. If we add two such σ's that use the same y and z blocks, we get $\tau = \sigma \oplus \sigma' = L(x, F(x)) \oplus L(x', F(x')) \oplus L(W[i], W[i+1]) \oplus L(W[j], W[j+1])$. The last bit does not depend on y, z anymore. We can repeat this process, adding two such τ's that use the same masks, we end up with a bit

$$\mu = \tau \oplus \tau' = L(x, F(x)) \oplus L(x', F(x')) \oplus L(x'', F(x'')) \oplus L(x''', F(x'''))$$

---

[4] In a separate paper [1], we show that these two types of attacks can be viewed as two special cases of a generalized distinguishing attack.

Since $L(x, F(x))$ has bias of 2^{-9}, the bit μ has bias of 2^{-36}, so after seeing about 2^{72} such bits, we can distinguish the cipher from random.

Since each of the masks is used sixteen times before it is modified, we have about $\binom{16}{2}$ choices for the pairs of σ's to add (still ignoring the rotation of y), and about $\binom{16}{2}$ choices for the pairs of τ's to add. Hence, 256 steps of the cipher gives us about $\binom{16}{2}^2 \approx 2^{14}$ bits μ. After seeing roughly $256 \cdot 2^{58} = 2^{66}$ steps of the cipher (i.e., 2^{70} bytes of output), we can to collect the needed 2^{72} samples of μ's to distinguish the cipher from random.

The rotation of y. The rotation of y makes it harder to devise attacks as above. To cancel both the y and the z blocks, one would have to use two different approximations with the same output bit pattern, but where the input bit patterns are rotated accordingly. We do not know if it possible to devise such approximation with bias of 2^{-9}.

The secret S-boxes. The introduction of key-dependent S-boxes in Scream does not significantly alter the analysis from above. Since the S-boxes are key-dependent, an attacker cannot pick "the best approximations" for them, but on the other hand these S-boxes have better approximations than the Rijndael S-box. Thus, the attacker can use a random approximation, and it will likely to be roughly as good as the best approximation for the fixed S-boxes.

4.2 Low-Diffusion Attacks

A low-diffusion attack exploits the fact that not every byte of $F(x)$ is influenced by every byte of x (and vise versa). For example, there are output bytes that only depend on six input bytes. In fact, in the full version of this report we show that knowing two bytes of x and one byte of (linear combination of bytes in) $F(x)$, we can compute another byte of (linear combination of bytes in) $F(x)$. Namely, we have a (non-degenerate) linear function L with output length of four bytes, so that we can write $L(X, F(x))_3 = g(L(X, F(x))_{0..2})$, where g is an known deterministic function (with three bytes of input and one byte of output).

As for the linear attacks, here too we need to eliminate the linear masking, introduced by the y, z and the $W[i]$'s. This is done in very much the same way. Again, we ignore for now the rotation of the block y. For each step of the cipher the attacker sees the four bytes $L(x \oplus y \oplus W[i], F(x) \oplus z \oplus W[i+1])$. We eliminate the dependence on y, z by adding two such quantities that use the same y, z blocks. This gives a four-byte quantity $L(x, F(x)) \oplus L(x', F(x')) \oplus L(W[i], W[i+1]) \oplus L(W[j], W[j+1])$. Adding two of those with the same i, j, we then obtain the four byte quantity

$$L(x, F(x)) \oplus L(x', F(x')) \oplus L(x'', F(x'')) \oplus L(x''', F(x'''))$$

We can write this last quantity in terms of the function g, as a pair $(r_1 \oplus r_2 \oplus r_3 \oplus r_4, \ g(r_1) \oplus g(r_2) \oplus g(r_3) \oplus g(r_4))$, where the r_i's are three-byte long, and the $g(r_i)$'s are one-byte long. In a separate paper [1], we analyze the statistical properties of such expressions, and calculate the number of samples that needs to be seen to distinguish them from random.

The rotation of y. Again, the rotation of y makes it harder to devise attacks as above. In the full paper we show, however, that we can still use a low-diffusion attack on the F function, in which guessing six bytes of $(x, F(x))$ yields the value of four other bytes. Applying tools from our paper [1] to this relation, we can compute that the amount of output text that is needed to distinguish the cipher from random along the lines above, is merely 2^{44} bytes. However, the procedure for distinguishing is quite expensive. The most efficient way that we know how to use these 2^{44} bytes would require roughly 2^{50} space and 2^{80} time.

The secret S-boxes. At present, we do not know how to extend low-diffusion attacks such as above to deal with secret S-boxes. Although we can still write the same expression $L(X, F(x))_3 = g(L(X, F(x))_{0..2})$, the function g now depends on the key, so it is not known to the attacker. Although it is likely that some variant of these attacks can be devised for this case too, we strongly believe that such variants would require significantly more text than the 2^{64} bytes that we "allow" the attacker to see.

5   The Cipher Scream-F

In Scream, we used key-dependent S-boxes to defend against "low-diffusion attacks". A different approach is to keep the S-box fixed, but to add to the main body of the cipher some "key dependent operation" before outputting each block. This approach was taken in Scream-F, where we added one round of Feistel ladder after the round function, using a key-dependent table. However, since the only key-dependent table that we have is the mask table W, we let W double also as an "S-box". Specifically, we add the following lines 3a-3e between lines 3 and 4 in the main-loop routine from Figure 3.

3a.   view the table W as an array of 64 4-byte words $\hat{W}[0..63]$

3b. $x_{0..3}\ :=\ x_{0..3}\ \oplus \hat{W}[1 + (x_4 \wedge 00111110)]$

3c. $x_{4..7}\ :=\ x_{4..7}\ \oplus \hat{W}[x_8 \wedge 00111110]$

3d. $x_{8..11} := x_{8..11} \oplus \hat{W}[1 + (x_{12} \wedge 00111110)]$

3e. $x_{12..15} := x_{12..15} \oplus \hat{W}[x_0 \wedge 00111110]$

We note that the operation $x_i \wedge 00111110$ in these lines returns an even number between 0 and 62, so we only use odd entries of W to modify $x_{0..3}$ and $x_{8..11}$, and even entries to modify $x_{4..7}$ and $x_{12..15}$. The reason is that to form the output block, the words $x_{0..3}, x_{8..11}$ will be masked with even entries of W, and the words $x_{4..7}, x_{12..15}$ will be masked by odd entries. The odd/even indexing is meant to avoid the possibility that these masks cancel with the entries that were used in the Feistel operation.[5]

---

[5] It is still possible that two words, say $x_{0..3}$ and $x_{4..7}$, are masked with the same mask, but it seems less harmful.

5.1   Conclusions

We presented Scream, a new stream cipher with the same design style as SEAL. The new cipher is roughly as fast as SEAL, but we believe that it is more secure. It has some practical advantages over SEAL, in flexibility of implementation, and also in the fact that it can take a full 128-bit nonce (vs. 32 bits in SEAL). In the process of designing Scream, we studied the advantages and pitfalls of the SEAL design style. We hope that the experience from this work would be beneficial also for future ciphers that uses this style of design.

Acknowledgments. This design grew out of a study group in IBM, T.J. Watson during the summer and fall of 2000. Other than the authors, the study group also included Ran Canetti, Rosario Gennaro, Nick Howgrave-Graham, Tal Rabin and J.R. Rao. The motivation for this work was partly due to the NESSIE "call for cryptographic primitives" (although we missed their deadline by more than one year).

References

1. D. Coppersmith, S. Halevi, and C. Jutla. Cryptanalysis of stream ciphers with linear masking. manuscript, 2002.
2. J. Daemen and V. Rijmen. AES proposal: Rijndael. Available on-line from NIST at http://csrc.nist.gov/encryption/aes/rijndael/, 1998.
3. S. Fluhrer. Cryptanalysis of the SEAL 3.0 pseudorandom function family. In *Proceedings of the Fast Software Encryption Workshop (FSE'01)*, 2001.
4. H. Handschuh and H. Gilbert. χ^2 cryptanalysis of the SEAL encryption algorithm. In *Proceedings of the 4th Workshop on Fast Software Encryption*, volume 1267 of *Lecture Notes in Computer Science*, pages 1–12. Springer-Verlag, 1997.
5. P. Rogaway and D. Coppersmith. A software optimized encryption algorithm. *Journal of Cryptology*, 11(4):273–287, 1998.

A   Constants and Test Vectors

```
The Rijndael S-box, S[0..255] = [
  63 7c 77 7b f2 6b 6f c5 30 01 67 2b fe d7 ab 76 ca 82 c9 7d fa 59 47 f0
  ad d4 a2 af 9c a4 72 c0 b7 fd 93 26 36 3f f7 cc 34 a5 e5 f1 71 d8 31 15
  04 c7 23 c3 18 96 05 9a 07 12 80 e2 eb 27 b2 75 09 83 2c 1a 1b 6e 5a a0
  52 3b d6 b3 29 e3 2f 84 53 d1 00 ed 20 fc b1 5b 6a cb be 39 4a 4c 58 cf
  d0 ef aa fb 43 4d 33 85 45 f9 02 7f 50 3c 9f a8 51 a3 40 8f 92 9d 38 f5
  bc b6 da 21 10 ff f3 d2 cd 0c 13 ec 5f 97 44 17 c4 a7 7e 3d 64 5d 19 73
  60 81 4f dc 22 2a 90 88 46 ee b8 14 de 5e 0b db e0 32 3a 0a 49 06 24 5c
  c2 d3 ac 62 91 95 e4 79 e7 c8 37 6d 8d d5 4e a9 6c 56 f4 ea 65 7a ae 08
  ba 78 25 2e 1c a6 b4 c6 e8 dd 74 1f 4b bd 8b 8a 70 3e b5 66 48 03 f6 0e
  61 35 57 b9 86 c1 1d 9e e1 f8 98 11 69 d9 8e 94 9b 1e 87 e9 ce 55 28 df
  8c a1 89 0d bf e6 42 68 41 99 2d 0f b0 54 bb 16 ]
```

The constant pi (for key-setup)

```
pi        = [24 3f 6a 88 85 a3 08 d3 13 19 8a 2e 03 70 73 44]
```

Test vectors for Scream-S

*** key-setup test vectors ***

```
key       = [00 00 00 00 00 00 00 00 00 00 00 00 00 00 00 00]
W0[0]     = [b6 a5 0b bf f3 9b 9e 99 28 b0 35 18 7b 7d 9c 7b]
W0[15]    = [83 32 53 22 db 10 00 31 49 3a a4 80 3a 41 8c b3]
```

*** nonce-setup test vectors ***

```
key       = [00 00 00 00 00 00 00 00 00 00 00 00 00 00 00 00]
nonce     = [00 00 00 00 00 00 00 00 00 00 00 00 00 00 00 00]
X         = [b4 b7 7e 35 6a 24 0c c8 a7 41 b8 c7 d7 29 68 82]
Y         = [e4 f4 1d 3b fd 07 d4 3c cb df a9 bb 25 df 65 6c]
Z         = [87 de 72 cd 96 5a 96 24 b4 eb 79 66 57 26 fd f9]
W[0]      = [66 d4 35 4d 2c 90 5f 0e 7f cc 25 59 43 ba d2 22]
W[15]     = [a8 0e b6 56 be aa 5d d2 8d ca fe 07 1b f9 9c 7a]
```

*** stream test vectors ***

```
key       = [01 23 45 67 89 ab cd ef fe dc ba 98 76 54 32 10]

nonce     = [00 00 00 00 00 00 00 00 00 00 00 00 00 00 00 00]
out[0]    = [74 8c 59 f2 0d 76 9e a8 7a 6d c1 87 46 e6 4a c0]
out[1]    = [bd 3b 39 cd 12 18 43 0f 80 fa e0 1b 2e 60 f1 74]
out[4]    = [15 21 8a 46 fb ee 26 54 98 8d 2b 80 8a 87 f4 5e]
out[16]   = [cb 32 f4 d6 f7 ce 57 69 e2 a3 ac d8 37 e1 37 82]
out[1023] = [97 ec 87 f0 a0 6c e7 0b 75 e6 12 25 50 1f 82 e3]

nonce     = [01 01 01 01 01 01 01 01 01 01 01 01 01 01 01 01]
out[0]    = [47 68 06 37 83 85 99 af d2 8f fb 2e dd fc 9d 2e]
out[1]    = [7b d3 0b e4 7a a6 3b 5f 4f 5f 05 06 66 17 d5 a2]
out[4]    = [98 aa 20 75 73 c7 fa fc 1c 4c 27 61 46 14 3c 1d]
out[16]   = [b3 33 a4 8e 17 50 8e ab b2 68 fb 60 67 56 46 1e]
out[1023] = [a5 41 b3 37 c6 bd 8a 4b 41 a1 40 5f ea c5 a3 f5]
```

Test vectors for Scream-F

*** key-setup test vectors ***

```
key       = [00 00 00 00 00 00 00 00 00 00 00 00 00 00 00 00]
W0[0]     = [be a0 cd 9a 5d f6 85 59 c0 3f a9 c5 53 fd ad e1]
W0[15]    = [eb 2e ab 45 26 ee 49 e1 34 db 97 87 62 d1 3b 25]
```

*** nonce-setup test vectors ***

```
key       = [00 00 00 00 00 00 00 00 00 00 00 00 00 00 00 00]
nonce     = [00 00 00 00 00 00 00 00 00 00 00 00 00 00 00 00]
X         = [d4 10 c5 bf bd 7b fd 81 37 4e e3 b0 c1 bf 8b a6]
Y         = [51 a6 7f 38 3d 0d 95 26 bf b5 b0 e8 26 b5 e4 93]
Z         = [53 50 b7 d6 87 3d df 8c 7f 9b 10 7c e0 92 d0 02]
W[0]      = [cb ad d5 c2 b0 85 af 77 6c d8 ef ce 7b 36 65 3a]
```

```
W[15]     = [19 14 5e 0a 4d 23 1c d5 f9 6f 85 8a 39 38 81 a1]

*** stream test vectors ***
key       = [01 23 45 67 89 ab cd ef fe dc ba 98 76 54 32 10]

nonce     = [00 00 00 00 00 00 00 00 00 00 00 00 00 00 00 00]
out[0]    = [39 ec 4a 06 45 4d c3 cd 96 dd ef 0c f0 c2 67 40]
out[1]    = [a0 ea 56 e7 e3 c8 f5 df 34 ea 35 ee 77 ed da 66]
out[4]    = [8a c8 93 af 83 ed 0a 53 6b e9 f4 7c b6 6d 21 67]
out[16]   = [e0 8c fe 31 34 a7 48 ca 14 10 f9 58 50 71 49 20]
out[1023] = [a4 e2 fc be 0a 47 53 9a 23 e0 79 25 5c be ea e7]

nonce     = [01 01 01 01 01 01 01 01 01 01 01 01 01 01 01 01]
out[0]    = [2e 70 fb 8c d5 d8 50 a8 94 38 0e 85 46 9d 33 fc]
out[1]    = [33 39 da 86 9c a1 f7 1b 3a d0 16 16 ea 42 24 1a]
out[4]    = [1a 79 cf 13 01 67 2c 52 25 13 8c c8 89 fb 50 72]
out[16]   = [c8 f2 3f ca 4e 0c 47 46 1a b3 7b 34 1b 57 c7 96]
out[1023] = [6e 63 21 c1 9b 49 08 57 84 87 14 ea 4f 08 4b 7d]
```

Distinguishing Attacks on SOBER-t16 and t32

Patrik Ekdahl and Thomas Johansson

Dept. of Information Technology, Lund University,
P.O. Box 118, 221 00 Lund, Sweden.
{Patrik,Thomas}@it.lth.se

Abstract. Two ways of mounting distinguishing attacks on two similar stream ciphers, SOBER-t16 and SOBER-t32, are proposed. It results in distinguishing attacks faster than exhaustive key search on full SOBER-t16 and on SOBER-t32 without stuttering.

1 Introduction

In the design of symmetric ciphers, security and performance are of outmost importance. For example, in the recent AES process we have seen a number of block ciphers competing in security and performance.

When choosing a symmetric encryption algorithm, the first choice is whether to choose a block cipher or a stream cipher. Most known block ciphers offer a sufficient security and a reasonably good performance. But a block cipher must usually be used in a "stream cipher" mode, which suggests that using a pure stream cipher primitive might be beneficial.

Modern stream ciphers will indeed offer an improved performance compared with block ciphers (typically a factor 4-5 if measured in speed). However, the security of stream ciphers is not as well understood as for block ciphers. Most proposed stream ciphers such as (alleged) RC4, A5/1, have security weaknesses [7,1].

In the recent call for primitives in the NESSIE project, two similar stream ciphers were submitted from Qualcomm Australia, called SOBER-t16 and SOBER-t32, respectively. These are two shift register based stream ciphers developed from previous versions of stream ciphers under the name of SOBER. There has been no known attacks better than exhaustive key search on these two stream ciphers, which means that they have offered full security. By full security we roughly mean that there is no attack that is better than an exhaustive key search attack. It should be noted that not many proposed stream ciphers offer full security.

A stream cipher consists of a keyed generator, producing a pseudo-random sequence that is added to the plaintext. In cryptanalysis, we consider the pseudo-random sequence to be known (known plaintext attack) and try to either recover the key, called a *key recovery attack*, or we try to distinguish the pseudo-random sequence from a truly random sequence, called a *distinguishing attack*.

J. Daemen and V. Rijmen (Eds.): FSE 2002, LNCS 2365, pp. 210–224, 2002.

The SOBER-t16 and SOBER-t32 generators can roughly be described as being nonlinear filter generators with an additional "stuttering" step before producing the output. Because of the stuttering step, the output will be irregularly produced. It is known that because of this irregularity, one can use a power analysis attack or a timing attack to recover the input to the stuttering step [9]. However, the authors claim that the generator is secure even without the stuttering step [9].

In this paper we consider several new ways of mounting distinguishing attacks on SOBER-t16 and SOBER-t32. The attacks are based on combining linear approximations of the nonlinear filter with the linear recurrence, defined through the feedback polynomial. Linear approximations have previously been used in e.g. the BAA attack on stream ciphers [3] and in linear cryptanalysis on block ciphers [8]. In our case we mainly derive the distribution of the noise introduced through the linear approximations by simulations. We consider attacks on the ciphers both including and excluding the stuttering step.

The final results are as follows. For SOBER-t16 without stuttering, which uses a 128 bit key, the output can be distinguished from a random sequence using at most 2^{92} output words and the same complexity. For the full SOBER-t16 with stuttering, we need at most 2^{111} output words and the same complexity. For SOBER-t32, without stuttering, which uses a 256 bit key, the output can be distinguished from a random sequence using at most 2^{87} output words and the same complexity. For the full SOBER-t32 with stuttering we could not find an exact complexity expression, but the proposed methods indicate a strong attack also here.

We should also mention that the proposed methods are applicable to the stream cipher SNOW [4], another candidate in the NESSIE project. The strength of such an attack on SNOW is considered in a subsequent paper.

The paper is organized as follows. In Section 2 we shortly describe the stream ciphers SOBER-t16 and SOBER-t32. Then we start by explaining the attack on SOBER-t16 without stuttering in Section 3. This is generalized to an attack on the full SOBER-t16 in Section 4. In Section 5 we describe a simple attack on SOBER-t32 without stuttering. In Section 6 we then elaborate on different possibilities for mounting an attack on the full SOBER-t32. Finally, we give some concluding remarks.

2   A Brief Description of SOBER-t16 and t32

Both SOBER-t16 and SOBER-t32 are word oriented stream ciphers. The word size is 16 bits for t16 and 32 bits for t32. The structure of t16 and t32 are very similar and we will here describe them as one cipher. The specific parameters for both t16 and t32 will be given alongside. To simplify the description of the common parts of t16 and t32, we will use the notation W to denote the word size. Thus, W is either 16 or 32 bits, depending on which cipher we are looking at. The operations in the ciphers include both addition in an extension field $\mathbb{F}_{2^W}$ and addition modulo 2^W, and we will denote the field addition by $\oplus$ (also called

XOR) and the ring addition by ⊞. In case there is no risk of confusion we will simply use the addition symbol +.

There are three main building blocks for the SOBER stream ciphers. The first is a word oriented linear feedback shift register (LFSR) which produces a LFSR sequence denoted $\{s_t, t \geq 0\}$. Secondly, a non-linear filter (NLF) takes some of these symbols as inputs and produces a new sequence $\{v_t, t \geq 0\}$. Finally, there is a so called stuttering unit. The stuttering unit takes $\{v_t, t \geq 0\}$ as input and produces an irregular output $\{z_n, n \geq 0\}$. The overall structure is pictured in Figure 1.

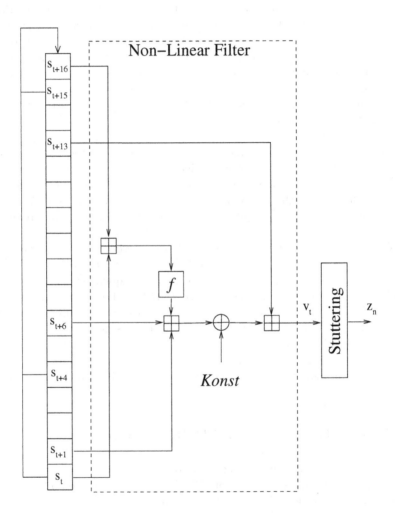

Fig. 1. Overall structure of SOBER-t16 and SOBER-t32.

2.1   The LFSR

The LFSR is a length 17 shift register, where each register element contains one word. Each word is considered as an element in an extension field ($\mathbb{F}_{2^w}$). The contents of the LFSR at time t is called the *state* of the LFSR at time t and will be denoted by a vector $\bar{S}_t = (s_t, s_{t+1}, \ldots, s_{t+16})$. The next state of the LFSR is obtained by shifting the previous state one step, and calculating a new word s_{t+17}. The new word is calculated as a certain linear combination of the contents of the previous state. Thus the next state will be $\bar{S}_{t+1} = (s_{t+1}, s_{t+2}, \ldots, s_{t+17})$ where

$$s_{t+17} = \sum_{i=0}^{16} c_i s_{t+i}, \tag{1}$$

for some known constants $c_i \in \mathbb{F}_{2^w}, i = 0, 1 \ldots, 16$. The arithmetics in Eq. (1) is performed in the extension field $\mathbb{F}_{2^w}$. Equation (1) is called the linear recurrence equation. The specific extension fields and recurrence equations for t16 and t32 are summarized below:

SOBER-t16
Defining polynomial for $\mathbb{F}_{2^{16}}$: $x^{16} + x^{14} + x^7 + x^6 + x^4 + x^2 + x + 1$
Linear recurrence equation: $s_{t+17} \oplus \alpha s_{t+15} \oplus s_{t+4} \oplus \beta s_t = 0$
where $\alpha = 0xE382$ and $\beta = 0x67C3$.

SOBER-t32
Defining polynomial for $\mathbb{F}_{2^{32}}$: $x^{32} + (x^{24} + x^{16} + x^8 + 1)(x^6 + x^5 + x^2 + 1)$
Linear recurrence equation: $s_{t+17} \oplus s_{t+15} \oplus s_{t+4} \oplus \alpha s_t = 0$
where $\alpha = 0xC2DB2AA3$.

The field elements α and β have been given in a hexadecimal form, corresponding to a polynomial basis. See [5,6] for more details.

2.2   The NLF Function

At time t, the NLF function takes five words from the LFSR state, $(s_t, s_{t+1}, s_{t+6}, s_{t+13}, s_{t+16})$ and one constant value ($Konst$) as input, and produces through a nonlinear function an output word, denoted by v_t. The value of $Konst \in \mathbb{F}_{2^w}$ is determined during the initialization phase of the LFSR and is kept constant throughout the entire session. The operations involved in the NLF function are XOR (denoted $\oplus$), addition modulo 2^W (denoted $\boxplus$) and application of a substitution box (denoted SBOX).

The output of the NLF function, v_t, at time t, can be written as:

$$v_t = ((s_{t+1} \boxplus s_{t+6} \boxplus f(s_t \boxplus s_{t+16})) \oplus Konst) \boxplus s_{t+13}, \tag{2}$$

where $f(x)$ is a function, different for SOBER-t16 and SOBER-t32, which in both cases involves an SBOX application. The interior design of the function f is pictured in Fig. 2. First the input is partitioned into a high part containing the 8 most significant bits, and a low part containing the remaining bits. The

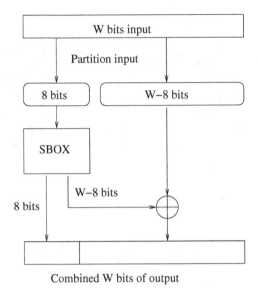

Fig. 2. The structure of the f-function in SOBER-t16 and SOBER-t32.

high part addresses an SBOX with W bits of output. The 8 most significant bits are directly taken as the f-function output, whereas the least significant part of the SBOX output is first XORed to the low part from the input, see Fig. 2.

2.3   Stuttering

Before producing the running key, the output from the NLF is passed through a stuttering unit. The stuttering decimates the NLF output, thus making e.g. a correlation attack harder. The first output from the NLF, v_0 is taken as the first *stutter control word* (SCW). The SCW is divided into pairs of bits (called dibits). Starting with the least significant dibit, the stuttering is determined from the value of these dibits. Actions are taken according to the value of the dibit, as listed in Table 1. The constant C has value 0x6996 for t16 and 0x6996C53A for t32, and $\sim C$ denotes the bitwise complement of C.

When all dibits in the SCW have been used, the LFSR is clocked once and a new SCW is read from the output of the NLF. This word determines the next 8 or 16 actions, depending on whether we are looking at SOBER-t16 or SOBER-t32. The resulting stream from the stuttering unit, denoted z_n, is the running key.

This concludes the brief description of SOBER-t16 and SOBER-t32. For a more detailed description, especially regarding the key initialization, we refer to [5] and [6].

Table 1. The possible actions taken in the stuttering unit depending on the value of the dibit.

Dibit	Action
00	1. Clock the LFSR, but do not output anything.
01	1. Clock the LFSR.
	2. Set the value of the next key stream word to be the XOR between C and the NLF output.
	3. Clock the LFSR again, but do not output anything.
10	1. Clock the LFSR, but do not output anything.
	2. Clock the LFSR.
	3. Set the value of the next key stream word to be the value of the NLF output.
11	1. Clock the LFSR.
	2. Set the value of the next key stream word to be the XOR between $\sim C$ and the NLF output.

3   A Distinguishing Attack on SOBER-t16 without Stuttering

We start by analyzing a version of SOBER-t16 where the stuttering unit has been removed. In this scenario each NLF output word is taken as a running key word. Thus we have $z_t = v_t$ for all $t \geq 0$. We also assume that we are given N words of the output key stream, so we have access to $z_0, z_1, \ldots, z_{N-1}$.

The first step in our attack is to approximate the NLF-function with a linear function and then argue that the noise introduced by the approximation possesses a nonuniform distribution. Recall the expression for the NLF output:

$$v_t = ((s_{t+1} \boxplus s_{t+6} \boxplus f(s_t \boxplus s_{t+16})) \oplus Konst) \boxplus s_{t+13}. \qquad (3)$$

We now approximate this function with a linear function by replacing $\boxplus$ with $\oplus$, and f by the identity map. When we do this approximation we introduce a noise (an error), which we denoted by w_t. We also move the value of $Konst$ into the noise w_t. I.e., we write

$$v_t = s_{t+1} \oplus s_{t+6} \oplus s_t \oplus s_{t+16} \oplus s_{t+13} \oplus w_t, \qquad (4)$$

where $w_t, t \geq 0$ denotes random variables that represent the error we get in the approximation at each time t. The distribution of w_t will be dependent on the value $Konst$. Also, w_t will have the same distribution for all t, and this distribution is denoted F.

Introduce the notation $\Omega_t = s_t \oplus s_{t+1} \oplus s_{t+6} \oplus s_{t+13} \oplus s_{t+16}$ for the XOR of the words from the LFSR that are inputs to the NLF function. Then we can write the output word v_t as

$$v_t = \Omega_t \oplus w_t. \qquad (5)$$

By looking at the running key at time $t, t+4, t+15$ and $t+17$ in combination with Eq. (5) we can express $z_{t+17} \oplus \alpha z_{t+15} \oplus z_{t+4} \oplus \beta z_t$ in the following way,

$$z_{t+17} \oplus \alpha z_{t+15} \oplus z_{t+4} \oplus \beta z_t =$$
$$v_{t+17} \oplus \alpha v_{t+15} \oplus v_{t+4} \oplus \beta v_t = (\Omega_{t+17} \oplus w_{t+17}) \oplus \alpha(\Omega_{t+15} \oplus w_{t+15}) \oplus \quad (6)$$
$$(\Omega_{t+4} \oplus w_{t+4}) \oplus \beta(\Omega_t \oplus w_t).$$

Rearranging the terms of the right hand side of (6) we get

$$z_{t+17} \oplus \alpha z_{t+15} \oplus z_{t+4} \oplus \beta z_t = \Omega_{t+17} \oplus \alpha \Omega_{t+15} \oplus \Omega_{t+4} \oplus \beta \Omega_t \oplus \quad (7)$$
$$w_{t+17} \oplus \alpha w_{t+15} \oplus w_{t+4} \oplus \beta w_t.$$

Recalling the linear recurrence relation for SOBER-t16,

$$s_{t+17} \oplus \alpha s_{t+15} \oplus s_{t+4} \oplus \beta s_t = 0, \quad (8)$$

we see that $\Omega_{t+17} \oplus \alpha \Omega_{t+15} \oplus \Omega_{t+4} \oplus \beta \Omega_t = 0$ and we can reduce (6) to

$$z_{t+17} \oplus \alpha z_{t+15} \oplus z_{t+4} \oplus \beta z_t = w_{t+17} \oplus \alpha w_{t+15} \oplus w_{t+4} \oplus \beta w_t, \quad (9)$$

where the multiplications with α and β are in the extension field $\mathbb{F}_{2^w}$.

3.1   Estimating the Distribution of w_t

The noise $w_t, t \geq 0$ are random variables taken from $\mathbb{F}_{2^{16}}$ with a nonuniform but unknown distribution F. Let us write the distribution F in the form

$$F = \begin{bmatrix} f_0 \\ f_1 \\ \vdots \\ f_{2^{16}-1} \end{bmatrix}$$

where $P(w_t = x) = f_x$. We can not hope to find a closed expression for the distribution F, since it is computationally too complex to derive. However, we can run the cipher and estimate the distribution F.

In the simulations, we measure the frequency of different values for the noise w_t, calculated as

$$w_t = (((s_{t+1} \boxplus s_{t+6} \boxplus f(s_t \boxplus s_{t+16})) \oplus Konst) \boxplus s_{t+13}) \oplus \Omega_t. \quad (10)$$

Assume that we sample 2^ν values of w_t according to (10), and denote the measured frequencies by the vector $\hat{F} = (\hat{f}_0, \hat{f}_1, \ldots, \hat{f}_{2^{16}-1})$. $\hat{F}$ is an estimation of F and we can write $F = \hat{F} + \bar{E}$, where $\bar{E}$ is a vector representing the error in the estimation. Focusing on one single component of $\bar{E}$, it will be approximately Gaussian distributed with zero mean and a standard deviation of $2^{-(\nu/2+8)}$. Simulations show that F is quite nonuniform. For example, in simulation with $Konst = 0$ and $\nu = 38$, the maximum value is $2^{-16} + 2^{-17.6}$. The error in this estimation is of order 2^{-28}. The distribution F has been tabulated for a number of different values of $Konst$.

3.2   Calculating the Full Noise Distribution

Let us define

$$W_t = w_{t+17} \oplus \alpha w_{t+15} \oplus w_{t+4} \oplus \beta w_t,$$

i.e, $W_t, t \geq 0$ are the random variables corresponding to the full noise that we can sample from the running key. Looking at Eq. (9) we see that we must combine four F distributions (as above) to get the overall noise distribution, denoted $P(W)$. Since the samples are taken at different positions in time, we assume $w_t, t \geq 0$ to be independent variables.

The distribution $H = [h_i]$ of the XOR of two random variables with distribution $F = [f_i]$ and $G = [g_i]$ respectively, is obtained by

$$h_l = \sum_{i \oplus j = l} f_i g_j. \tag{11}$$

The distribution of αw_t is simply a permutation of the distribution F.

It can be shown that when combining distributions as done in (11), we sustain significance in the resulting distribution. So by estimating the F distribution by simulation and then combining the probabilities according to (11), we can estimate the distribution $P(W)$, of the right hand side of (9) for different values of $Konst$.

To be able to distinguish the full noise distribution, $P(W)$, from the uniform distribution we need have some N different keystream symbols. The theory of hypothesis testing [2] gives us a bound on N.

Let $Z_t = z_{t+17} \oplus \alpha z_{t+15} \oplus z_{t+4} \oplus \beta z_t$. For short, the optimal test for distinguishing between the two possible distributions ($P(W)$ and uniform distribution) is according to the Neyman-Pearson lemma to check if the likelihood ratio $\sum_{t=0}^{N} \log[P(W_t = Z_t)/2^{-16}]$ is smaller or larger than 0.

The probability that we make an incorrect decision, denoted P_e, when trying to distinguish between two distributions P_1 and P_2, given N samples from one of the distributions is bounded by

$$P_e \leq 2^{-N \cdot C(P_1, P_2)}, \tag{12}$$

where $C(P_1, P_2)$ is the Chernoff information between the two distributions. The Chernoff information is defined as

$$C(P_1, P_2) = - \min_{0 \leq \lambda \leq 1} \log_2 (\sum_x P_1^{\lambda}(x) P_2^{1-\lambda}(x)). \tag{13}$$

We get a lower bound on $C(P_1, P_2)$ by using e.g. $\lambda = 1/2$. We fix a probability of error of $P_e = 2^{-32}$. Then we need to choose $N \geq 32 C(P_1, P_2)^{-1}$.

3.3   Summarizing the Results

The distribution $P(W)$ has been determined through simulation as previously described. The analysis in this section summarizes to the following attack:

> For $t = 1 \ldots N$ do
>
>   1. Calculate $Z_t = z_{t+17} \oplus \alpha z_{t+15} \oplus z_{t+4} \oplus \beta z_t$.
>   2. Let $\hat{f}_{Z_t} = \hat{f}_{Z_t} + 1$.
>
> end for.
>
> Calculate $I = \sum_{x \in \mathbb{F}_{2^{16}}} \hat{f}_x \log_2 \left[\frac{P(W=x)}{2^{-16}} \right]$.
>
> If $I > 0$ then output **SOBER** otherwise output **random**

We have calculated the combined $P(W)$ distribution using $2^{38}, (\nu = 38)$ outputs to generate the F distribution. Note that since F (and thus $P(W)$) is dependent on the unknown value of $Konst$, we actually need to determine the $P(W)$ distribution for all 2^{16} possible values of $Konst$.

The resulting Chernoff information between $P(W)$ and the uniform distribution, have been derived for 50 random values of $Konst$. They were all between 2^{-84} and 2^{-87}. We assume that calculating the Chernoff information for other values of $Konst$ will give similar results. In the worst case, we need at least $N = 32 \cdot 2^{87} = 2^{92}$ words from the running key to be able to distinguish a SOBER-t16 output sequence without stuttering from a uniform distribution with a probability of error $P_e = 2^{-32}$. The computational complexity of the attack is roughly 2^{92}.

Finally, the Neyman-Pearson test must be performed for each of the 2^{16} possible values of $Konst$. Still, the probability of error is smaller than 2^{-16}, which is small enough. Note that this step does not change the overall complexity.

4   A Distinguishing Attack on SOBER-t16 with Stuttering

When the stuttering unit is present, not every NLF output, v_t, is used to produce a keystream symbol. Recalling the functionality of the stuttering unit, we see that each v_t can be either discarded, used as a new SCW, or used (possible XORed with a constant) as a keystream symbol z_n. To be able to use the results from Section 3, we must have access to the NLF output quadruple $(v_t, v_{t+4}, v_{t+15}, v_{t+17})$.

Assume that we look at one output symbol $z_n = \mathcal{C}_0 \oplus v_t$, where $\mathcal{C}_0 \in \{0, C, \sim C\}$ is the constant that is XORed to v_t in the stuttering unit to form z_n. Simulations show that the most probable position in the key stream for v_{t+4} to appear in is z_{n+2}. Similar simulations to determine the most probable position for v_{t+15} and v_{t+17} give the following results

$$P(\mathcal{C}_1 \oplus v_{t+4} \to z_{n+2} | \mathcal{C}_0 \oplus v_t \to z_n) = 0.31,$$
$$P(\mathcal{C}_2 \oplus v_{t+15} \to z_{n+7} | \mathcal{C}_1 \oplus v_{t+4} \to z_{n+2}) = 0.19,$$
$$P(\mathcal{C}_3 \oplus v_{t+17} \to z_{n+8} | \mathcal{C}_2 \oplus v_{t+15} \to z_{n+7}) = 0.40.$$

Having established the most probable positions in the keystream for $(v_{t+4}, v_{t+15}, v_{t+17})$, given an output $z_n = C_0 \oplus v_t$, we are still faced with the problem of which constants $C_i, i = 0, 1, 2, 3$ each NLF output is XORed with.

Denote by $\mathcal{E}$ the event that, given $C_0 \oplus v_t \to z_n$, we have $C_1 \oplus v_{t+4} \to z_{n+2}$, $C_2 \oplus v_{t+15} \to z_{n+7}$, $C_3 \oplus v_{t+17} \to z_{n+8}$. The probability of event $\mathcal{E}$, denoted p_0, is $p_0 \approx 2^{-5.5}$.

By looking at Table 1 we note that certain combinations of (C_0, C_1, C_2, C_3) can not occur under the assumption $\mathcal{E}$. In general, the distribution is nonuniform and, for example, the five combinations, $(C_0, C_1, C_2, C_3) =$

$$(0, 0, 0, 0)$$
$$(C, C, C, \sim C)$$
$$(C, C, \sim C, 0)$$
$$(C, \sim C, 0, 0)$$
$$(\sim C, 0, 0, 0)$$

are more likely to occur than others.

From Eq. (6) and (9) in Section 3, we know that

$$v_{t+17} \oplus \alpha v_{t+15} \oplus v_{t+4} \oplus \beta v_t = w_{t+17} \oplus \alpha w_{t+15} \oplus w_{t+4} \oplus \beta w_t. \tag{14}$$

Given event $\mathcal{E}$, we can write

$$z_{n+8} \oplus \alpha z_{n+7} \oplus z_{n+2} \oplus \beta z_n = W_t \oplus C_3 \oplus \alpha C_2 \oplus C_1 \oplus \beta C_0, \tag{15}$$

where $W_t = w_{t+17} \oplus \alpha w_{t+15} \oplus w_{t+4} \oplus \beta w_t$ and has known distribution $P(W)$.

Again, we derive the distribution of the right hand side of (15) and denote this distribution by $P(W')$, assuming $W'_t = W_t \oplus C_3 \oplus \alpha C_2 \oplus C_1 \oplus \beta C_0$. The Chernoff information between $P(W')$ and the uniform distribution, is calculated to be $C(P(W'), P_U) \approx 2^{-95}$, where P_U is the uniform distribution.

Sampling the keystream output sequence at $(z_n, z_{n+2}, z_{n+7}, z_{n+8})$ will give us a sample of the noise from the distribution $P(W')$ with probability $p_0 = 2^{-5.5}$. With probability $1 - p_0$ the assumption was wrong and it is reasonable to assume that we then get a uniform distribution. Write the distribution $P(W')$ as a vector

$$P(W') = \begin{bmatrix} 2^{-16} + \xi_0 \\ 2^{-16} + \xi_1 \\ \vdots \\ 2^{-16} + \xi_{2^{16}-1} \end{bmatrix}, \tag{16}$$

where each element $2^{-16} + \xi_x$ represents $P(W' = x) = 2^{-16} + \xi_x$.

Let $Y = z_{n+8} \oplus \alpha z_{n+7} \oplus z_{n+2} \oplus \beta z_n$. The distribution of Y, denoted $P(Y)$, can then be calculated to be

$$P(Y) = \begin{bmatrix} 2^{-16} + \xi_0 p_0 \\ 2^{-16} + \xi_1 p_0 \\ \vdots \\ 2^{-16} + \xi_{2^{16}-1} p_0 \end{bmatrix}. \tag{17}$$

The resulting Chernoff information between Y and the uniform distribution, is finally calculated to be $C(P(Y), P_U) \approx p_0^2 C(P(W'), P_U)$, where P_U is the uniform distribution.

From the discussion in Section 3.3, we conclude that we need at most $N = 32 \cdot p_0^{-2} 2^{95} \approx 2^{111}$ keystream symbols to be able to distinguish the output of SOBER-t16 with stuttering from a uniform source. The complexity is of the same size. We summarize the results in this section in the following attack.

For $t = 1 \ldots N$ do

    1. Calculate $Z_n = z_{n+8} \oplus \alpha z_{n+7} \oplus z_{n+2} \oplus \beta z_n$.
    2. Let $\hat{f}_{Z_n} = \hat{f}_{Z_n} + 1$.

end for.

Calculate $I = \sum_{x \in \mathbb{F}_{2^{16}}} \hat{f}_x \log_2 \left[\frac{P(Y=x)}{2^{-16}} \right]$.

If $I > 0$, then output **SOBER** otherwise output **random**

Again, we should note that $P(Y)$ is dependent on *Konst*, and a full attack includes testing against 2^{16} different distributions.

5 A Distinguishing Attack on SOBER-t32 without Stuttering

The attack on SOBER-t16 was possible because we could compute the noise distribution F by simulation. From F we could derive $P(W)$.

Obtaining significance in simulation was possible because of the small word size of 16 bits. In SOBER-t32 we cannot directly use the same method to obtain a similar distribution F, due to our computational limitations. We note, however, that *if* we could simulate and find a noise distribution, then the attack on t32 would probably be strong. This is due to the fact that the linear recurrence relation in t32 has only one constant α different from one, whereas t16 has two, α and β. The multiplications by these constants tend to smooth out the distribution.

However, in this section we present another attack, based on a bitwise linear approximation through the NLF function. Using the same notation as before, we denote the XOR of the input words to the NLF at time t, by $\Omega_t = s_t \oplus s_{t+1} \oplus s_{t+6} \oplus s_{t+13} \oplus s_{t+16}$. The output from the NLF at time t, is denoted v_t. Since the stuttering unit is removed, we have $z_t = v_t$ for all $t \geq 0$. Each word is 32 bits, and we will denote a specific bit i, $0 \leq i \leq 31$, in a word x, with $x[i]$. Let k denote the value of *Konst*.

We start by considering the linear recurrence relation of t32 given by

$$s_{t+17} \oplus s_{t+15} \oplus s_{t+4} \oplus \alpha s_t = 0, \tag{18}$$

and the corresponding characteristic polynomial for the recurrence

$$x^{17} + x^{15} + x^4 + \alpha. \tag{19}$$

Repeated squaring of this polynomial will still yield a valid linear recurrence equation for the considered linear recurrence of t32. Specifically, exponentiation with 2^{32} gives

$$x^{17 \cdot 2^{32}} + x^{15 \cdot 2^{32}} + x^{4 \cdot 2^{32}} + \alpha^{2^{32}}. \tag{20}$$

Since $\alpha \in \mathbb{F}_{2^{32}}$ we have $\alpha^{2^{32}} = \alpha$ and addition of (19) and (20) gives

$$x^{17} + x^{15} + x^4 + x^{17 \cdot 2^{32}} + x^{15 \cdot 2^{32}} + x^{4 \cdot 2^{32}}. \tag{21}$$

Here we can divide with x^4, and the resulting linear recurrence is given by

$$s_{t+17 \cdot 2^{32} - 4} \oplus s_{t+15 \cdot 2^{32} - 4} \oplus s_{t+4 \cdot 2^{32} - 4} \oplus s_{t+13} \oplus s_{t+11} \oplus s_t = 0,$$

which is written

$$s_{t+\tau_5} \oplus s_{t+\tau_4} \oplus s_{t+\tau_3} \oplus s_{t+\tau_2} \oplus s_{t+\tau_1} \oplus s_t = 0, \tag{22}$$

by introducing the constants $\tau_1 = 11$, $\tau_2 = 13$, $\tau_3 = 4 \cdot 2^{32} - 4$, $\tau_4 = 15 \cdot 2^{32} - 4$ and $\tau_5 = 7 \cdot 2^{32} - 4$. Note that in Eq. (22) we have derived a linear recurrence equation that holds *for each single bit position*.

Consider the XOR between two adjacent bits, i and $i - 1$, $i \geq 1$, in the running key z_t. As before, we use a linear approximation of the NLF function, $z_t = \Omega_t \oplus w_t$, where the value of *Konst* is merged into the binary random variable w_t representing the noise. We can now write

$$z_t[i] \oplus z_t[i - 1] = \Omega_t[i] \oplus \Omega_t[i - 1] \oplus w_t[i]. \tag{23}$$

where $w_t[i]$ denotes the noise in bit position i introduced by the linear approximation. Let $F[i]$ be the distribution of $w_t[i]$. We can estimate the distribution $F[i]$ by simulation and the result shows that the distribution is quit nonuniform for many positions $0 \leq i \leq 31$. We can write the correlation between the XOR of bit i and $i - 1$ of the input and output as

$$P(z_t[i] \oplus z_t[i - 1] = \Omega_t[i] \oplus \Omega_t[i - 1]) =$$
$$P(F[i] = 0) = \frac{1}{2} + \varepsilon_i, \tag{24}$$

for each bit position $0 < i \leq 31$.

The largest correlation we have found is for the XOR of bit 29 and bit 30 (i.e. $F[30]$) in the input and output words. Simulations with 2^{30} samples for 100 random values of k, indicates that the correlation in (24) for $i = 30$ is only dependent on the two corresponding bits in k, i.e. $k[30]$ and $k[29]$. We have the following result,

$$\varepsilon_{30} \approx \begin{cases} -0.0086 & \text{if } k[30] = 0 \text{ and } k[29] = 0 \\ -0.0052 & \text{if } k[30] = 1 \text{ and } k[29] = 1 \\ +0.0086 & \text{if } k[30] = 1 \text{ and } k[29] = 0 \\ +0.0052 & \text{if } k[30] = 0 \text{ and } k[29] = 1. \end{cases} \tag{25}$$

Now, given a key stream output, $z_0, z_1, \ldots, z_{N-1}$, of length N, we can use the linear recurrence relation (22) to calculate

$$z_{t+\tau_5} \oplus z_{t+\tau_4} \oplus z_{t+\tau_3} \oplus z_{t+\tau_2} \oplus z_{t+\tau_1} \oplus z_t = \Omega_{t+\tau_5} \oplus w_{t+\tau_5} \oplus \Omega_{t+\tau_4} \oplus w_{t+\tau_4} \oplus$$
$$\Omega_{t+\tau_3} \oplus w_{t+\tau_3} \oplus \Omega_{t+\tau_2} \oplus w_{t+\tau_2} \oplus$$
$$\Omega_{t+\tau_1} \oplus w_{t+\tau_1} \oplus \Omega_t \oplus w_t \oplus$$

where the sum of all the Ω_j terms will equal zero because of Eq. (22). Thus, we have

$$z_{t+\tau_5} \oplus z_{t+\tau_4} \oplus z_{t+\tau_3} \oplus z_{t+\tau_2} \oplus z_{t+\tau_1} \oplus z_t = \bigoplus_{j=0}^{5} w_{t+\tau_j}. \qquad (26)$$

Introduce the notation $Z_t = z_{t+\tau_5} \oplus z_{t+\tau_4} \oplus z_{t+\tau_3} \oplus z_{t+\tau_2} \oplus z_{t+\tau_1} \oplus z_t$ for the left hand side of (26), and $W_t = \bigoplus_{j=0}^{5} w_{t+\tau_j}$ for the right hand side. We can calculate the probability that

$$P(Z_t[i] \oplus Z_t[i-1] = 0) =$$
$$P(W_t[i] \oplus W_t[i-1] = 0) = \frac{1}{2} + 2^5 \varepsilon_i^6, \qquad (27)$$

where the last equality comes from combining the six independent noise distributions of $w_t[i]$, each with probability $1/2 + \varepsilon_i$ of being zero.

Recalling the measured correlation for bits 29 XOR 30 from (25), we see that ε_{30} takes four possible values. If we want to distinguish the distribution of w from a uniform source, the worst case is the smallest value of ε_{30}. Thus, using $\varepsilon_{30} = 0.0052$ and combining the six noise distribution according to (27) we derive the final correlation probability for the six independent key stream positions as

$$p_0 = P(Z_t[i] \oplus Z_t[i-1] = 0) = \frac{1}{2} + 2^5(0.0052)^6 \approx \frac{1}{2} + 2^{-40.5}. \qquad (28)$$

5.1   Summarizing the Results

To be able to distinguish this nonuniform distribution, denoted P_0, from a uniform source, denoted P_U, we again calculate the Chernoff information between the two distributions,

$$C(P_0, P_U) = - \min_{0 \le \lambda \le 1} \log_2 \sum_x P_0^\lambda(x) P_U^{1-\lambda}(x) \approx 2^{-81.5}. \qquad (29)$$

Settling for an error probability of $P_e = 2^{-32}$ we see that we need $N = 2^{86.5}$ samples from the key stream. Each sample spans a distance of $\tau_5 = 17 \cdot 2^{32} - 4 \approx 2^{36}$ positions, so all in all we need $N + \tau_5 \le 2^{87}$ key stream output words, to distinguish an output sequence from SOBER-t32 without stuttering unit, from a uniform source. The attack presented in this section summarizes as follows.

For $t = 1 \ldots N$ do

   1. Calculate $Z_t = z_{t+\tau_5} \oplus z_{t+\tau_4} \oplus z_{t+\tau_3} \oplus z_{t+\tau_2} \oplus z_{t+\tau_1} \oplus z_t$.
   2. Let $\hat{f} = \hat{f} + (1 - (Z_t[i] \oplus Z_t[i-1]))$.

end for.
Calculate $I = \hat{f} \log \left[\frac{\frac{1}{2} + 2^{-40.5}}{1/2} \right] + (2^N - \hat{f}) \log \left[\frac{\frac{1}{2} - 2^{-40.5}}{1/2} \right]$.
If $I > 0$, then output **SOBER** otherwise output **random**

6   Some Remarks on SOBER-t32 with Stuttering

The obvious extension of the attack in the previous section would be to guess the most probable key stream positions for $v_{t+\tau_5}, \ldots, v_{t+\tau_1}$, given $z_n = v_t$. Since τ_3, τ_4, τ_5 are all in the order of 2^{32}, the probability of guessing the positions of $v_{t+\tau_5}, \ldots, v_{t+\tau_3}$ in the output will be very small. However, it might be possible to get an attack using $N < 2^{256}$ words, in this way.

    Another approach would be to repeat the attack on SOBER-t16 but consider only a specific subset of the bit positions of the words. Then we can simulate the distribution of the selected bit positions of w_t as well as the same bit positions of αw_t. If these distributions show a non-uniformity of similar magnitude as SOBER-t16, we can distinguish the full SOBER-t32 using about the same method as for t16.

7   Conclusions

We have derived a distinguishing attack, based on a linear approximation of the NLF function, on SOBER-t16 with and without stuttering unit. We can distinguish the output sequence from a random source using at most 2^{92} keystream words and same complexity in the case of no stuttering, and using at most 2^{111} key stream words and same complexity for full SOBER-t16. For SOBER-t32 without the stuttering unit we can, due to a fairly strong bit correlation in the NLF function, distinguish the output from a random source using 2^{87} key stream output words and same complexity.

References

1. A. Biryukov, A. Shamir, and D. Wagner. Real time crypanalysis of A5/1 on a PC. In *Proceeding of Fast Software Encryption Workshop*, LNCS vol.1978, pp 1–18, Springer-Verlag, 2001.
2. T. Cover and J.A. Thomas. *Elements of Information Theory*. Wiley series in Telecommunication. Wiley, 1991.
3. C. (Cunsheng) Ding, G. Xiao, and W. Shan. *The stability theory of stream ciphers*, volume 561. Springer-Verlag Inc., New York, NY, USA, 1991.
4. P. Ekdahl and T. Johansson. SNOW – a new stream cipher. In *Proceedings of the First Open NESSIE Workshop*, 13-14 November 2000, Heverlee, Belgium.

5. P. Hawkes and G. Rose. Primitive specification and supporting documentation for SOBER-t16 submission to NESSIE. In *Proceedings of the First Open NESSIE Workshop*, 13-14 November 2000, Heverlee, Belgium.

6. P. Hawkes and G. Rose. Primitive specification and supporting documentation for SOBER-t32 submission to NESSIE. In *Proceedings of the First Open NESSIE Workshop*, 13-14 November 2000, Heverlee, Belgium.

7. I. Mantin and A. Shamir. Practical attack on broadcast RC4. In *Preproceedings of Fast Software Encryption*, 2001.

8. M. Matsui. Linear cryptanalysis method for DES cipher. In T. Helleseth, editor, *Advances in Cryptology — Eurocrypt '93*, volume 765, pages 386–397, Berlin, 1994. Springer-Verlag.

9. M. Schafheutle. A first report on the stream ciphers SOBER-t16 and SOBER-t32. NESSIE document NES/DOC/SAG/WP3/025/2, NESSIE, 2001.

Linearity Properties of the **SOBER**-t32 Key Loading* **

Markus Dichtl and Marcus Schafheutle

Siemens AG, Corporate Technology, 81730 München, Germany,
`Markus.Dichtl,Marcus.Schafheutle@mchp.siemens.de`

Abstract. In the course of the evaluation of the stream cipher SOBER-t32 submitted to NESSIE, a correlation between initial states has been found for related keys. With high probability some sums of bits of the initial state after key loading do not change their value when a bit of the key is inverted. This holds also for the loading of frame keys. It is shown that the required condition for the frame keys is met very naturally when using counters as frame keys. The linearity properties of the SOBER-t32 key loading are caused by non-optimal diffusion of the non-linear filter function of the cipher.

1 Introduction

SOBER-t32 is a synchronous additive stream cipher designed for key sizes up to 256 bits. SOBER-t32 was submitted to NESSIE by Philip Hawkes and Gregory Rose at Qualcomm Australia. NESSIE (New European Schemes for Signatures, Integrity, and Encryption) is a project within the IST program of the European Commission. Its main purpose is to put forward a portfolio of strong cryptographic primitives that has been obtained after an open call and been evaluated using a transparent and open process.

2 Description of **SOBER**-t32

The stream cipher is constructed from a linear feedback shift register ($LFSR$), a non-linear filter (NLF), and a form of irregular decimation, called stuttering. SOBER-t32 outputs the key stream as 32-bit blocks. The $LFSR$ is of length 17 and operates over $GF(2^{32})$.

The NLF consists of XOR ($\oplus$), addition modulo 2^{32} ($\boxplus$), and a 32-to-32 bit transformation called f-function. The output of the non-linear filter at time t is described as

$$NLF(t) = ((f(s_t \boxplus s_{t+16}) \boxplus s_{t+1} \boxplus s_{t+6}) \oplus \text{const}) \boxplus s_{t+13}$$

---

* The work described in this paper has been supported by the Commission of the European Communities through the IST program under contract IST-1999-12324.
** The information in this document is provided as is, and no warranty is given or implied that the information is fit for any particular purpose. The user thereof uses the information at its sole risk and liability.

J. Daemen and V. Rijmen (Eds.): FSE 2002, LNCS 2365, pp. 225–230, 2002.

where s_{t+k} is the content of the k'th shift register cell at time t, const is a session key dependent constant value, derived during the key loading phase.

The function f uses the 8 most significant bits of its input as an input to a lookup table S-box with 32 bits of output. The 24 least significant bits of the input are just XOR-ed to the output of the S-box in order to obtain the result of the function f.

The stuttering decimates the output of the *NLF* in an irregular fashion. For the stuttering of SOBER-t32 it can be shown that there is an average of $\frac{12}{25}$ key stream output per clock of the *LFSR*.

With the size of the *LFSR* and the size of the key dependent parameter const, it is obvious that the initial state size of SOBER-t32 is $2^{17\cdot32+32}$.

A detailed specification of SOBER-t32 can be found at the NESSIE Web site [NES].

3 SOBER-t32 Key Loading

The key loading determines the initial state of the *LFSR* and the value of const from the key. It relies on the operations "Include()" and "Diffuse()".

Include(X) adds the 32-bit word X to the *LFSR* cell r_{15}. Diffuse() clocks the register, computes the output of the *NLF*, and XORs this output to the *LFSR* cell r_4.

For key loading, the key is divided into 32-bit words. The Include() operator is applied to each of these words, and each Include() operation is followed by a Diffuse() operation. As an immediate consequence one observes that the key words included last are diffused less than those included first. The Include() operation is also applied to the key length in bytes. Then the Diffuse() operation is applied 17 more times.

4 Diffusion Properties of the *NLF*

A closer inspection of *NLF* shows that its diffusion properties are not ideal. Modifications in the most significant 8 bits of the input of f are very efficiently diffused over the output word, whereas for modifications in the 24 least significant bits of the input of f no diffusion by f occurs. The only way of diffusion for these bit positions is by carry propagation. However, carry chains tend to be short. Burks, Goldstine, and von Neumann found out in 1946 [BGvN46] that on average the longest carry chain in adding k-bit numbers is of length $\log_2(k)$. Hence carry badly propagates bit modifications to bit positions far away from the bit position of the modification.

Nevertheless long carry chains occur from time to time. This explains why the linearity properties of the SOBER-t32 key loading described in the next section do not hold always, but only with a very high probability. The low probability long carry chains provide enough diffusion to disturb the linear relationships occasionally.

A lot of diffusion occurs by the clocking of the *LFSR*. However, this linear operation in $GF(2^{32})$ is also linear in $GF(2)$.

The linear recurrence over $GF(2^{32})$ of the SOBER-t32 *LFSR* can be shown, see [Her85], to be equivalent to implementing 32 parallel bit-wise *LFSRs*, each of length $17 \cdot 32 = 544$. These linear recurrences are identical, represented by the primitive polynomial $p_{32}(x)$ over $GF(2)$:

$$
\begin{aligned}
p_{32}(x) = {} & 1 + x^{17} + x^{19} + x^{21} + x^{23} + x^{25} + x^{27} + x^{29} + x^{30} + x^{31} + x^{33} + x^{34} + x^{35} + x^{37} + x^{38} \\
& + x^{39} + x^{41} + x^{42} + x^{46} + x^{47} + x^{49} + x^{50} + x^{51} + x^{53} + x^{54} + x^{55} + x^{56} + x^{57} + x^{58} \\
& + x^{59} + x^{61} + x^{62} + x^{63} + x^{64} + x^{65} + x^{66} + x^{67} + x^{68} + x^{70} + x^{74} + x^{76} + x^{77} + x^{78} \\
& + x^{79} + x^{84} + x^{85} + x^{87} + x^{89} + x^{90} + x^{91} + x^{92} + x^{95} + x^{97} + x^{98} + x^{100} + x^{101} + x^{102} \\
& + x^{109} + x^{111} + x^{113} + x^{114} + x^{117} + x^{118} + x^{121} + x^{125} + x^{130} + x^{131} + x^{132} + x^{133} \\
& + x^{137} + x^{138} + x^{140} + x^{142} + x^{143} + x^{145} + x^{146} + x^{147} + x^{148} + x^{149} + x^{151} + x^{153} \\
& + x^{156} + x^{160} + x^{163} + x^{164} + x^{165} + x^{172} + x^{173} + x^{175} + x^{176} + x^{177} + x^{179} + x^{180} \\
& + x^{184} + x^{185} + x^{186} + x^{190} + x^{191} + x^{193} + x^{198} + x^{200} + x^{201} + x^{202} + x^{206} + x^{207} \\
& + x^{208} + x^{209} + x^{210} + x^{211} + x^{212} + x^{213} + x^{219} + x^{220} + x^{221} + x^{225} + x^{227} + x^{229} \\
& + x^{231} + x^{232} + x^{233} + x^{235} + x^{236} + x^{238} + x^{239} + x^{240} + x^{241} + x^{242} + x^{244} + x^{245} \\
& + x^{246} + x^{247} + x^{249} + x^{252} + x^{255} + x^{258} + x^{262} + x^{263} + x^{264} + x^{265} + x^{266} + x^{277} \\
& + x^{279} + x^{281} + x^{284} + x^{285} + x^{288} + x^{289} + x^{290} + x^{291} + x^{292} + x^{294} + x^{296} + x^{300} \\
& + x^{301} + x^{302} + x^{304} + x^{306} + x^{307} + x^{309} + x^{310} + x^{316} + x^{321} + x^{323} + x^{324} + x^{325} \\
& + x^{327} + x^{334} + x^{335} + x^{336} + x^{337} + x^{340} + x^{341} + x^{342} + x^{344} + x^{345} + x^{346} + x^{347} \\
& + x^{350} + x^{352} + x^{355} + x^{357} + x^{360} + x^{361} + x^{362} + x^{363} + x^{364} + x^{365} + x^{368} + x^{373} \\
& + x^{377} + x^{379} + x^{381} + x^{382} + x^{383} + x^{385} + x^{388} + x^{389} + x^{390} + x^{391} + x^{392} + x^{394} \\
& + x^{398} + x^{403} + x^{404} + x^{405} + x^{406} + x^{407} + x^{413} + x^{416} + x^{420} + x^{421} + x^{422} + x^{425} \\
& + x^{426} + x^{428} + x^{430} + x^{431} + x^{433} + x^{435} + x^{436} + x^{437} + x^{438} + x^{440} + x^{441} + x^{442} \\
& + x^{445} + x^{446} + x^{447} + x^{448} + x^{449} + x^{450} + x^{453} + x^{458} + x^{461} + x^{463} + x^{465} + x^{466} \\
& + x^{469} + x^{471} + x^{473} + x^{474} + x^{477} + x^{478} + x^{479} + x^{481} + x^{483} + x^{484} + x^{487} + x^{488} \\
& + x^{489} + x^{490} + x^{493} + x^{494} + x^{496} + x^{499} + x^{500} + x^{503} + x^{505} + x^{506} + x^{508} + x^{511} \\
& + x^{513} + x^{514} + x^{516} + x^{519} + x^{521} + x^{524} + x^{527} + x^{529} + x^{532} + x^{536} + x^{540} + x^{544}
\end{aligned}
$$

5   Linearity Properties of the SOBER-t32 Key Loading

The insufficient diffusion explained in the previous section is the reason for the existence of sums of bits from the initial state of the shift register which keep their value if some bit of the key is inverted. We denote the bits of the initial state of the shift register by $b_1, b_2, \ldots$ where b_1 is the least significant bit of the 17th *LFSR* cell, b_{32} the most significant bit of this word, b_{33} the least significant bit of the 16th *LFSR* cell, $\ldots$. We computed the sum

$b_{542} + b_{537} + b_{531} + b_{530} + b_{529} + b_{528} + b_{527} + b_{525} + b_{524} + b_{520} + b_{519} + b_{518} + b_{516} + b_{514} + b_{513} + b_{478} + b_{477} + b_{474} + b_{473} + b_{471} + b_{470} + b_{469} + b_{466} + b_{465} + b_{383} + b_{382} + b_{381} + b_{380} + b_{379} + b_{377} + b_{371} + b_{370} + b_{369} + b_{363} + b_{362} + b_{361} + b_{360} + b_{356} + b_{256} + b_{254} + b_{250} + b_{249} + b_{248} + b_{247} + b_{241} + b_{238} + b_{235} + b_{234} + b_{232} + b_{231} + b_{229} + b_{228} + b_{125} + b_{122} + b_{119} + b_{118} + b_{117} + b_{115} + b_{112} + b_{111} + b_{109} + b_{108} + b_{104} + b_{103} + b_{102} + b_{100} + b_{98} + b_{97} + b_{62} + b_{61} + b_{58} + b_{57} + b_{55} + b_{54} + b_{53} + b_{50} + b_{49}$

in $GF(2)$ for 100000 keys chosen randomly. In 99957 cases the value of this sum remained the same when the least significant bit of the last key word was inverted.

We also found 16 other sums of this kind and 7 bit sums whose values change with high probability when the least significant bit of the last key word is toggled. In all cases the success probability determined from 100000 trials was at least 99.4 percent.

Of course, only linearly independent solutions were considered. By forming linear combinations, many more sums of this kind could be found, which do not provide additional information.

We also identified sums whose values remain invariant under the inversion of other key bits with high probability or change their value with high probability. In total, we found 249 such equations with a success probability of at least 98.6 percent. Again, these probabilities were determined by using 100000 random keys.

For each of the 11 least significant bits of the last key word such sums were found. Apparently from more significant bit positions, the carry chains reach the 8 most significant bit positions with sufficient probability to provide enough non-linear diffusion to destroy such linearity properties. For 8 of the 9 least significant bit positions of the second to last key word such sums exist as well.

For earlier key words, the number of applications of the `Diffuse()` operation seems to be sufficiently high in order to prevent the existence of such sums.

One way to strengthen SOBER-t32 against the linearity properties described is to increase the number of final `Diffuse()` steps for key loading. Our experiments showed 21 final steps instead of 17 to be sufficient.

6   Linearity Properties of the **SOBER**-t32 Rekeying

For some applications it is convenient to be able to generate more than one key stream from one key. To make the streams different, SOBER-t32 can process an initialization vector, called frame key, which can be assumed to be public. Ideally, the streams generated with the same key but different frame keys should be completely independent. We are not able to show correlations between the streams generated, but between the initial states derived from different frame keys but the same key. Since the loading of the frame key is very similar to the key loading, it does not come as a big surprise that sums of the kind described also exist for the frame key loading.

First the cipher key is loaded, then the frame key. The only difference is that for key loading the value of const in the NLF is zero. For frame key loading the const value determined in the key loading phase is used. (This values is also used for the actual generation of the stream.)

The following sum of bits of the initial $LFSR$ state almost never changed its value when the second to least significant bit of the to last key frame word was toggled:

$b_{511} + b_{508} + b_{507} + b_{506} + b_{505} + b_{501} + b_{494} + b_{492} + b_{491} + b_{488} + b_{487} + b_{486} +$
$b_{448} + b_{447} + b_{445} + b_{444} + b_{441} + b_{440} + b_{437} + b_{436} + b_{434} + b_{433} + b_{432} + b_{431} +$
$b_{428} + b_{427} + b_{425} + b_{422} + b_{417} + b_{351} + b_{348} + b_{346} + b_{345} + b_{344} + b_{341} + b_{339} +$
$b_{336} + b_{335} + b_{334} + b_{333} + b_{331} + b_{330} + b_{325} + b_{224} + b_{222} + b_{221} + b_{218} + b_{217} +$
$b_{215} + b_{213} + b_{211} + b_{209} + b_{205} + b_{203} + b_{202} + b_{201} + b_{199} + b_{198} + b_{195} + b_{193} + b_{96} +$
$b_{93} + b_{91} + b_{90} + b_{88} + b_{84} + b_{82} + b_{81} + b_{80} + b_{79} + b_{78} + b_{73} + b_{72} + b_{71} + b_{65} + b_{32} +$
$b_{31} + b_{29} + b_{28} + b_{25} + b_{24} + b_{21} + b_{20} + b_{18} + b_{17} + b_{16} + b_{15} + b_{12} + b_{11} + b_9 + b_6 + b_1$

The value of the sum did not change in 99806 cases of 100000 where both key and frame key were chosen randomly. In total, 28 sums with probabilities above 99 percent were identified.

Our experiments suggest that the non-zero values of the variable const cause the diffusion for the key frame loading to be a little better than for the key loading, where const is zero.

Whereas it might look quite artificial that single key bits should be inverted as required for the linearity properties of the key loading described in the previous section, the inversion of individual bits of key frame bits occurs in real life. Most commonly, the key frame is just a binary counter. Counter states where the only bit difference is at a bit position of low significance occur very frequently. We have seen in the case of the key loading that only bit positions of low significance can be inverted with good probabilities for the linearity properties, and this holds also for the key frame loading. So when using the counter method for key frames, invariant sums of initial state bits occur frequently.

7   Relation to Other Attacks

The key loading of previous versions of SOBER has been the base for earlier attacks. In the original version of SOBER, the key loading was linear. This was exploited by the attack of Bleichenbacher and Patel [BP99].

In the description of the NESSIE submission of SOBER-t32 [NES] a paper by Bleichenbacher, Patel, and Meier [BPM] is quoted in which a correlation between initial states of SOBER-II (an attempt to fix the problems of SOBER) for different key frames but the same initial key material was found. The updated key and frame loading used in SOBER-t32 is claimed by the authors of the cipher to destroy this correlation. Above, we have shown another correlation that they did not succeed to destroy.

8   Applicability of the Linearity Properties to SOBER-t16

The linearity properties we identified for SOBER-t32, do not exist in SOBER-t16. SOBER-t16 is very similar to SOBER-t32, but based on 16-bit words. As we pointed out above, the non-linear diffusion of SOBER-t32 relies on carry propagation. This also holds for SOBER-t16, but shorter carry chains, which occur with higher probability, are sufficient for the non-linear diffusion within the 16-bit words. If the key loading of SOBER-t16 is reduced to 15 final Diffuse() operations instead of 17, the linearity properties appear.

9   Conclusion

We have found high probability correlations of sums of initial state bits of SOBER-t32 for related keys and also for related key frames. Such correlations are undesirable for a stream cipher, even when it is not clear how to exploit them for an attack. As we have identified the non-optimal diffusion of the *NLF* as the main source of the problem, we suggest not to rely on carry propagation as a means of diffusion in the next version of SOBER.

References

[BGvN46] A.W. Burkes, H.H. Goldstine, and J. von Neumann, *Preliminary discussion of the logical design of an electronic computing instrument*, Tech. report, Institute for Advanced Study Report, Princeton, NJ, 1946.

[BP99]   D. Bleichenbacher and S. Patel, *SOBER cryptanalysis*, Proceedings of Fast Software Encryption '99, Lecture Notes in Computer Science, Springer Verlag, 1999, pp. 305–316.

[BPM]   D. Bleichenbacher, S. Patel, and W. Meier, *Analysis of the SOBER stream cipher*, Tech. report, TIA contribution TR45.AHAG/99.08.30.12.

[Her85]   T. Herlestam, *On functions of linear shift register sequences*, Proceedings of EUROCRYPT '85, Lecture Notes in Computer Science, Springer Verlag, 1985, pp. 119–129.

[NES]   *NESSIE web site*, http://www.cryptonessie.org.

A Time-Memory Tradeoff Attack Against LILI-128

Markku-Juhani Olavi Saarinen

Helsinki University of Technology
Laboratory for Theoretical Computer Science
P.O. Box 5400, FIN-02015 HUT, Finland
mjos@tcs.hut.fi

Abstract. In this note we discuss a novel and simple time-memory tradeoff attack against the stream cipher LILI-128. The attack defeats the security advantage of having an irregular stepping function. The attack requires 2^{46} bits of keystream, a lookup table of 2^{45} 89-bit words and computational effort which is roughly equivalent to 2^{48} DES operations.

1 Introduction

The LILI-128 keystream generator [5] is a LFSR-based synchronous stream cipher with a 128 bit key. It was accepted as one of six candidate stream ciphers for NESSIE, but was rejected from the second round.

In the original LILI-128 specification, the authors conjecture that the complexity of divide and conquer attacks is "at least 2^{112} operations, requiring knowledge of at least 1700 known keystream bits".

In this paper, we use the approximate number of "equivalent DES operations" as a measure of computational efficiency. While the number of bit operations is an useful measure in asymptotic analysis of algorithms, we feel that our approach is more appropriate for the purposes of this paper, since it allows easy comparison of security level to other ciphers (esp. block ciphers).

1.1 Previous Work

After its initial release, some cryptanalytic results on LILI-128 has been published [6,9]. The best known attacks are:

- In [8], Jönsson and Johansson describe an attack requiring 2^{71} bit operations (2^{79} in precomputation phase), 2^{30} keystream bits and an off-line precomputed table with 2^{40} entries.
- In [1], Babbage discusses a rekeying attack and generic time-memory tradeoff attacks.

J. Daemen and V. Rijmen (Eds.): FSE 2002, LNCS 2365, pp. 231–236, 2002.
© Springer-Verlag Berlin Heidelberg 2002

1.2   Time/Memory/Data Tradeoffs

In 1980 Hellman introduced a technique for breaking block ciphers using time-memory tradeoffs [7]. An analogous (but very different) technique for stream ciphers was proposed by Babbage in 1995 [2], although the underlying idea is not algorithmically new. More recently, Biryukov, Shamir and Wagner combined these approaches in work related to the A5/1 cipher [3,4].

The basic idea of Time/Memory/Data tradeoff attacks against stream ciphers is as follows. Most stream ciphers can be described in terms of a state x_i (which is characterized by N, its size) and two functions, *Step* and *Output*. The initial state x_0 is derived from a secret key. To generate one bit of keystream $z(i)$, *Output* is invoked, followed by an update of the internal state using *Step*:

$$z(i) = Output(x_{i-1})$$
$$x_i = Step(x_{i-1})$$

The attack consists of two stages:

1. **Off-line preprocessing stage.** We pick random x_i states and compute the bit sequence $z(i), z(i+1), \dots, z(i+O(\log N))$. This output bit sequence is stored in a table together with the random state x_i. We sort the list in increasing order of the output bit sequence.
2. **On-line computation phase.** Each $O(\log N)$-bit window of the known keystream is considered. For each window we perform a table lookup to determine whether or not this state was one of the states considered in the preprocessing stage. If the state is found, we can compute keystream bits after or (if *Step* is invertible) before the known keystream segment.

It can be shown that $O(\sqrt{N})$ bits of known keystream, $O(\log N \sqrt{N})$ bits of memory, and $O(\log N \sqrt{N})$ time is sufficient to find one internal state x_i. From this internal state we can derive future bits or possibly even the original session key. It turns out that in case of LILI-128, further improvements are possible over this generic attack.

2   Description of LILI-128

LILI-128 uses two LFSRs, $LFSR_c$ and $LFSR_d$. $LFSR_c$ has an internal state of 39 bits and is clocked once for each output bit. $LFSR_d$ has an internal state of 89 bits and is clocked 1 to 4 times, depending on two bits in $LFSR_c$. During key setup phase a $128 = 39 + 89$ bit cryptovariable is directly loaded into these two registers.[1]

In the following, we let $t_0, t_1, \dots, t_{38}$ denote the individual bits of $LFSR_c$, t_0 being the most significant bit in the register and t_{38} being the least significant

---

[1] In [6] the authors also discuss other keying methods for LILI-128.

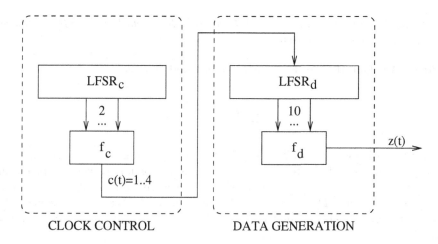

Fig. 1. Overview of the LILI-128 keystream generator.

bit. Similarly we use $u_0, u_1, \ldots, u_{88}$ to denote the individual bits of $LFSR_d$. The primitive polynomial for $LFSR_c$ is

$$x^{39} + x^{35} + x^{33} + x^{31} + x^{17} + x^{15} + x^{14} + x^2 + 1$$

while $LFSR_d$ uses the primitive polynomial

$$x^{89} + x^{83} + x^{80} + x^{55} + x^{53} + x^{42} + x^{39} + x + 1.$$

The procedure for generating keystream is as follows:

1. Ten bits from $LFSR_d$ are fed to a highly nonlinear function f_d, $f_d : \mathbb{F}_2^{10} \to \mathbb{F}_2$ to generate one output bit $z(t)$. [2]

$$z(t) = f_d(u_0, u_1, u_3, u_7, u_{12}, u_{20}, u_{30}, u_{44}, u_{65}, u_{80}).$$

2. Two bits from $LFSR_c$ are fed to a "linear" clock control function $f_c : \mathbb{F}_2^2 \to \mathbb{F}_2$ to form the clocking amount $c(t)$:

$$c(t) = f_c(t_{12}, t_{20}) = 2t_{12} + t_{20} + 1$$

3. The clock control register $LFSR_c$ is clocked once and the data generation register $LFSR_d$ is clocked $c(t)$ times (i.e. 1, 2, 3, or 4 times).

Note that the output bit is indeed generated *before* the LFSRs are clocked, hence effectively halving the key search effort in some applications.

---

[2] The f_d function is specified as a 1024-entry table in the original specification [5], and is excluded from this paper since it is irrelevant to the present attack.

Lemma 1. *For each $\Delta_c = 2^{39} - 1$ times $LFSR_c$ is clocked, $LFSR_d$ is clocked exactly $\Delta_d = 5 * 2^{38} - 1$ times.* [3]

Proof. We claim that for all cryptovariables and all values t:

$$\sum_{i=1}^{2^{39}-1} c(t+i) = \Delta_d$$

Since the polynomial of $LFSR_c$ is primitive, it's period is $2^{39} - 1 = \Delta_c$ and thus the internal state of the register goes through all possible values except the all zero state $t_0 = t_1 = \ldots = t_{38} = 0$. During this cycle the control bits (t_{12}, t_{20}) have value $(0, 0)$ exactly $2^{37} - 1$ times and values $(0, 1)$, $(1, 0)$, and $(1, 1)$ exactly 2^{37} times, bringing the total sum to $1*(2^{37}-1)+(2+3+4)*2^{37} = 1374389534719 = \Delta_d$.

Lemma 2. *$LFSR_d$ can be stepped by Δ_d number of positions forward or backward by performing a vector-matrix multiplication with a precomputed 89×89 bit matrix over $GF(2)$. The matrix can be constructed with roughly 2^{28} bit operations using a binary matrix exponentiation algorithm.*

Proof. Trivial.

This could also be achieved using a multiplication algorithm in $GF(2^{89})$, but for a constant Δ_d, vector-matrix multiplication actually appears to be slightly faster (and functionally equivalent). The Hamming weight of the matrix is $3949 \approx 2^{11.4}$.

3   The Attack

Although many tradeoffs are possible, we will present a concrete version of the attack where we have tried to minimize the amount of keystream required. The amount of off-line and on-line computation are approximately the same.

3.1   Constructing the Lookup Table

A table of 2^{45} 89-bit words is set up by iterating the following procedure 2^{46} times. A pseudorandom 89-bit value is loaded into $LFSR_d$ and 45 bits from f_d are sampled Δ_d steps apart (see Lemma 2). This 45-bit vector is used as an index in the table to store the original 89-bit register value.

Analysis. The expected proportion of filled slots in the table is $1 - e^{-2} = 0.8647$. The table size is $2^{51.48}$ bits. The computational effort required to construct the table is roughly equivalent to 2^{48} DES operations.

---

[3] This lemma follows implicitly from Theorem 2 in [5]

3.2 Lookup Stage

We have 2^{46} bits of keystream $z(0), z(1), \ldots, z(2^{46} - 1)$.

1.   For i = 0 to $2^{46} - 44\Delta_c - 1$ Do:
2.   idx $= z(i) \mid z(i + \Delta_c) \mid \ldots \mid z(i + 44\Delta_c)$
3.   Load Table[idx] to $LFSR_d$.
4.   Rewind $LFSR_d$ back $\Delta_d \lfloor \frac{i}{\Delta_c} \rfloor$ positions.
5.   For j = 0 to 127 Do:
6.   If $(f_d(LFSR_d) \neq z(j\Delta_c + (i \bmod \Delta_c))$ break loop.
7.   Advance $LFSR_d$ by Δ_d positions.
8.   If previous loop was not broken, return $LFSR_d$.

In line 2, a 45-bit index to the lookup table is constructed. In line 3, this index is used to fetch a 89-bit candidate value for $LFSR_d$. In line 4, this guess for $LFSR_d$ is rewinded back a multiple of Δ_d steps so that its output bit should match with $z(i \bmod \Delta_d)$ (see Lemma 1). The loop in lines 5 to 7 compares 128 bits sampled from $LFSR_d$ Δ_d steps apart to keystream bits sampled Δ_c bits apart. If all 128 bits match, the correct value for $LFSR_d$ has been found with high probability and is returned on line 8. This guess can be furthermore verified by performing an exhaustive search for the 39-bit value of $LFSR_c$.

Analysis. The probability of finding the correct value for $LFSR_d$ at least once in the main loop is

$$1 - \left(1 - \frac{0.8647 * 2^{45}}{2^{89}}\right)^{2^{46} - 44\Delta_c} \approx 90\%.$$

The inner loop on lines 5 to 7 breaks with high probability after only few tries. The main loop runs for about 2^{45} iterations (before a correct value is found). Therefore we claim that the computational effort is roughly equivalent to 2^{48} DES operations.

4   Conclusions

We have presented a novel time-memory tradeoff attack against LILI-128, which greatly improves the time complexity required to break this cipher over previous published results. The attack requires 2^{46} bits (8 terabytes) of keystream, and therefore is not usable in many applications. We conjecture that LILI-128 can be broken using a lookup table of 2^{45} 89-bit words ($2^{51.48}$ bits) and computational effort equivalent to 2^{48} DES operations.

We therefore feel that the security of LILI-128 is not as high as suggested by the designers. We do not recommend the use of this encryption algorithm for high volumes of data, or as a general-purpose standard for high security applications.

Acknowledgments. The author would like to thank Steve Babbage, Kaisa Nyberg, and anonymous program committee members for very helpful input.

References

1. S. Babbage. *Cryptanalysis of LILI-128*. NESSIE Public Report, https://www.cosic.esat.kuleuven.ac.be/nessie/reports, 2001.
2. S. Babbage. *A Space/Time Tradeoff in Exhaustive Search Attacks on Stream Ciphers*, European Convention on Security and Detection, IEE Conference Publication No. 408, 1995.
3. A. Biryukov and A. Shamir, *Cryptanalytic Time/Memory/Data Tradeoffs for Stream Ciphers*, Proceedings of ASIACRYPT 2000, LNCS 1976, pp. 1–13, Springer-Verlag, 2000.
4. A. Biryukov, A. Shamir, and D. Wagner, *Real Time Cryptanalysis of A5/1 on a PC*, Proceedings of FSE '2000, LNCS 1978, pp. 1–18, Springer-Verlag, 2001.
5. E. Dawson, J. Golić, W. Millan and L. Simpson, *The LILI-128 Keystream Generator*, Proceedings of the Seventh Annual Workshop on Selected Areas in Cryptology – SAC 2000, LNCS 2012, Springer-Verlag, 2000.
6. E. Dawson, J. Golić, W. Millan and L. Simpson, *Response to Initial Report on LILI-128*, Submitted to Second NESSIE Workshop, 2001.
7. M. E. Hellmab, *A Cryptanalytic Time-Memory Trade-Off*, IEEE Transactions on Information Theory, Vol. IT-26, N 4, pp. 401–406, 1980.
8. F. Jönsson and T. Johansson, *A Fast Correlation Attack on LILI-128.*, Information Processing Letters Vol 81, N. 3, Pages 127–132, 2001.
9. J. White, *Initial Report on the LILI-128 Stream Cipher*, NESSIE Public Report, https://www.cosic.esat.kuleuven.ac.be/nessie/reports, 2001.

On the Security of Randomized CBC–MAC Beyond the Birthday Paradox Limit A New Construction

Éliane Jaulmes, Antoine Joux, and Frédéric Valette

DCSSI Crypto Lab
18, rue du Dr. Zamenhof
F-92131 Issy-Les-Moulineaux.
{eliane.jaulmes,fred.valette}@wanadoo.fr
antoine.joux@m4x.org

Abstract. In this paper, we study the security of randomized CBC–MACs and propose a new construction that resists birthday paradox attacks and provably reaches full security. The size of the MAC tags in this construction is optimal, i.e., exactly twice the size of the block cipher. Up to a constant, the security of the proposed randomized CBC–MAC using an n–bit block cipher is the same as the security of the usual encrypted CBC–MAC using a $2n$–bit block cipher. Moreover, this construction adds a negligible computational overhead compared to the cost of a plain, non-randomized CBC–MAC. We give a full standard proof of our construction using one pass of a block-cipher with $2n$-bit keys but there also is a proof for n-bit keys block-ciphers in the random oracle model.

1 Introduction

The message authentication code (MAC) is a well-known and widely used cryptographic primitive whose goal is to authenticate messages and to check their integrity in a secret key setting. For historical and efficiency reasons, MACs are often based on block ciphers. Of course, other constructions are possible. A well-known method to build MACs is for example to start from a hash function and transform it into a secure MAC. The idea first appeared in the work of Wegman and Carter [15]. Other existing constructions are for example XOR-MACs [3], HMAC [1] and UMAC [6]. However, in low end cryptographic devices, the ability to reuse an existing primitive is an extremely nice property. In practice, a simple construction called CBC–MAC is frequently encountered. Several variants of the CBC–MAC are described in normative documents [9,14]. The simplest of those works as follows: let E be a block cipher using a key K to encrypt n–bit blocks. To compute the CBC–MAC of the message M with the key K, we split M into a sequence of n–bit blocks $M_1, \ldots, M_l$ and compute

$$C_0 = 0^n,$$
$$C_i = E_K(M_i \oplus C_{i-1}) \text{ for } i \text{ in } 1 \cdots l.$$

J. Daemen and V. Rijmen (Eds.): FSE 2002, LNCS 2365, pp. 237–251, 2002.
© Springer-Verlag Berlin Heidelberg 2002

After this computation, the value of the CBC–MAC is $\mathbf{CBC}_{E_K}(M) = C_l$. Note that the length of the message M has to be a multiple of the block size n, however several padding techniques have been proposed to remove this constraint [9].

This simple CBC–MAC has been proved secure in [4] for messages of fixed (non zero) length. However, when the length is no longer fixed, forgery attacks exist. The simplest of those uses two messages of one block each M and M', and queries their MACs C and C'. Then it can forge the MAC of $M\|(M' \oplus C)$, namely C'.

In order to remove this limitation, it is shown in [11] and [7] that it suffices to encrypt the plain CBC–MAC of a message with another key. However, the security level offered by these CBC–MACs is not optimal, since they all suffer from a common weakness: birthday paradox based attacks. In fact, all iterated MACs suffer from this kind of attacks, as has been shown in [12]. The basic idea beyond the birthday paradox attacks is to find two different messages with the same MAC value. Due to the birthday paradox, this search can be done in $2^{n/2}$ MAC computations, where n is the size of the MAC tag. Then one just need to append any fixed string to these messages and the MAC values of the extended messages are again the same. Thus forgery is easy since it suffices to query the MAC of one of the extended messages and use it as a forged MAC of the other extended message.

In order to protect MACs from birthday paradox attacks, it is suggested in [9] to add to each message a unique identifier, leading to a stateful MAC, or some kind of randomization, leading to a randomized MAC. These ideas have been studied in deeper details by some recent papers. In [3], a stateful construction based on XOR-MAC is given. It turns out that this leads to a reasonably simple and efficient construction. However, this approach has a major drawback, since it forces the MAC generation device to maintain an internal state from one generation to the next, which is extremely inconvenient when several MAC generation devices share the same key. On the other hand, randomization is much easier to deal with in practice. However, building a randomized MAC provably secure against birthday paradox attacks is not a simple matter. Indeed, the best currently known solution, called MACRX [2], is not CBC–MAC based and it expands the size of the MAC values by a factor of 3 instead of the expected 2. Indeed since with a MAC of size kn, an adversary can always obtain collisions in $2^{\frac{kn}{2}}$, in order to have a security against birthday paradox, the size of the MAC must be at least $2n$ bits. So MACRX is not optimal. Moreover it proposes to use hash functions and requires the use of a pseudorandom function family which itself is secure beyond the birthday paradox limit. Up to now, the only known solutions for designing pseudorandom functions with security beyond the birthday limit using block-ciphers (pseudorandom permutations) are counter-based [5,8]. Moreover, it was shown in [13] that the simple and arguably reasonable approach of adding a random value at the beginning of a message before computing its CBC–MAC does not give full security. Indeed, this construction suffers from the so-called L-collision attack and forgery is possible after $2^{\alpha n}$ queries, where $\alpha = 2/3$.

Our paper is organized as follows. In section 2 we recall the standard deterministic CBC–MAC algorithm, **DMAC**, and explain how we construct our randomized CBC-MAC **RMAC** from **DMAC**. We also present our security model and recall a few notations. The section 3 contains the theorems stating the security of our construction as well as some sketches of proof. In section 4 we show how to instantiate our construction with a block-cipher. Two proofs are given. The first one is in the standard model but make use of block-ciphers with $2n$-bit keys, the second one is in the random oracle model and uses only block-ciphers with n-bit keys. Then section 5 proposes a detailed instantiation using the AES block-cipher and we conclude in section 6.

2   Preliminaries

2.1   Standard Deterministic CBC–MAC

According to [11] and [7], we know that encrypted CBC–MAC has a security level of $O(2^{n/2})$. In particular, in [11] the security of a CBC–MAC named **DMAC** is analyzed. We briefly recall the definition of **DMAC**. Given two random permutations f_1 and f_2 on n bits, $\mathbf{DMAC}_{f_1,f_2}$ is defined on messages whose length is a multiple of n. Given $M = (M_1, M_2, \cdots, M_l)$, we compute:

$$C_0 = 0^n,$$
$$C_i = f_1(M_i \oplus C_{i-1}) \quad \text{for } i \text{ in } 1 \cdots l,$$
$$\mathbf{CBC}_{f_1}(M) = C_l$$
$$\mathbf{DMAC}_{f_1,f_2}(M) = f_2(\mathbf{CBC}_{f_1}(M)).$$

The first block appearing in the computation C_0 is called the initial value, it can safely be chosen as the all-zero block 0^n. The resulting algorithm may be seen on figure 1.

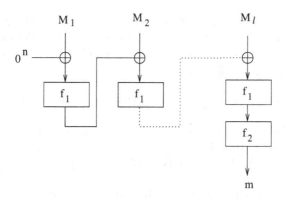

Fig. 1. The **DMAC** algorithm.

In order to deal with messages of arbitrary size, it suffices to define a padding process **Pad** such that for any pair of distinct messages M and M', we have **Pad**$(M) \neq$ **Pad**(M'). Such a padding can be obtained by simply adding a '1' bit at the end of the message followed by enough '0' bits to turn the length of the padded message into a multiple of n. Note that in order to ensure that **Pad**$(M) \neq$ **Pad**(**Pad**(M)), we need to pad messages whose length is already a multiple of n. In that case one full block is added.

Another approach for dealing with messages of arbitrary length was proposed in [7]. This approach nicely avoids the padding of messages which already contain an integral number of blocks. This is achieved by taking one permutation f_2 for messages that need to be padded and a different permutation f'_2 for others. In fact, this is a first step towards randomizing the function f_2 and it neatly fits into the construction we propose in this paper. However, to avoid cumbersome details, we ignore this variation in the proofs.

An advantage of [7] is that the security proof it gives for **DMAC** is much simpler than the proof from [11]. However, the result stated in [7] is slightly weaker. Indeed, in [11] the probability for an adversary of attacking **DMAC** is bounded by a function of the form $O(L^2/2^n)$, where L is the sum of the length (in blocks) of the messages whose MAC are computed during the attack. In [7], the result is expressed in terms of the number of messages q and of the length l of the longer message as a function of the form $O(l^2 q^2/2^n)$. When all the messages are roughly of the same length, the two are equivalent. However, if the adversary queries $2^{n/4} - 1$ messages of one block and a single message of $2^{n/4}$ blocks, then $L = 2^{n/4+1} - 1$, $q = 2^{n/4}$ and $l = 2^{n/4}$, we see that the result from [11] bounds the advantage of the adversary as $O(2^{-n/2})$ while the bound from [7] is $O(1)$. In truth, it seems that the authors of [7] chose to present a weaker result for the sake of clarity. In the security proof we present in this paper, we closely mimic the proof from [7], however we bound the advantage of the adversary as a function of L instead of using q and l.

2.2   Randomizing CBC-MACs

The above definition can easily be turned into a randomized CBC–MAC. Let f_1 be a random permutation on n bits and F_2 be a set of random permutations or functions $f_2^{(R)}$ on n bits, indexed by R a r–bit number. A randomized CBC–MAC is built on the following function:

$$\mathbf{RMAC}_{f_1, F_2}(M, R) = \left(\mathbf{DMAC}_{f_1, f_2^{(R)}}(M), R \right).$$

To compute the MAC of a message, we proceed as follows: we choose a random r–bit value R and returns **RMAC**$_{f_1, F_2}(M, R)$. To verify a given MAC (m, R) of a message M, we check whether **RMAC**$_{f_1, F_2}(M, R) = (m, R)$. The algorithm may be seen on figure 2.

When dealing with messages of arbitrary length, we can pad all messages as in [11]. Alternatively, we can also follow the approach from [7] (see section 2.1) to avoid padding messages whose length is already a multiple of n. This is simply

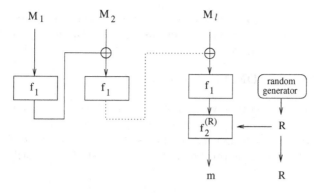

Fig. 2. The **RMAC** algorithm.

done by adding one bit to R, thus turning it into a $(r+1)$–bit number. The added bit is set to '0' when computing or verifying the MAC tag of a padded message and is set to '1' for an unpadded message. This ensures that a padded and a non-padded message never share the same R. In the boundary (non-randomized) case $r = 0$, we are clearly back to the proposal from [7], i.e. using f_2 in one case and f_2' in the other.

2.3 Security Model

The main goal of the paper is to prove that **RMAC** achieves full security. In order to make this statement precise, we need to define a new security model.

Perfect MACs. A perfect (ordinary) MAC is usually seen as a random function f from the set of messages $\{0,1\}^*$ to the set of possible MAC tags $\{0,1\}^n$. Thus to each message the function associates a random MAC tag. Similarly, a **perfect randomized MAC** is a family of independent random functions $f^{(R)}$ indexed by R, a r–bit number. Each function in the family goes from $\{0,1\}^*$ to $\{0,1\}^n$ and is randomly and independently chosen for each $R \in \{0,1\}^r$ among all possible such functions. This family of functions can be accessed through two oracles, a MAC generation oracle G_f and a MAC verification oracle V_f. The generation oracle takes a message M, chooses a random r–bit value R and returns $(f^{(R)}(M), R)$. The verification oracle takes a message M and a MAC tag (m, R), checks whether $f^{(R)}(M) = m$ and accordingly returns valid MAC or invalid MAC.

When there is no randomness, i.e. when $r = 0$, we get a perfect MAC as special case. In that case, the verification oracle becomes redundant since verification can be achieved by generating a MAC for M and testing equality with m.

Information theoretic model. The classical approach in proving the security of **DMAC** is to show the security of an information theoretic version of the construction and then come to the computational result (see [11] or [7]). Recall that **DMAC** uses two functions f_1 and f_2. For a padded message $M = (M_1, M_2, \cdots, M_l)$, we compute:

$$\mathbf{DMAC}_{f_1, f_2}(M) = f_2(\mathbf{CBC}_{f_1}(M)).$$

In the information theoretic version of the construction, it is first assumed that the functions f_1 and f_2 are randomly chosen among all possible functions and the security of the resulting construction is shown. Then f_1 and f_2 are replaced by block ciphers and it is proved that such an instantiation still offers a good security.

Now if we look at **RMAC**, we see that

$$\mathbf{RMAC}_{f_1, F_2}(M, R) = \left(\mathbf{DMAC}_{f_1, f_2^{(R)}}(M), R \right).$$

Here we assume that f_1 is a random permutation and that F_2 is a family of independent random permutations indexed by R and we are going to prove the security of **RMAC** under these assumptions. But before proceeding further, we need to define a few notations.

Notations. Let $\mathbf{Rand}(A, B)$ be the set of all functions from A to B. When A or B is replaced by a positive number n, then the corresponding set is $\{0, 1\}^n$. Let $\mathbf{Perm}(n)$ be the set of all permutations on $\{0, 1\}^n$. By $x \xleftarrow{R} A$ we denote the choice of an element x uniformly at random in A.

A function family F is a set of functions from A to B where A and B are subsets of $\{0, 1\}^*$. Each element in F is indexed by a key K. A block cipher is a function family from A to A that contains permutations only.

Adversaries against ordinary MACs. When dealing with ordinary MACs, an adversary is an algorithm given access to an oracle that computes some function. Adversaries are assumed to never ask queries outside the domain of the oracle and to never repeat the same query.

Let F be a function family from A to B, f be a function randomly chosen in F and $\mathcal{A}$ be an adversary. We say that $\mathcal{A}^f$ forges, if $\mathcal{A}$ outputs $(x, f(x))$ and $\mathcal{A}$ never queried its oracle f at x. We denote:

$$\mathbf{Adv}_F^{\mathbf{mac}}(\mathcal{A}) = \mathbf{Pr}[f \xleftarrow{R} F | \mathcal{A}^f \text{ forges}],$$

$$\mathbf{Adv}_F^{\mathbf{prf}}(\mathcal{A}) = \left| \mathbf{Pr}[f \xleftarrow{R} F | \mathcal{A}^f = 1] - \mathbf{Pr}[f \xleftarrow{R} \mathbf{Rand}(A, B) | \mathcal{A}^f = 1] \right|,$$

and when $A = B = \{0, 1\}^n$:

$$\mathbf{Adv}_F^{\mathbf{prp}}(\mathcal{A}) = \left| \mathbf{Pr}[f \xleftarrow{R} F | \mathcal{A}^f = 1] - \mathbf{Pr}[f \xleftarrow{R} \mathbf{Perm}(n) | \mathcal{A}^f = 1] \right|.$$

$\mathbf{Adv}_F^{\mathbf{mac}}(\mathcal{A})$ represents the probability for the adversary $\mathcal{A}$ of forging a valid MAC knowing that the MAC function f is not a true random function but is randomly chosen among the family F. $\mathbf{Adv}_F^{\mathbf{prf}}(\mathcal{A})$ represents the advantage for adversary $\mathcal{A}$ of distinguishing a function f randomly chosen from one chosen in the family F. $\mathbf{Adv}_F^{\mathbf{prp}}(\mathcal{A})$ is the same as above but with permutations instead of functions.

We also write $\mathbf{Adv}^{\mathbf{mac}}(t,\mu)$ for the maximal value of $\mathbf{Adv}^{\mathbf{mac}}(\mathcal{A})$ among adversaries that are bounded as follows: the running time should be less than t, and the sum of the bit length of all the oracle queries should be less than μ. We likewise define $\mathbf{Adv}^{\mathbf{prf}}(t,\mu)$ and $\mathbf{Adv}^{\mathbf{prp}}(t,\mu)$. In the case of $\mathbf{Adv}^{\mathbf{mac}}(t,\mu)$, μ also counts the length of an additional query to verify if the adversary's output is a valid forgery.

Adversaries against randomized MACs. When dealing with randomized MACs, an adversary is an algorithm given access to the generation and to the verification oracles for some randomized MAC. Adversaries are assumed to never ask queries outside the domain of the oracle, however, they may repeat the same query. Indeed, it might be useful to obtain several different MAC tags for the same message. Without loss of generality, since the adversary can always get rid of duplicates, we may assume that when MAC generation is queried several times with the same message, the generation oracle always chooses a different random value (among a total of 2^r possibilities). In that case, the adversary should not be allowed to query a given message more than 2^r times from the generation oracle. Moreover, we may assume that the adversary never repeats verifications, and never verifies previously generated MAC tags or obviously false tags. This means that when a tag (m, R) was generated for a message M, the adversary never verifies $(M, (m', R))$. Indeed, the answer is obviously valid when $m = m'$ and invalid otherwise.

Let $\mathcal{P}$ be the family of all perfect randomized MAC from a set A to a set B, let $\mathcal{F}$ be a given family of randomized MAC from A to B and let f be a randomized MAC randomly chosen in $\mathcal{F}$. We say that $\mathcal{A}^{G_f, V_f}$ forges, if $\mathcal{A}$ obtains the answer valid MAC from the oracle V_f for a tag $(x, (f^{(R)}(x), R))$ where $\mathcal{A}$ never got this MAC tag from its generation oracle G_f. We let:

$$\mathbf{Adv}_{\mathcal{F}}^{\mathbf{Rmac}}(\mathcal{A}) = \mathbf{Pr}[f \xleftarrow{R} \mathcal{F} | \mathcal{A}^{G_f, V_f} \text{ forges}],$$
$$\mathbf{Adv}_{\mathcal{F}}^{\mathbf{Rprf}}(\mathcal{A}) = \left| \mathbf{Pr}[f \xleftarrow{R} \mathcal{F} | \mathcal{A}^{G_f, V_f} = 1] - \mathbf{Pr}[f \xleftarrow{R} \mathcal{P} | \mathcal{A}^{G_f, V_f} = 1] \right|.$$

$\mathbf{Adv}_{\mathcal{F}}^{\mathbf{Rmac}}(\mathcal{A})$ represents the probability for an adversary $\mathcal{A}$ of forging a valid MAC knowing that the family f is not a perfect randomized MAC but is randomly chosen among the set $\mathcal{F}$. $\mathbf{Adv}_{\mathcal{F}}^{\mathbf{Rprf}}(\mathcal{A})$ represents the advantage for an adversary $\mathcal{A}$ of distinguishing a family f randomly chosen among all possible perfect randomized MACs from one chosen in the set $\mathcal{F}$.

As before, we write $\mathbf{Adv}^{\mathbf{Rmac}}(t,\mu)$ for the maximal value of $\mathbf{Adv}^{\mathbf{Rmac}}(\mathcal{A})$ among adversaries that are bounded as follows: the running time should be less than t, and the sum of the bit length of all the oracle queries should be less than μ. We likewise define $\mathbf{Adv^{Rprf}}(t, \mu)$. In the case of $\mathbf{Adv^{Rmac}}(t, \mu)$, no additional queries are necessary, since the adversary has access to a verification oracle and can test its forgery by itself. This differs from $\mathbf{Adv^{mac}}(t, \mu)$ in the case of non randomized MAC.

3   Security of RMAC

We are now going to state the security reached by **RMAC** in the information-theoretic model. We evaluate this security in terms of $\mathbf{Adv}_{\mathcal{G}}^{\mathbf{Rmac}}(\mathcal{A})$ when $\mathcal{G}$ is the family described in 2.3, i.e. $\mathcal{G}$ is the family of all couples (f_1, F_2) where f_1 is a random permutation and F_2 is a family of independent random **permutations** indexed by R.

Theorem 1 states that the advantage of a forging adversary against $\mathbf{RMAC}_{f_1, F_2}$ with f_1 and F_2 as above increases as a linear function of L, the total length of messages.

Theorem 1. *[Forging **RMAC** is hard] Fix $n \geq 2, r = n$ and let $N = 2^n$. Let $\mathcal{G}$ denotes the family of randomized MAC $\mathbf{RMAC}_{f_1, F_2}$ built from the couple (f_1, F_2) where f_1 is a random permutation and F_2 a family of random permutations $f_2^{(R)}$. Let $\mathcal{A}$ be an adversary which asks queries of total length at most L n–bit blocks. Assume $L \leq N/4$, then:*

$$Adv_{\mathcal{G}}^{Rmac}(\mathcal{A}) \leq \frac{4nL + 4L + 2}{N}.$$

Proof of theorem 1. If an adversary A is able to forge, then he is able to distinguish between **RMAC** and a **Rprf**. Indeed recall that $\mathcal{A}$ forges when he has verified his forgery through the verification oracle, thus forgery leads directly to distinction. We have:

$$\mathbf{Adv}_{\mathcal{G}}^{\mathbf{Rmac}}(\mathcal{A}) \leq \mathbf{Adv}_{\mathcal{G}}^{\mathbf{Rprf}}(\mathcal{A}) + \frac{1}{N}.$$

We just need to prove an indistinguishability theorem in the information-theoretic model. Theorem 2 states that the advantage for distinguishing $\mathbf{RMAC}_{f_1, F_2}$ from a perfect randomized MAC increases as a linear function of L.

Theorem 2. *[**RMAC** $\approx$ **Rand**] Fix $n \geq 1, r = n$ and let $N = 2^n$. Let $\mathcal{G}$ denotes the family of randomized MAC $\mathbf{RMAC}_{f_1, F_2}$ built from the couple (f_1, F_2) where f_1 is a random permutation and F_2 a family of random permutations. Let $\mathcal{A}$ be an adversary which asks queries of total length at most L n–bit blocks. Assume $L \leq N/4$, then:*

$$Adv_{\mathcal{G}}^{Rprf}(\mathcal{A}) \leq \frac{4nL + 4L + 1}{N}.$$

In order to prove this theorem, we are first going to prove a lemma where the family F_2 of random permutations has been replaced by a family of random functions.

Lemma 1. *Fix $n \geq 1, r = n$ and let $N = 2^n$. Let $\mathcal{F}$ denotes the family of randomized MAC $\mathbf{RMAC}_{f_1, F_2}$ built from the couple (f_1, F_2) where f_1 is a random permutation and F_2 a family of random functions. Let $\mathcal{A}$ be an adversary which asks queries of total length at most L n–bit blocks. Assume $L \leq N/4$, then:*

$$Adv_{\mathcal{F}}^{Rprf}(\mathcal{A}) \leq \frac{3nL + 3L + 1}{N}.$$

Sketch of proof of Lemma 1. [1] Here f_1 is a random permutation and F_2 a family of random functions. The proof of the theorem is close to the proof given in [7] but there are some fundamental differences. The adversary $\mathcal{A}$ has access to the two oracles described in section 2.3, the generation and the verification oracles. The total length of the queries it may ask is bounded by L. An adaptive adversary can always be replaced by a non-adaptive adversary that performs as well, so we are going to separately bound the advantage $\mathbf{Adv}_G(\mathcal{A})$ gained by $\mathcal{A}$ by means of the generation queries and the advantage $\mathbf{Adv}_V(\mathcal{A})$ gained by means of the verification queries, those two advantages being independent from each other.

In order to bound $\mathbf{Adv}_G(\mathcal{A})$, we observe that only a small number of messages will be processed with the same R^2. Moreover since all the functions $f_2^{(R)}$ for different Rs are independent, the adversary only learns information from MACs generated with the same R (else he only sees outputs of independent functions). Within such a group, the adversary only learns information when the CBC output of two messages is the same (else he only sees the outputs of a random function on different inputs). So we need to evaluate the probability of collision within a group of messages at the end of the CBC computation. The collision probability is defined as follows:

$$V_n(M, M') = \mathbf{Pr}[\pi \overset{R}{\leftarrow} \mathbf{Perm}(n) | \mathbf{CBC}_\pi(M) = \mathbf{CBC}_\pi(M')].$$

We improve a lemma from [7][3] and obtain that the probability of collision among q_R messages M_i of size m_i blocks is

$$\mathbf{Pr}[\pi \overset{R}{\leftarrow} \mathbf{Perm}(n) | \exists i \neq j \text{ such that } \mathbf{CBC}_\pi(M_i) = \mathbf{CBC}_\pi(M_j)] \leq \frac{3q_R \sum_{i=1}^{q_R} m_i}{2^n},$$

with $\sum_{i=1}^{q_R} m_i \leq N/4$.

So the advantage $\mathbf{Adv}_G(\mathcal{A})$ is bounded by the sum of the probability of collision within the different groups plus the probability of existence of a group larger than n:

---

[1] The full proof is given in [10].
[2] Less than n messages with probability $1 - 1/2^n$, see in the full paper [10].
[3] See in the full paper [10].

$$\mathbf{Adv}_G(\mathcal{A}) \leq \sum_R \frac{3q_R \sum_{i=1}^{q_R} m_i}{2^n} + \frac{1}{2^n} \leq \frac{3n}{2^n} \sum_R \sum_{i=1}^{q_R} m_i + \frac{1}{2^n} \leq \frac{3nL}{2^n} + \frac{1}{2^n}.$$

In order to bound $\mathbf{Adv}_V(\mathcal{A})$, we observe that the adversary learns information only when he checks a previously received MAC with a new message (else he just guesses at random). The adversary succeeds if the new message collides with the reference message at the end of the CBC computation. We thus need to evaluate the probability of collision of messages with a reference message[4]. We find that

$$\mathbf{Pr}[\pi \xleftarrow{R} \mathbf{Perm}(n) | \exists i \in [1, q] \text{ such that } \mathbf{CBC}_\pi(M_i) = \mathbf{CBC}_\pi(M_0)] \leq \frac{3 \sum_{i=0}^{q} m_i}{2^n}.$$

Summing over all reference messages of total length L we get:

$$\mathbf{Adv}_V(\mathcal{A}) \leq \frac{3L}{2^n}.$$

Finally, adding $\mathbf{Adv}_G(\mathcal{A})$ and $\mathbf{Adv}_V(\mathcal{A})$, we conclude the proof of lemma 1.

$$\mathbf{Adv}_{\mathcal{F}}^{\mathbf{Rprf}}(\mathcal{A}) \leq \frac{3nL + 3L + 1}{N}.$$

Sketch of proof of Theorem 2. In theorem 2 we replace the family F_2 of random functions by a family of random permutations. We evaluate the advantages $\mathbf{Adv}_G^{(2)}(\mathcal{A})$ and $\mathbf{Adv}_V^{(2)}(\mathcal{A})$ obtained by $\mathcal{A}$ respectively with generation and verification queries when we do this modification.

We use the well-known PRF/PRP switching lemma [4] on each permutation $f_2^{(R)}$. Indeed the adversary tries to separately distinguish the different permutations from functions. If q_R denotes the number of calls made to $f_2^{(R)}$, we recall from the proof of theorem 2 that with probability $1/2^n$ we have $q_R \leq n$. So we obtain

$$\mathbf{Adv}_G^{(2)}(\mathcal{A}) \leq \sum_R \frac{q_R^2}{2^{n+1}} \leq n \sum_R \frac{q_R}{2^{n+1}} \leq \frac{nL}{2^{n+1}}.$$

During the verification phase, the adversary wins when he distinguishes the random permutations $f_2^{(R)}$ from random functions. This happens when the verification oracle answers `valid MAC` for either a guessed MAC or a MAC obtained for another message. We find that:

$$\mathbf{Adv}_V^{(2)}(\mathcal{A}) \leq \frac{L}{2^n - n}.$$

Finally, adding $\mathbf{Adv}_G^{(2)}(\mathcal{A})$ and $\mathbf{Adv}_V^{(2)}(\mathcal{A})$ with $\mathbf{Adv}_{\mathcal{F}}^{\mathbf{Rprf}}(\mathcal{A})$, we conclude the proof of theorem 2.

---

[4] See in the full paper [10].

4 Instantiation of the RMAC Construction with a Block Cipher

4.1 The Computational Model

Proof in the standard model. When proving the security of a MAC construction, it is customary to first show that their information-theoretic versions approximate random functions. Then, we need to transport the result from the information-theoretic model to the computational complexity model. This improves the advantage of the adversary since he can now try to distinguish the pseudo-random functions or permutations from truly random ones. It is a general principle that the advantage in the computational-complexity model is the sum of the advantage in the information-theoretic model and of the advantages to distinguish each component of the construction from its idealized version with the number of calls made in the construction. An example of this principle appears in section 4 of [4].

To go from the information theoretic model to the computational model, we replace the random permutation f_1 with a block-cipher B. The adversary gains that way an advantage $\mathbf{Adv}_B^{\mathbf{prf}}$ of distinguishing the block-cipher B from a random function.

The random family of permutations F_2 indexed by a n-bit key can be viewed as a function $f_2(R, X) = f_2^{(R)}(X)$ of $2n$-bit to n-bit where $f_2^{(R)}$ is a permutation. We want to replace the family F_2 by a construction based on a block-cipher B with keys of $2n$ bits. We propose to choose a random $2n$-bit key K and to let $f_2^{(R)}(X) = B_{K \oplus R}(X)$, where R has been padded with n zeroes for the XOR. When such a construction is used, the adversary gains some new advantage. This advantage comes either from a weakness in the block cipher or from a weakness in the construction itself. In order to separate the two kinds of weaknesses, we would like to assume that the block cipher is "perfect". In order to do this, we use the following model. Assume that the block cipher is replaced by a family F_3 containing 2^{2n} random permutations together with a numbering. Given access to F_3, can an adversary distinguish the case where F_2 is built from F_3 as above from the case where F_2 itself is a family of random permutations ? Clearly, the adversary gains no advantage unless in the former case he manages to query the same function once through F_2 and once through F_3. In order to bound the probability that this event occurs and since the adversary is computationally bounded, we assume that he makes less than 2^n calls to F_3. Thus he queries at mots 2^n permutations among a total of 2^{2n}. On the other hand, with q queries, at most q permutations can be seen on the F_2 side. Unless the two sets collide, the adversary sees nothing. Thus, the probability for an adversary to detect that F_2 is a subfamily of F_3 is at most $\dfrac{2^n q}{2^{2n}} = \dfrac{q}{2^n}$.

Now, when using a real block-cipher, some new weaknesses may arise. In that case, this leads to a correlated key attack against the block-cipher. Indeed, the MAC construction allows us to distinguish B from F_3, when given access to $B_{K \oplus R}$ (and the corresponding decryption oracle).

The advantage gained by the adversary from the information-theoretic model to the computational complexity model is thus equal to $\frac{q}{2^n} + \mathbf{Adv}_B^{F_3}$, where $\mathbf{Adv}_B^{F_3}$ is the advantage for an adversary of distinguishing the block-cipher B from a "perfect" block-cipher.

Now that no attack other than exhaustive search is possible against B, we can express the advantage of an adversary is the standard model. We can bound q by L, the total number of queries done by the adversary.

$$\mathbf{Adv}_{\mathbf{RMAC}_B} \leq \mathbf{Adv}_{\mathcal{G}}^{\mathbf{RMAC}} + \mathbf{Adv}_B^{\mathrm{prf}} \leq \frac{4nL + 5L + 2}{2^n} + \frac{t}{2^n}.$$

Going further with the random oracle model. Instead of a $2n$-bit key block cipher, it would be more satisfying to use a standard n-bit key. We see in this section that this can be done if we accept a weaker security proof. Indeed with n-bit keys, we only prove security in the random oracle model. As above, we define $f_2(R, X) = f_3^{(R \oplus K)}(X)$. However, F_3 is now a smaller family made of 2^n permutations "only". The adversary trying to forge the MAC in this model has still access to the two oracles G_f and V_f and to F_3 through two other oracles C_f and C_f^{-1}. These computation oracles work as follows. In C_f, the adversary queries a chosen function $f_3^{(S)}$ of the family F_3, indexed by some n-bit integer S, with some input X and the oracle returns $f_3^{(S)}(X)$. In C_f^{-1}, the adversary also queries a chosen instance of the block-cipher F_3, indexed by S and asks for the value of X corresponding to the output U; the oracle returns $\left(f_3^{(S)}\right)^{-1}(U)$.

Let $\mathcal{H}$ be the family of all triplets (f_1, F_2, F_3) as described above. We want to bound the probability of forging for the adversary $\mathcal{A}$:

$$\mathbf{Adv}_{\mathcal{H}}^{\mathbf{Rmac}}(\mathcal{A}) = \mathbf{Pr}\left[f \xleftarrow{R} \mathcal{H} | \mathcal{A}^{G_f, V_f, C_f, C_f^{-1}} \text{ forges}\right].$$

Theorem 3. *[Forging **RMAC** with idealized block-cipher] Fix $n \geq 2, r = n$ and let $N = 2^n$. Let $\mathcal{H}$ denotes the family of randomized MAC $\mathbf{RMAC}_{f_1, F_2, F_3}$ built from the triplet (f_1, F_2, F_3) where F_3 is a random family of 2^n permutations, f_1 is a random permutation and F_2 is a permuted copy of the family F_3 determined by a key K. Let $\mathcal{A}$ be an adversary which asks queries of total length at most L n–bit blocks. Assume $L \leq N/4$, then:*

$$Adv_{\mathcal{H}}^{Rmac}(\mathcal{A}) \leq \frac{4nL + 6L + 2}{N}.$$

Sketch of proof of theorem 3. [5] Let $\mathcal{A}$ be an adversary trying to forge. $\mathcal{A}$ has access to the four oracles G_f, V_f, C_f and C_f^{-1}. Against $\mathcal{A}$, we play a simulator $\mathcal{S}$ that works as follows. When $\mathcal{A}$ queries G_f or V_f on a value of a permutation of the family F_2, $\mathcal{S}$ chooses this value randomly under the condition that the

---

[5] The full formal proof is given in the full paper [10].

underlying function $f_2^{(R)}$ is a permutation. Of course this implies that when asked twice the value of some permutation it answers twice the same result. When $\mathcal{A}$ queries C_f or C_f^{-1} on a value of a permutation of the family F_3, $\mathcal{S}$ also chooses this value randomly under the same conditions as above. The important fact here is that the simulator answers questions about F_3 independently from questions about F_2. When the attacker $\mathcal{A}$ terminates, the simulator chooses an n-bit key K uniformly at random. Then he tries to redefine F_2 using the formula $f_2^{(R)} = f_3^{(K \oplus R)}$. Unless two incompatible answers were given while the attacker asks questions, this can be done easily. Indeed, all the answers define some $f_3^{(S)}$ and thus $f_2^{(S \oplus K)}$ or some $f_2^{(R)}$ and thus $f_3^{(R \oplus K)}$. The rest of F_3 (and F_2) can be chosen randomly (under the condition that all $f_3^{(S)}$ are permutations). When two incompatible answers were given, we assume that the simulator has lost, i.e. that the adversary wins.

We want to evaluate the probability $\mathbf{Pr}[F_2$ and F_3 incompatible$]$. The answers to F_2 and F_3 match if the tables of the answers for F_2 and those for F_3 are compatible for the chosen key K. The probability that F_2 and F_3 are not compatible is less than the probability that $f_2^{(R)}$ and $f_3^{(R \oplus K)}$ have been evaluated on one common point or that $f_2^{(R)}$ and $f_3^{(R \oplus K)}$ have one common output. Since the simulator independently answers questions on permutations of F_3 and F_2, the adversary cannot adapt its queries to one family from the answers to the queries of the other family. Moreover, when K is chosen, $\mathcal{A}$ has already terminated and it is too late for him to be adaptive. Since $\mathcal{A}$ cannot be adaptive on K, we can compute an upper bound on the probability that F_2 and F_3 mismatch. We find:

$$\mathbf{Pr}\left[F_2 \text{ and } F_3 \text{ incompatible}\right] \leq \frac{2L}{2^n}.$$

Since $\mathbf{Adv}_{\mathcal{H}}^{\mathbf{Rmac}}(\mathcal{A}) \leq \mathbf{Adv}_{\mathcal{G}}^{\mathbf{Rmac}}(\mathcal{A}) + \dfrac{2L}{2^n}$, we have $\mathbf{Adv}_{\mathcal{H}}^{\mathbf{Rmac}}(\mathcal{A}) \leq \dfrac{4nL + 6L + 2}{2^n}$.

This concludes the sketch of proof of theorem 3.

5   Detailed Instantiations with the AES

Up to now, the only known attack against the AES is exhaustive search. We propose two different instantiations of **RMAC** with the AES. The first one assumes that all messages are padded. The second instantiation takes advantage of the technique from [7] that allows not to pad messages which are formed from an integral number of blocks (see section 2.2). We describe here the instantiations with the AES using n-bit keys and $2n$-bit keys. In these instantiations, the longest key size for K_2 gives security in the standard model, while the shortest key size restricts us to security in the random oracle model.

First instantiation. Let K_1 be a 128-bit key and K_2 a 128 or 256-bit key. Let $f_1 = \mathbf{AES}_{K_1}$ and $f_2^{(R)} = \mathbf{AES}_{K_2 \oplus R}$. Here R is a 128-bit integer padded with zeros for the XOR if necessary. The proposed instantiation is simply $\mathbf{RMAC}_{f_1, F_2}$.

Second instantiation. Let K_1 be a 128-bit key and K_2 be a 192 or 256-bit key. Let $f_1 = \mathbf{AES}_{K_1}$ and $f_2^{(R)} = \mathbf{AES}_{K_2 \oplus R}$. Here R is a 129-bit number padded with zeros for the XOR. The 128 low order bits of R are randomly chosen by the generation oracle. The 129-th bit is a '0' when the message needs to be padded and a '1' otherwise. This additional bit is never included as part of the MAC tags, it should be set by the verification oracle according to the properties of the message being verified.

Security of the instantiations. Glueing together theorem 1 and theorem 3 with the known attacks against the AES, we claim that the advantage of an adversary making queries of total length at most L and with runtime t – including the run time of the generation and verification queries themselves – is at most:

$$\mathbf{Adv}_{\mathbf{RMAC}_{\mathbf{AES}}} \leq \frac{4 \cdot 128L + 5L + 2}{2^{128}} + \frac{t}{2^{128}} \leq \frac{518L + t}{2^{128}},$$

using a key K_2 of 256 bits and

$$\mathbf{Adv}_{\mathbf{RMAC}_{\mathbf{AES}}} \leq \frac{4 \cdot 128L + 6L + 2}{2^{128}} + \frac{t}{2^{128}} \leq \frac{519L + t}{2^{128}},$$

using a key K_2 of 128 bits.
This should be compared with the security of the traditional $\mathbf{DMAC}$:

$$\mathbf{Adv}_{\mathbf{DMAC}_{\mathbf{AES}}} \leq \frac{2L^2 + t}{2^{128}}.$$

In other words, $\mathbf{RMAC}_{\mathbf{AES}}$ is secure as long as the total length of the queries is smaller than 2^{118}, while $\mathbf{DMAC}_{\mathbf{AES}}$ is secure as long as the total length of the queries is smaller than 2^{63}. In fact, the security of $\mathbf{RMAC}_{\mathbf{AES}}$ is almost as good as the security of $\mathbf{DMAC}$ with a good 256–bit block cipher.

6   Conclusion

The $\mathbf{RMAC}$ construction proposed in this paper gives an efficient solution to the problem of constructing a randomized CBC–MAC provably secure against birthday paradox attacks. The only previously known example of a birthday paradox resistant MAC was given in [2] and called MACRX. Compared to MACRX, $\mathbf{RMAC}$ has two main advantages. Firstly, its output has twice the length of the underlying block-cipher instead of three times for MACRX. Secondly, being a CBC–MAC variant, $\mathbf{RMAC}$ does not require any special functions other than the block cipher.

Moreover, $\mathbf{RMAC}$ unleashes the full power of the AES in MAC computation, thus making the need for 256-bit block ciphers a very remote perspective. Quite interestingly, the proof is stronger when using 256-bit keys in AES.

References

1. M. Bellare, R. Canetti, and H. Krawczyk. Keying hash functions for message authentication. In *CRYPTO'96*, volume 1109 of *LNCS*. Springer, 1996.
2. M. Bellare, O. Goldreich, and H. Krawczyk. Stateless evaluation of pseudorandom functions: Security beyond the birthday barrier. In *CRYPTO'99*, volume 1666 of *LNCS*, pages 270–287. Springer, 1999.
3. M. Bellare, R. Guerin, and P. Rogaway. XOR MACs: New methods for message authentication using finite pseudorandom functions. In *CRYPTO'95*, volume 963 of *LNCS*, pages 15–28. Springer-Verlag, 1995.
4. M. Bellare, J. Killian, and P. Rogaway. The security of the cipher block chaining message authentication code. In *CRYPTO'94*, volume 839 of *LNCS*, pages 341–358. Springer, 1994. See new version at http://www.cs.ucdavis.edu/~rogaway/.
5. M. Bellare, T. Krovetz, and P. Rogaway. Luby-rackoff backwards: increasing security by making block-ciphers non-invertible. In *EUROCRYPT'98*, volume 1403 of *LNCS*, pages 266–280. Springer, 1998.
6. J. Black, S. Halevi, H. Krawczyk, T. Krovetz, and P. Rogaway. UMAC: Fast and secure message authentication. In *CRYPTO'99*, volume 1666 of *LNCS*, pages 216–233. Springer-Verlag, 1999.
7. J. Black and P. Rogaway. CBC MACs for arbitrary-length messages: The three-key constructions. In *CRYPTO 2000*, volume 1880 of *LNCS*, pages 197–215. Springer, 2000.
8. C. Hall, D. Wagner, J. Kelsey, and B. Schneier. Building PRFs from PRPs. In *CRYPTO'98*, volume 1462 of *LNCS*, pages 370–389. Springer, 1998.
9. International Organization for Standards, Geneva, Switzerland. *ISO/IEC 9797-1. Information Technology – Security Techniques – Data integrity mechanism using a cryptographic check function employing a block cipher algorithm*, second edition edition, 1999.
10. É. Jaulmes, A. Joux, and F. Valette. On the security of randomized cbc–mac beyond the birthday paradox limit: A new construction. Available at http://eprint.iacr.org, 2002. Full version of this paper.
11. E. Petrank and C. Rackoff. CBC-MAC for real-time data sources. Technical Report 97-10, Dimacs, 1997.
12. B. Preneel and P. van Oorschot. MDx-MAC and building fast MACs from hash functions. In *CRYPTO'95*, volume 963 of *LNCS*, pages 1–14. Springer, 1995.
13. M. Semanko. L-collision attacks against randomized MACs. In *CRYPTO 2000*, volume 1880 of *LNCS*, pages 216–228. Springer, 2000.
14. U.S. Department of Commerce/National Bureau of Standards, National Technical Information Service, Springfield, Virginia. *FIPS 113. Computer Data Authentication. Federal Information Processing Standards Publication 113*, 1994.
15. M. Wegman and J. Carter. New hash functions and their use in authentication and set equality. *Journal of Computer and System Sciences*, 22(3):265–279, 1981.

Cryptanalysis of the Modified Version of the Hash Function Proposed at PKC'98

Daewan Han, Sangwoo Park, and Seongtaek Chee

National Security Research Institute
161 Gajeong-dong, Yuseong-gu, Daejeon, 305-350, Korea
{dwh,psw,chee}@etri.re.kr

Abstract. In the conference PKC'98, Shin *et al.* proposed a dedicated hash function of the MD family. In this paper, we study the security of Shin's hash function. We analyze the property of the Boolean functions, the message expansion, and the data dependent rotations of the hash function. We propose a method for finding the collisions of the modified Shin's hash function and show that we can find collisions with probability 2^{-30}.

1 Introduction

Hash functions are used for many cryptographic applications, such as message authentication and digital signature. A hash function is a computationally efficient function which maps binary strings of arbitrary length to binary strings of some fixed length. Cryptographic hash functions should satisfy the following properties [3]:

- *pre-image resistance*: given a y in the image of a hash function h, it is computationally infeasible to find any pre-image x such that $h(x) = y$.
- *2nd pre-image resistance*: given x and $h(x)$, it is computationally infeasible to find a $x' \neq x$ such that $h(x) = h(x')$.
- *collision resistance*: it is computationally infeasible to find any two distinct inputs x, x' which hash to the same output.

Since the hash function MD4 [6] was introduced by R. Rivest, many dedicated hash functions based on design principles of MD4 have been proposed. MD5 [7], HAVAL [10], RIPEMD [5], RIPEMD-160 [2], and SHA-1 [4] are the dedicated hash functions of the MD family.

In the conference PKC'98, Shin *et al.* proposed a dedicated hash function of the MD family [8]. We call it Shin's hash function. The compression function of Shin's hash function processes a message block of 512 bits and consists of 4 rounds. Each of the rounds consists of 24 steps. Shin's hash function employs the message expansion similar to SHA-1, and Boolean functions similar to HAVAL. Another feature of Shin's hash function is to adopt the data-dependent rotations: rotations are processed by variable amounts determined by message words.

J. Daemen and V. Rijmen (Eds.): FSE 2002, LNCS 2365, pp. 252–262, 2002.

In this paper, we study the security of Shin's hash function. We analyze the property of the Boolean functions, the message expansion, and the data dependent rotations of Shin's hash function. We indicate that, unlike the designer's intention, some of the Boolean functions of Shin's hash function fail to satisfy the Strict Avalanche Criterion(SAC) [9]. Also, we point out that there can be some weakness of the message expansion and the data dependent rotations. We consider the modified Shin's hash function which is Shin's hash function whose Boolean functions all satisfy the SAC, and propose a method for finding the collisions for the modified Shin's hash function.

2   The Compression Function of Shin's Hash Function

Throughout this paper, the symbol $+$ represents a modulo 2^{32} addition, $X \oplus Y$, $X \wedge Y$ and $X \vee Y$ represent the bitwise exclusive OR, AND, and OR of X and Y, respectively. The symbol $X^{\lll s}$ denotes the left cyclic shift of X by s bit positions to the left.

The compression function of Shin's hash function processes a 16-word message block of 512 bits, $(X_0, X_1, \ldots, X_{15})$, and consists of 4 rounds. The 16-word message block is expanded to a 24-word message block, $(X_0, X_1, \ldots, X_{23})$. In the 24-word message block, $X_i(i = 0, 1, \ldots, 15)$ are the same as the message words of the original 16-word message blocks and the additional 8 message words, $X_i(i = 16, 17, \cdots, 23)$ are determined by the 16-word message blocks as follows:

$$X_{16+i} = (X_{0+i} \oplus X_{2+i} \oplus X_{7+i} \oplus X_{12+i})^{\lll 1}, i = 0, 1, \cdots, 7. \tag{1}$$

With the expanded 24-word message block, the compression function transforms a 5-word(160 bits) initial value (A, B, C, D, E) into a 160-bit output value. The 5-word initial values are the followings:

$$A = \text{0x67452301}, \ B = \text{0x}efcdab89, \ C = \text{0x98}badcfe,$$

$$D = \text{0x10325476}, \ E = \text{0x}c3d2e1f0.$$

Each round of the compression function consists of 24 steps and each step processes a different word. The orders in which the words are processed differ from round to round. The word processing orders are determined by the following:

Round 1	Round 2	Round 3	Round 4
id	ρ	ρ^2	ρ^3

and the permutation ρ is as follows:

i	0	1	2	3	4	5	6	7	8	9	10	11
$\rho(i)$	4	21	17	1	23	18	12	10	5	16	8	0

i	12	13	14	15	16	17	18	19	20	21	22	23
$\rho(i)$	20	3	22	6	11	19	15	2	7	14	9	13

In addition, each round employs a different constant. The constant $K_i(i = 1, 2, 3, 4)$ is adopted by i-th round.

$$K_1 = 0\text{x}0, \ K_2 = 0\text{x}5a827999, \ K_3 = 0\text{x}6ed9eba1, \ K_4 = 0\text{x}8f1bbcdc.$$

In each round, one of the following Boolean functions is employed.

$$f_0(x_1, x_2, x_3, x_4, x_5) = (x_1 \wedge x_2) \oplus (x_3 \wedge x_4) \oplus (x_2 \wedge x_3 \wedge x_4) \oplus x_5$$
$$f_1(x_1, x_2, x_3, x_4, x_5) = x_2 \oplus ((x_4 \wedge x_5) \vee (x_1 \wedge x_3))$$
$$f_2(x_1, x_2, x_3, x_4, x_5) = x_1 \oplus (x_2 \wedge (x_1 \oplus x_4)) \oplus (((x_1 \wedge x_4) \oplus x_3) \vee x_5)$$

The Boolean functions perform bitwise operations on words. f_0, f_1, f_2, f_0 are adopted by the 1st, 2nd, 3rd, and 4th round, respectively.

Now, we describe the step function of Shin's hash function. Let $T_{i,j}(j = 0, 1, \cdots, 4)$ be the input of the step function at step i. Then, the step function of Shin's hash function has a transformation of the form

$$T_{i,0} = (f(T_{i,0}, T_{i,1}, T_{i,2}, T_{i,3}, T_{i,4}) + X_i + K)^{\lll s}, \quad T_{i,1} = T_{i,1}^{\lll 10}$$
$$T_{i+1,1} = T_{i,0}, T_{i+1,2} = T_{i,1}, T_{i+1,3} = T_{i,2}, T_{i+1,4} = T_{i,3}, T_{i+1,0} = T_{i,4}.$$

The rotation amount s_i at the step i is determined by the following:

$$s_i = X_{ord(i)} \bmod 32,$$

where $ord(i)$ is determined by the following permutations:

Round 1	Round 2	Round 3	Round 4
ρ^3	ρ^2	ρ	id

3   Some Properties of Shin's Hash Function

In this section, we analyze the property of the Boolean functions, the message expansion, and the data dependent rotations of Shin's hash function.

3.1   The Property of the Boolean Functions

The designers of Shin's hash function claimed that each of the Boolean functions of the hash function is 0-1 balanced, has a high nonlinearity, and satisfies the SAC [8]. Yet, it is easy to find out that some of the Boolean functions of Shin's hash function fail to satisfy the SAC.

We define the Boolean function f as satisfying the SAC if whenever one input bit of f is changed, each output bit is changed with probability 1/2[9]. In case of Shin's hash function, it is easy to know that the Boolean function f_0 and f_1 do not satisfy the SAC. In case of f_0, we can know that, whenever the input bit, x_5, is changed, the output bit is changed with probability 1. Similarly, in case of f_1, whenever the input bit, x_2, is changed, the output bit is changed with probability 1.

$$f_0(x_1, x_2, x_3, x_4, x_5) = (x_1 \wedge x_2) \oplus (x_3 \wedge x_4) \oplus (x_2 \wedge x_3 \wedge x_4) \oplus x_5,$$
$$f_1(x_1, x_2, x_3, x_4, x_5) = x_2 \oplus ((x_4 \wedge x_5) \vee (x_1 \wedge x_3)).$$

Since the designers of Shin's hash function intended that each of the Boolean functions satisfies the SAC, it can be adequate that we consider the Shin's hash function whose Boolean functions all satisfy the SAC. We call it the modified Shin's hash function.

3.2   The Property of the Message Expansion

The additional 8 message words $X_{16}, X_{17}, \cdots, X_{23}$ are determined by the 16-word message block by the equation (1). The equation (1) can be restated as follows:

$$X_{16} = (X_0 \oplus X_2 \oplus X_7 \oplus X_{12})^{\lll 1}$$
$$X_{17} = (X_1 \oplus X_3 \oplus X_8 \oplus X_{13})^{\lll 1}$$
$$X_{18} = (X_2 \oplus X_4 \oplus X_9 \oplus X_{14})^{\lll 1}$$
$$X_{19} = (X_3 \oplus X_5 \oplus X_{10} \oplus X_{15})^{\lll 1}$$
$$X_{20} = (X_4 \oplus X_6 \oplus X_{11} \oplus X_{16})^{\lll 1}$$
$$X_{21} = (X_5 \oplus X_7 \oplus X_{12} \oplus X_{17})^{\lll 1}$$
$$X_{22} = (X_6 \oplus X_8 \oplus X_{13} \oplus X_{18})^{\lll 1}$$
$$X_{23} = (X_7 \oplus X_9 \oplus X_{14} \oplus X_{19})^{\lll 1}$$

For two 32-bit words X and $\tilde{X}$, we will define the difference of X and $\tilde{X}$ as follows:

$$\Delta X = X - \tilde{X} \pmod{2^{32}}.$$

To find a collision for Shin's hash function, we should find two distinct message blocks X and $\tilde{X}$ which have the same hash value. In the two distinct message blocks X and $\tilde{X}$, if non-zero difference occurs between X_i and $\tilde{X}_i$ ($0 \leq i \leq 15$), non-zero difference can occur between some of the additional 8 message words which are generated from X_i and $\tilde{X}_i$. For example, if non-zero difference occurs between X_0 and $\tilde{X}_0$, then non-zero difference can occur between X_{16} and $\tilde{X}_{16}$. Furthermore, non-zero difference between X_{16} and $\tilde{X}_{16}$ can make non-zero difference between X_{20} and $\tilde{X}_{20}$. Thus, the message expansion can increase the difficulty of finding collisions for the hash function.

Table 1 shows the property of the message expansion of Shin's hash function. It shows that X_{10} and X_{15} affect X_{19} and X_{23} simultaneously. It means that, although ΔX_{10} and ΔX_{15} are non-zero, we can have $\Delta X_{19} = \Delta X_{23} = 0$ in the case $X_{10} = X_{15}$ and $\tilde{X}_{10} = \tilde{X}_{15}$. Similarly, although ΔX_{17}, ΔX_{21}, and ΔX_{22} are non-zero, we can have $\Delta X_8 = \Delta X_{13} = 0$ in the case $X_{17} \oplus X_{21} \oplus X_{22} = 0$ and $\tilde{X}_{17} \oplus \tilde{X}_{21} \oplus \tilde{X}_{22} = 0$.

Table 1. The effect of message expansion of the SHF

X_0	X_1	X_2	X_3	X_4	X_5	X_6	X_7	X_8	X_9	X_{10}	X_{11}	X_{12}	X_{13}	X_{14}	X_{15}
X_{16}	X_{17}	X_{16}	X_{17}	X_{18}	X_{19}	X_{20}	X_{16}	X_{17}	X_{18}	X_{19}	X_{20}	X_{16}	X_{17}	X_{18}	X_{19}
X_{20}	X_{21}	X_{18}	X_{19}	X_{20}	X_{21}	X_{22}	X_{20}	X_{21}	X_{22}	X_{23}		X_{20}	X_{21}	X_{22}	X_{23}
		X_{20}	X_{21}	X_{22}	X_{23}		X_{21}	X_{22}	X_{23}			X_{21}	X_{22}	X_{23}	
		X_{22}	X_{23}				X_{23}								

Now, we know that there are instances in Shin's hash function where non-zero difference between some original message words cannot be diffused to non-zero difference between some additional message words. We can use this property of the message expansion to find the collisions for the modified Shin's hash function.

3.3   The Property of the Data Dependent Rotations

In Shin's hash function, the data dependent rotations are adopted by the equation

$$s_i = X_{ord(i)} \bmod 32,$$

at step i. We can know that in the 1st round, $ord(i)$ is determined by the word processing orders of the 4th round, and $ord(i)$ of the 2nd round is determined by the word processing orders of the 3rd round, and so on.

For example, at step 1, s_1 is determined by the word X_{13}(see Appendix A). So, if we have the word X_{13} such that $X_{13} = 0 \bmod 32$, then, $s_1 = 0$. It means that we can make the data dependent rotations ineffective by choosing the appropriate message block, i.e. if we have the message block $(X_i, i = 0, 1, \dots, 15)$ such that $X_i = 0 \bmod 32$, we can have all s_i equal to 0.

Also, if we have X_i and $\tilde{X}_i$ such that $\Delta X_i \neq 0$ and $X_i = \tilde{X}_i \bmod 32$, then we know that the shift amounts determined by X_i and $\tilde{X}_i$ are the same although X_i and $\tilde{X}_i$ are different.

4   Attack on the Modified Shin's Hash Function

Although some of the Boolean functions of Shin's hash function do not satisfy the SAC, since the designers of Shin's hash function intended that each of the Boolean functions satisfies the SAC, it seems to be adequate that we study the security of the modified Shin's hash function. In this section, we propose a method for finding the collisions for the modified Shin's hash function.

We define some notations. A_i, B_i, C_i, D_i, E_i represent the chaining variables after step i for a message block $X = (X_0, \dots, X_{23})$, and $\tilde{A}_i, \tilde{B}_i, \tilde{C}_i, \tilde{D}_i, \tilde{E}_i$ represent the chaining variables after step i for a message block $\tilde{X} = (\tilde{X}_0, \dots, \tilde{X}_{23})$. s_i and $\tilde{s}_i$ represent the shift value used in step i for X and $\tilde{X}$, respectively.

4.1 Attack on the 6 Consecutive Steps of the Modified Shin's Hash Function

We analyze the 6 consecutive steps of the modified Shin's hash function and show how to find the collisions for 6 consecutive steps. For convenience, we consider the 6 consecutive steps from step 1 to step 6.

To find the collisions for 6 consecutive steps, we should find two distinct message blocks $X = (X_0, X_1, \ldots, X_5)$ and $\tilde{X} = (\tilde{X}_0, \tilde{X}_1, \ldots, \tilde{X}_5)$ which have the same chaining variables after step 6, i.e. $A_6 = \tilde{A}_6$, $B_6 = \tilde{B}_6$, $C_6 = \tilde{C}_6$, $D_6 = \tilde{D}_6$, and, $E_6 = \tilde{E}_6$.

We consider two distinct message blocks X and $\tilde{X}$ such that $\Delta X_0 = 1^{\lll 31}$ and $\Delta X_1 = \Delta X_2 = \Delta X_3 = \Delta X_4 = \Delta X_5 = 0$. Then, we can have a situation that $s_i = \tilde{s}_i (i = 1, 2, \ldots, 6)$. We assume that $A_0 = \tilde{A}_0$, $B_0 = \tilde{B}_0$, $C_0 = \tilde{C}_0$, $D_0 = \tilde{D}_0$, and, $E_0 = \tilde{E}_0$. Note that the Boolean function f satisfies the SAC.

Now, we analyze each step from step 1 to step 6 and find the probability with which $(X, \tilde{X})$ can be a collision of the 6 consecutive steps. Table 2 shows the updated chaining variables at each step of the 6 consecutive steps. The boxed variables represent the updated chaining variables at each step.

Table 2. Chaining variables updated in each step

Step	A_0	B_0	C_0	D_0	E_0	Input message
1	$\boxed{A_1}$	B_1	C_1	D_1	E_1	X_0
2	A_2	B_2	C_2	D_2	$\boxed{E_2}$	X_1
3	A_3	B_3	C_3	$\boxed{D_3}$	E_3	X_2
4	A_4	B_4	$\boxed{C_4}$	D_4	E_4	X_3
5	A_5	$\boxed{B_5}$	C_5	D_5	E_5	X_4
6	$\boxed{A_6}$	B_6	C_6	D_6	E_6	X_5

A_1 and $\tilde{A}_1$ are updated at step 1 as follows:

$$A_1 = (f(A_0, B_0, C_0, D_0, E_0) + X_0 + K)^{\lll s_1}, \quad B_1 = B_0^{\lll 10}$$
$$\tilde{A}_1 = (f(\tilde{A}_0, \tilde{B}_0, \tilde{C}_0, \tilde{D}_0, \tilde{E}_0) + \tilde{X}_0 + K)^{\lll \tilde{s}_1}, \quad \tilde{B}_1 = \tilde{B}_0^{\lll 10}$$

Since $\Delta X_0 = 1^{\lll 31}$, we can have a situation that $X_0 \oplus \tilde{X}_0 = 1^{\lll 31}$ with probability 1. Also, we know that $\Delta A_0 = \Delta B_0 = \Delta C_0 = \Delta D_0 = \Delta E_0$ and $s_1 = \tilde{s}_1$. So, we have the following:

$$A_1 \oplus \tilde{A}_1 = 1^{\lll s_1(\text{or } \tilde{s}_1) - 1}.$$

At step 2, E_2 and $\tilde{E}_2$ are updated as follows:

$$E_2 = (f(E_1, A_1, B_1, C_1, D_1) + X_1 + K)^{\lll s_2}, \quad A_2 = A_1^{\lll 10},$$
$$\tilde{E}_2 = (f(\tilde{E}_1, \tilde{A}_1, \tilde{B}_1, \tilde{C}_1, \tilde{D}_1) + \tilde{X}_1 + K)^{\lll \tilde{s}_2}, \quad \tilde{A}_2 = \tilde{A}_1^{\lll 10}.$$

Since $\Delta E_1 = \Delta B_1 = \Delta C_1 = \Delta D_1 = 0$, $\Delta X_1 = 0$, $s_2 = \tilde{s}_2$, we can have the following equation:

$$\Delta E_2 = 0 \iff f(E_1, A_1, B_1, C_1, D_1) = f(E_1, \tilde{A}_1, B_1, C_1, D_1).$$

Note that $A_1 \oplus \tilde{A}_1 = 1^{\lll s_1(\text{or } \tilde{s}_1)-1}$. Since f satisfies the SAC, we can have a result such that

$$f(E_1, A_1, B_1, C_1, D_1) = f(E_1, \tilde{A}_1, B_1, C_1, D_1)$$

with probability $1/2$, i.e. $\Delta E_2 = 0$ with probability $1/2$.

We assume that we have $\Delta E_2 = 0$ at step 2. At the next step, D_3 and $\tilde{D}_3$ are updated as follows:

$$D_3 = (f(D_2, E_2, A_2, B_2, C_2) + X_2 + K)^{\lll s_3}, \qquad E_3 = E_2^{\lll 10},$$
$$\tilde{D}_3 = (f(\tilde{D}_2, \tilde{E}_2, \tilde{A}_2, \tilde{B}_2, \tilde{C}_2) + \tilde{X}_2 + K)^{\lll s_3}, \qquad \tilde{E}_3 = \tilde{E}_2^{\lll 10}.$$

Since $\Delta D_2 = \Delta E_2 = \Delta B_2 = \Delta C_2$, $\Delta X_2 = 0$, and $s_3 = \tilde{s}_3$, we can have the following equation:

$$\Delta D_3 = 0 \iff f(D_2, E_2, A_2, B_2, C_2) = f(D_2, E_2, \tilde{A}_2, B_2, C_2).$$

Since f satisfies the SAC, we can have a result such that

$$f(D_2, E_2, A_2, B_2, C_2) = f(D_2, E_2, \tilde{A}_2, B_2, C_2)$$

with probability $1/2$, i.e. $\Delta D_3 = 0$ with probability $1/2$.

Similarly, we can have that $\Delta C_4 = 0$ with probability $1/2$ at step 4. Also, we can have that $\Delta B_5 = 0$ with probability $1/2$ at step 5, and $\Delta A_6 = 0$ with probability $1/2$ at step 6.

As a result, for two distinct message blocks $X = (X_0, X_1, \dots, X_5)$ and $\tilde{X} = (\tilde{X}_0, \tilde{X}_1, \dots, \tilde{X}_5)$ such that $\Delta X_0 = 1^{\lll 31}$ and $\Delta X_1 = \Delta X_2 = \Delta X_3 = \Delta X_4 = \Delta X_5 = 0$, we can have that $\Delta A_6 = 0$, $\Delta B_6 = 0$, $\Delta C_6 = 0$, $\Delta D_6 = 0$, $\Delta E_6 = 0$ with probability 2^{-5}. So, we can find a collision for the 6 consecutive steps of the modified Shin's hash function with about 2^5 operations.

4.2   Attack on the Full Steps of the Modified Shin's Hash Function

Now, we propose a method for finding a collision for the full steps of the modified Shin's hash function. We consider two distinct message blocks $X = (X_0, X_1, \dots, X_{15})$ and $\tilde{X} = (\tilde{X}_0, \tilde{X}_1, \dots, \tilde{X}_{15})$ which satisfy the following conditions.

- Condition 1 : X_i is arbitrary for $i \neq 8, 9$, and 10.
- Condition 2 : $X_8 = \text{0x00000016}$ and $X_9 = X_2 \oplus X_4 \oplus X_{14} \oplus \text{0x0000000b}$
- Condition 3 : $X_{10} = X_{15}$
- Condition 4 : $\tilde{X}_i = X_i$ for $i \neq 10, 15$
- Condition 5 : $\tilde{X}_{10} = \tilde{X}_{15} = X_{15} + 1^{\lll 31}$

Condition 2 implies that $X_8 = 22 \bmod 32$ and $X_{18} = 22 \bmod 32$. So, the shift amounts determined by X_8 and the shift amounts determined by X_{18} are equal to 22. From Condition 3, 4, and 5, we can have that $X_i = \tilde{X}_i (i = 16, 17, \cdots, 23)$ from the property of the message expansion. Also, since $X_i = \tilde{X}_i \bmod 32(0 \le i \le 23)$, it is easy to know that $s_i = \tilde{s}_i$ $(1 \le i \le 96)$. Finally, we notice that $\Delta X_{10} = \Delta X_{15} = 1^{\lll 31}$.

We denote I_i as the section of the 6 consecutive steps from step i to step $i + 5$ and consider the sections $I_{11}, I_{32}, I_{43}, I_{54}, I_{69}$, and, I_{81}. Note that the first input message word of the sections is X_{10} or X_{15}, and $\Delta X_{10} = \Delta X_{15} = 1^{\lll 31}$(see Appendix A).

First, we consider the section I_{32}. We have two distinct message blocks (X_{10}, $X_5, X_{16}, X_8, X_0, X_{20}$) and ($\tilde{X}_{10}, \tilde{X}_5, \tilde{X}_{16}, \tilde{X}_8, \tilde{X}_0, \tilde{X}_{20}$) such that $\Delta X_{10} = 1^{\lll 31}$ and $\Delta X_5 = \Delta X_{16} = \Delta X_8 = \Delta X_0 = \Delta X_{20} = 0$. So, to the section I_{32}, we can apply the attack on the 6 consecutive steps of the modified Shin's hash function, i.e. if $\Delta A_{31} = \Delta B_{31} = \Delta C_{31} = \Delta D_{31} = \Delta E_{31} = 0$, we can have $\Delta A_{37} = \Delta B_{37} = \Delta C_{37} = \Delta D_{37} = \Delta E_{37} = 0$ with probability 2^{-5}. Similarly, the attack on the 6 consecutive steps can be applied to I_{43}, I_{54}, and I_{69} with the same probability.

However, to the section I_{11}, the attack on the 6 consecutive steps cannot be directly applied because $\Delta X_{15} \ne 0$, where X_{15} and $\tilde{X}_{15}$ are the input message words of the last step of I_{11}. However, by using the property of the data dependent rotations, this problem can be solved. Note that, at step 16, A_{16} and $\tilde{A}_{16}$ are updated as follows:

$$A_{16} = (f_0(A_{15}, B_{15}, C_{15}, D_{15}, E_{15}) + X_{15} + K_1)^{\lll s_{16}}, \quad B_{16} = B_{15}^{\lll 10}$$
$$\tilde{A}_{16} = (f_0(\tilde{A}_{15}, \tilde{B}_{15}, \tilde{C}_{15}, \tilde{D}_{15}, \tilde{E}_{15}) + \tilde{X}_{15} + K_1)^{\lll s_{16}}, \quad \tilde{B}_{15} = \tilde{B}_{15}^{\lll 10}$$

We can have that $\Delta B_{15} = \Delta C_{15} = \Delta D_{15} = \Delta E_{15} = 0$ with probability 2^{-4}. Note that at step 11, A_{11} and $\tilde{A}_{11}$ are updated by the message words X_{10} and $\tilde{X}_{10}$ such that $\Delta X_{10} = 1^{\lll 31}$, and s_{11} and $\tilde{s}_{11}$ are determined by X_{18} and $\tilde{X}_{18}$, respectively. Since we have $X_{18} = \tilde{X}_{18}$ and $X_{18} = 22 \bmod 32$ from Condition 2, s_{11} and $\tilde{s}_{11}$ are equal to 22. Thus, we know that $A_{11} \oplus \tilde{A}_{11} = 1^{\lll 21}$. At step 12, A_{11} and $\tilde{A}_{11}$ are left-rotated by 10 bit positions, i.e. $A_{12} \oplus \tilde{A}_{12} = 1^{\lll 31}$. Since, from step 13 to step 15, $A_{13} = A_{14} = A_{15}$ and $\tilde{A}_{13} = \tilde{A}_{14} = \tilde{A}_{15}$, the equation $A_{15} \oplus \tilde{A}_{15} = 1^{\lll 31}$ holds. Now, we know that $\Delta X_{15} = \Delta A_{15} = 1^{\lll 31}$, so we can have the following equation:

$$\Delta A_{16} = 0$$
$$\Longleftrightarrow f_0(A_{15}, B_{15}, C_{15}, D_{15}, E_{15}) \oplus f_0(\tilde{A}_{15}, B_{15}, C_{15}, D_{15}, E_{15}) = 1^{\lll 31}$$

Since f_0 satisfies the SAC, we can have that

$$f_0(A_{15}, B_{15}, C_{15}, D_{15}, E_{15}) \oplus f_0(\tilde{A}_{15}, B_{15}, C_{15}, D_{15}, E_{15}) = 1^{\lll 31}$$

with probability $1/2$. So, $\Delta A_{16} = 0$ with probability 2^{-5}. Similarly, by using the property of the data dependent rotations, this attack can be applied to the section I_{81}. In this case, by the value of X_8 in Condition 2, we have that $s_{81} = \tilde{s}_{81} = 22$.

As our main result, if we have the two distinct message blocks X and $\tilde{X}$ which satisfy the Condition 1,2,3,4, and 5, we can find a collision of the modified Shin's hash function with probability $(2^{-5})^6 = 2^{-30}$.

Now, we find a collision of the modified Shin's hash function by computer simulation. We employ the Boolean functions $f_i (i = 0, 1, 2, 3)$ which satisfy the SAC as follows:

$$f_0(x_1, x_2, x_3, x_4, x_5) = x_2 \oplus (x_3 \wedge (x_2 \oplus x_5)) \oplus (((x_2 \wedge x_5) \oplus x_4) \vee x_1)$$

$$f_1(x_1, x_2, x_3, x_4, x_5) = x_3 \oplus (x_4 \wedge (x_3 \oplus x_1)) \oplus (((x_3 \wedge x_1) \oplus x_5) \vee x_2)$$

$$f_2(x_1, x_2, x_3, x_4, x_5) = x_1 \oplus (x_2 \wedge (x_1 \oplus x_4)) \oplus (((x_1 \wedge x_4) \oplus x_3) \vee x_5)$$

$$f_3(x_1, x_2, x_3, x_4, x_5) = x_4 \oplus (x_5 \wedge (x_4 \oplus x_2)) \oplus (((x_4 \wedge x_2) \oplus x_1) \vee x_3)$$

$f_i(i = 0, 1, 2, 3)$ is the modified version of the Boolean function f_2 of Shin's hash function, which satisfies the SAC. Note that our attack does not use the specific properties of the Boolean functions except the SAC.

As a result of computer simulation, we give a collision for the modified Shin's hash function in Table 3 which has the following hash value:

0xdfe4e58f, 0x1f21fb34, 0x9956457f, 0x8726dff2, 0x0a45bef3

Table 3. A collision for the modified Shin's hash function

X_0 = 0xe64ec066	$\tilde{X}_0$ = 0xe64ec066	
X_1 = 0xfd126b95	$\tilde{X}_1$ = 0xfd126b95	
X_2 = 0x6d80c03e	$\tilde{X}_2$ = 0x6d80c03e	
X_3 = 0x09d32e0c	$\tilde{X}_3$ = 0x09d32e0c	
X_4 = 0x767d3ff5	$\tilde{X}_4$ = 0x767d3ff5	
X_5 = 0x2bc1b633	$\tilde{X}_5$ = 0x2bc1b633	
X_6 = 0x40727b94	$\tilde{X}_6$ = 0x40727b94	
X_7 = 0xd7e17540	$\tilde{X}_7$ = 0xd7e17540	
X_8 = 0x00000016	$\tilde{X}_8$ = 0x00000016	
X_9 = 0x278364e1	$\tilde{X}_9$ = 0x278364e1	
X_{10} = 0xe7e7d228	$\tilde{X}_{10}$ = 0x67e7d228	
X_{11} = 0x8014bf7d	$\tilde{X}_{11}$ = 0x8014bf7d	
X_{12} = 0xd5a3b0de	$\tilde{X}_{12}$ = 0xd5a3b0de	
X_{13} = 0x5a70ffd6	$\tilde{X}_{13}$ = 0x5a70ffd6	
X_{14} = 0x3c7e9b21	$\tilde{X}_{14}$ = 0x3c7e9b21	
X_{15} = 0xe7e7d228	$\tilde{X}_{15}$ = 0x67e7d228	

5   Conclusion

In this paper, we have studied the security of Shin's hash function proposed by Shin *et al.* in the conference PKC'98. We have pointed out that, unlike the designer's intention, some of the Boolean functions of Shin's hash function do not satisfy the SAC. Also, we have indicated that there are instances in Shin's hash function that the message expansion is not effective. We have proposed a method for finding the collisions with probability 2^{-30} for the modified Shin's hash function. Furthermore, we have found a collision of the modified Shin's hash function by computer simulation.

Recently, the collisions of the original Shin's hash function have been found by Chang et al.. They have extended our attack, and the complexity of the attack is 2^{37} hashing operations [1].

This paper has provided a good example that, although it is known that the SAC is one of the important properties of the cryptographic Boolean functions, it can be absolutely irrelevant for the dedicated hash functions. So, we recommend that the Boolean functions of the dedicated hash function of the MD family be carefully chosen. Also, the message expansion should be carefully designed, and we conjecture that the data dependent rotations seem to be inadequate for the dedicated hash functions.

References

1. Donghoon Chang, Jaechul Sung, Soo Hak Sung, Sangjin Lee, and Jongin Lim. Full-Round Differential Attack on the Hash Function Proposed at PKC'98. Proceedings of Koreacrypt'01, pages 24–35, 2002.
2. Hans Dobbertin, Antoon Bosselaers, and Bart Preneel. RIPEMD-160: A strengthened version of RIPEMD. *ftp.esat.kuleuven.ac.be/pub/COSIC/bosselae/ripemd*, April 1996.
3. Alfred J. Menezes, Paul C. van Oorschot, and Scott A. Vanstone. *Handbook of Applied Cryptography*. CRC Press, 1996.
4. National Institute of Standards and Technology. FIPS PUB 180-1 : Secure Hash Standard, April 1995.
5. Research and Development in Advanced Communications Technologies in Europe. RIPE: Integrity primitives for secure information systems. Final Report of RACE Integrity Primitives Evaluation(R1040),RACE, 1995.
6. Ronald L. Rivest. The MD4 message digest algorithm. In Alfred J. Menezes and Scott A. Vanstone, editors, *Advances in Cryptology - Crypto'90*, volume 537 of *Lecture Notes in Computer Science*, pages 303–311. Springer-Verlag, 1991.
7. Ronald L. Rivest. The MD5 message digest algorithm. In *Request for Comments(RFC) 1321*, April. Internet Activities Board, Internet Privacy Task Force, 1992.
8. Sang Uk Shin, Kyung Hyune Rhee, Dae Hyun Ryu, and Sang Jin Lee. A new hash function based on MDx-family and its application to MAC. In Hideki Imai and Yuliang Zheng, editors, *Public Key Cryptography - PKC'98*, volume 1431 of *Lecture Notes in Computer Science*, pages 234–246. Springer, 1998.
9. A. F. Webster and Stafford E. Tavares. On the design of S-boxes. In Hugh C. Williams, editor, *Advances in Cryptology - Crypto'85*, volume 218 of *Lecture Notes in Computer Science*, pages 523–534. Springer-Verlag, New York, 1986.
10. Yuliang Zheng, Josef Pieprzyk, and Jennifer Seberry. HAVAL-A One-Way Hashing Algorithm with Variable Length of Output. In Jennifer Seberry and Yuliang Zheng, editors, *Advances in Cryptology - Auscrypt'92*, volume 718 of *Lecture Notes in Computer Science*, pages 83–104. Springer, 1992.

A   Message Processing Orders of Shin's Hash Function

Step	Word	Step	Word	Step	Word	Step	Word
1	X_0	25	X_4	49	X_{23}	73	X_{13}
2	X_1	26	X_{21}	50	X_{14}	74	X_{22}
3	X_2	27	X_{17}	51	X_{19}	75	X_2
4	X_3	28	X_1	52	X_{21}	76	X_{14}
5	X_4	29	X_{23}	53	X_{13}	77	X_3
6	X_5	30	X_{18}	54	$\boxed{X_{15}}$	78	X_6
7	X_6	31	X_{12}	55	X_{20}	79	X_7
8	X_7	32	$\boxed{X_{10}}$	56	X_8	80	X_5
9	X_8	33	X_5	57	X_{18}	81	$\boxed{X_{15}}$
10	X_9	34	X_{16}	58	X_{11}	82	X_0
11	$\boxed{X_{10}}$	35	X_8	59	X_5	83	X_{18}
12	X_{11}	36	X_0	60	X_4	84	X_{23}
13	X_{12}	37	X_{20}	61	X_7	85	$\boxed{X_{10}}$
14	X_{13}	38	X_3	62	X_1	86	X_{21}
15	X_{14}	39	X_{22}	63	X_9	87	X_{16}
16	$\boxed{X_{15}}$	40	X_6	64	X_{12}	88	X_{20}
17	X_{16}	41	X_{11}	65	X_0	89	X_4
18	X_{17}	42	X_{19}	66	X_2	90	X_{17}
19	X_{18}	43	$\boxed{X_{15}}$	67	X_6	91	X_{12}
20	X_{19}	44	X_2	68	X_{17}	92	X_{19}
21	X_{20}	45	X_7	69	$\boxed{X_{10}}$	93	X_8
22	X_{21}	46	X_{14}	70	X_{22}	94	X_9
23	X_{22}	47	X_9	71	X_{16}	95	X_{14}
24	X_{23}	48	X_{13}	72	X_3	96	X_1

Compression and Information Leakage of Plaintext

John Kelsey

Certicom
kelsey.j@ix.netcom.com

1 Introduction

Cryptosystems like AES and triple-DES are designed to encrypt a sequence of input bytes (the plaintext) into a sequence of output bytes (the ciphertext) in such a way that the output carries no information about that plaintext except its length. In recent years, concerns have been raised about "side-channel" attacks on various cryptosystems–attacks that make use of some kind of leaked information about the cryptographic operations (e.g., power consumption or timing) to defeat them. In this paper, we describe a somewhat different kind of side-channel provided by data compression algorithms, yielding information about their inputs by the size of their outputs. The existence of some information about a compressor's input in the size of its output is obvious; here, we discuss ways to use this apparently very small leak of information in surprisingly powerful ways.

The compression side-channel differs from side-channels described in [Koc96] [KSHW00] [KJY00] in two important ways:

1. It reveals information about plaintext, rather than key material.
2. It is a property of the *algorithm*, not the implementation. That is, any implementation of the compression algorithm will be equally vulnerable.

1.1 Summary of Results

Our results are as follows:

1. Commonly-used lossless compression algorithms leak information about the data being compressed, in the size of the compressor output. While this would seem like a very small information leak, it can be exploited in surprisingly powerful ways, by exploiting the ability of many compression algorithms to adapt to the statistics of their previously-processed input data.
2. We consider the "stateless compression side-channel," based on the compression ratio of an unknown string without reference to the rest of the message's contents. We also consider the much more powerful "stateful compression side-channel," based on the compression ratio of an unknown string, given information about the rest of the message.
3. We describe a number of simple attacks based mainly on the stateless side-channel.

J. Daemen and V. Rijmen (Eds.): FSE 2002, LNCS 2365, pp. 263–276, 2002.
© Springer-Verlag Berlin Heidelberg 2002

4. We describe attacks to determine whether some string S appears often in a set of messages, using the stateful side-channel.
5. We describe attacks to extract a secret string S that is repeated in many compressed messages, under partial chosen plaintext assumptions, using the stateful side-channel.
6. We consider countermeasures that can make both the stateless and the stateful side-channels substantially harder to exploit, and which may thus block some of these attacks.
7. We discuss the implications of these results, in light of the widespread use of compression with encryption, and the "folk wisdom" suggesting that adding compression to an encryption application will increase security.

1.2 Practical Impact of Results

Compression algorithms are widely used in real-world applications, and have a large impact on those applications' performance in terms of speed, bandwidth requirements, and storage requirements. For example, PGP and GPG compress using the Zip Deflate algorithm before encrypting, IPSec can use IPComp to compress packets before encrypting them, and both the SSH and TLS protocols support an option for on-the-fly compression.

Potential security implications of using compression algorithms are of practical importance to people designing systems that might use both compression and encryption.

The side-channel attacks described in this paper can have a practical impact on security in many situations. However, it is important to note that these attacks have little security impact on, say, a bulk encryption application which compresses data before encrypting. To a first-order approximation, the attacks in this paper are described in decreasing order of practicality. The string-extraction attacks are not likely to be practical against many systems, since they require such a specialized kind of partial chosen-plaintext access. The string-detection attacks have less stringent requirements, and so are likely to be useful against more systems. The passive information leakage attacks are likely practical to use against any system that uses compression and encryption together, and for which some information about input size is available.

In a broader sense, the results in this paper point to the need to consider the impact of any pre- or post-processing done along with encryption and authenticaton. For example, we have not considered timing channels from compression algorithms in this paper, but such channels will clearly exist for some compression algorithms, and must also exist for many other kinds of processing done on plaintext before it is sent, or ciphertext after it is received and decrypted. Similarly, anything done to the decrypted ciphertext of a message before authenticating the result is subject to reaction attacks: attacks in which changes in the ciphertext can cause different error messages or other behavior on the part of the receiver, depending on some secret information that the attacker seeks to reveal. (For decompressors which must terminate decompression with an error for some possible inputs, for example, there are serious dangers with respect to reaction attacks, or even with buffer-overrun or other related attacks.)

1.3   Previous Work

Although existence of the stateless compression side channel is obvious, we have seen very little reference to it in the literature. Nearly all published works discussing compression and encryption describe how compression *improves* the security of encryption.

One of the attendants of FSE2002 brought [BCL02] to our attention; in this article, researchers had noticed that they could use the compression ratio of a file to determine the language in which it was written in. This is the same phenomenon on which is based one of our stateless side channels.

1.4   Guide to the Paper

The remainder of this paper is arranged as follows: First, we discuss commonly-used compression methods, and how they interact with encryption. Next, we describe the side-channel which we will use in our attacks. We then consider several kinds of attack, making use of this side channel. We conclude with a discussion of various complications to the attacks, and possible defenses against them.

2   Lossless Compression Methods and the Compression Side-Channels

The goal of any compression algorithm (note: in this paper, we consider only *lossless* compression algorithms) is to reduce the redundancy of some block of data, so that an input that required R bits to encode can be written as an output with fewer than R bits. All lossless compression algorithms work by taking advantage of the fact that not all messages of R bits are equally likely to be sent. These compression algorithms make a trade-off: they effectively encode the higher probability messages with fewer bits, while encoding the lower probability messages with more bits. The compression algorithms in widespread use today typically use two assumptions to remove redundancy: They assume that characters and strings that have appeared recently in the input are likely to recur, and that some values (strings, lengths, and characters) are more likely to occur than others. Using these two assumptions, these algorithms are effective at compressing a wide variety of commonly-used data formats.

Many compression algorithms (and specifically, the main one we will consider here) make use of a "sliding window" of recently-seen text. Strings that appear in the window are encoded by reference to their position in the window. Other compression algorithms keep recently-seen strings in an easily-searched data structure; strings that appear in that structure are encoded in an efficient way.

Essentially all compression algorithms make use of ways to efficiently encode symbols (characters, strings, lengths, dictionary entries) of unequal frequency, so that commonly-occurring symbols are encoded using fewer bits than rarely-occurring symbols.

For the purposes of this paper, it is necessary to understand three things about these compression functions:

1. At any given point in the process of compressing a message, there are generally many different input strings of the same length which will compress to different lengths. This inherently leaks information about these input strings.
2. The most generally useful compression algorithms encode the next few bytes of input in different ways (and to different lengths), depending on recently-seen inputs.
3. While a single "pass" of a compression algorithm over a string can leak only a small amount of data about that string, multiple "passes" with different data appearing before that string can leak a great deal of data about that string.

This summary necessarily omits a lot of detail about how compression algorithms work. For a more complete introduction to the techniques used in compression algorithms, see [Sal97a] or [CCF01a].

2.1 Interactions with Encryption

Essentially all real-world ciphers output data with no detectable redundancy. This means that ciphertext won't compress, and so if a system is to benefit from compression, it must compress the information before it is encrypted.

The "folk wisdom" in the cryptographic community is that adding compression to a system that does encryption adds to the security of the system, e.g., makes it less likely that an attacker might learn anything about the data being encrypted. This belief is generally based on concerns about unicity distance, keysearch difficulty, or ability of known- or chosen-plaintext attacks. We believe that this folk wisdom, though often repeated in a variety of sources, is not generally true; adding compression to a competently designed encryption system has little real impact on its security. We base this on three observations:

Unicity distance is irrelevant. The unicity distance of an encryption system is the number of bits of ciphertext an attacker must see before he has enough information that it is even theoretically possible to determine the key. Compression algorithms, decreasing the redundancy of plaintexts, clearly increase unicity distance. However, this is irrelevant for practical encryption systems, where a single 128-bit key can be expected to encrypt millions of bytes of plaintext.

Keysearch difficulty is only slightly increased. Since most export restrictions on key lengths have gone away, we can expect this to become less and less relevant over time, as existing fielded algorithms with 40- and 56-bit key lengths are replaced with triple-DES or AES. At any rate, for systems with keys short enough for brute force searching, adding general-purpose compression algorithms to the system seems like a singularly unhelpful way to fix the problem. Standard compression algorithms usually include fixed headers, and tend to be pretty predictable in their first few bytes of output. It seems unlikely that adding such a compression algorithm, even with fixed headers removed, increases the difficulty of keysearch by more than a factor of 10 to 100. Switching to a stronger cipher is a far cheaper solution that actually solves the problem.

Standard algorithms not that helpful. Compression with some additional features to support security (such as a randomized initial state) can make known-plaintext attacks against block ciphers much harder. However, off-the-shelf compression algorithms provide little help against known-plaintext attacks (since an attacker who knows the compression algorithm and the plaintext knows the compressor output). And while chosen-plaintext attacks can be made much harder by specially designed compression algorithms, they are also made much harder, at far lower cost, by the use of standard chaining modes.

In summary, compression algorithms add very little security to well-designed encryption systems. Such systems use keys long enough to resist keysearch attack and chaining modes that resist chosen-plaintext attack. The real reason for using compression algorithms isn't to increase security, but rather to save on bandwidth and storage. As we will disucuss below, this real advantage needs to be balanced against a (mostly academic) risk of attacks on the system, such as those described below, based on information leakage from the compression algorithm.

3   The Compression Side-Channel and Our Attack Model

In this section, we describe the compression side channel in some detail. We also consider some situations in which this side channel might leak important data.

Any lossless compression algorithm must compress different messages by different amounts, and indeed must expand some possible messages. The compression side channel we consider in this paper is simply the different amount by which different messages are compressed. When an unknown string S is compressed, and an attacker sees the input and output sizes, he has almost certainly learned only a very small amount about S. For almost any S, there will be a large set of alternative messages of the same length, which would also have had the same size of compressor output. Even so, *some* small amount of information is leaked by even this minimal side-channel. For example, an attacker informed that a file of 1MB had compressed to 1KB has learned that the original file must have been *extremely* redundant.

Fortunately (for cryptanalysts, at least), compression algorithms such as LZW and Zip Deflate adapt to the data they are fed. (The same is true of many other compression algorithms, such as adaptive markov coding and Burrows-Wheeler coding, and even adaptive Huffman coding of symbols.) As a message is processed, the state of the compressor is altered in a predictable way, so that strings of symbols that have appeared earlier in the message will be encoded more efficiently than strings of symbols that have not yet appeared in the message. This allows an enormously more powerful side-channel when the unknown string S is compressed with many known or chosen prefix strings, $P_0, P_1, ..., P_{n-1}$. Each prefix can put the compressor into a different state, allowing new information to be extracted from the compressor output size in each case. Similarly, if a known or chosen set of suffixes, $Q_0, Q_1, ..., Q_{n-1}$ is appended to the unknown string S before compression, the compressor output sizes that result will each carry a slightly different piece of information about S, because those suffixes with many strings in common with S will compress better than other suffixes, with fewer strings in common with S. This can allow an attacker to reconstruct all of S with reasonably high probability, even when the compressor output sizes for different prefixes or suffixes differ only by a few bytes. In this situation, it is quite possible for an attacker to rule out all incorrect values of S given enough input and output sizes for various prefixes, along with knowledge or control over the prefix values. Further, an attacker can build information about S gradually, refining a partial guess when the results of each successive compressor output are seen.

A related idea can be used against a system that compresses and encrypts, but does not strongly authenticate its messages. The effect of altering a few bytes of plaintext (through a ciphertext alteration) will be very much dependent on the state of the decompressor both before and after the altered plaintext bytes are processed. The kind of control exerted over the compressor state is different, but the impact is similar. However, we do not consider this class of attack in this paper.

3.1   Assumptions and Models

We will make the following assumptions in the remainder of this paper:

1. Each message is processed by first compressing it, then encrypting it.
2. The attacker can learn the precise compressor output length from the ciphertext.
3. The attacker somehow knows the precise input length, or (in some cases) at least the approximate input length.

In the sections that follow, we will consider three basic classes of attacks: First, we will consider purely passive attacks, where the attacker simply observes the ciphertext length and compression ratio, and learns information that should have been concealed by the encryption mechanism. Second, we will consider a kind of limited chosen-plaintext attack, in which the attacker attempts to determine whether and approximately how often some string appears in a set of messages. Third, we will consider a much more demanding kind of chosen-plaintext attack, in which the attacker must make large numbers of chosen or adaptive-chosen plaintext queries, in hopes of extracting a whole secret string.

4   Data Information Leakage

In this section, we consider purely passive attacks; ways that an attacker can learn some information he should not be able to learn, by merely observing the ciphertexts and corresponding compression ratio. One general property of these attacks is that they are quite hard to avoid, without simply eliminating compression from the system. However, it is also worth noting that most of these attacks are not particularly devastating under most circumstances.

4.1   Highly Redundant Data

Consider a large file full of binary zeros or some other very repetitive contents. Encrypting this under a block cipher in ECB-mode would reveal a lot of redundancy; this is one reason why well-designed encryption systems use block ciphers in one of the chaining modes. Using CBC- or CFB-mode, the encrypted file would reveal nothing about the redundancy of the plaintext file.

Compressing before encryption changes this behavior. Now, a highly-redundant file will compress extremely well. The very small ciphertext will be sufficient, given knowledge of the original input size, to inform an attacker that the plaintext was highly redundant.

We note that this information leak is not likely to be very important for most systems. However:

1. Chaining modes prevent this kind of information leakage, and this is, in fact, one very good reason to use chaining modes with block ciphers.
2. In some situations, leaking the fact that highly-redundant data is being transmitted may leak some very important information. (An example might be a compressed, encrypted video feed from a surveillance camera–an attacker could watch the bandwidth consumed by the feed, and determine whether the motion of his assistant trying to get past the camera had been detected.)

4.2   Leaking File or Data Types

Different data formats compress at different ratios. A large file containing ASCII-encoded English text will compress at a very different ratio from a large file containing a Windows executable file. Given knowledge only of the compression ratio, an attacker can thus infer something about the kind of data being transmitted. This is not so trivial, and may be relevant in some special circumstances.

This may be resisted by encoding the data to be transmitted in some other format, at the cost of losing some of the advantage of compression.

4.3   Compression Ratio as a Checksum

Consider a situation where an attacker knows that one of two different known messages of equal length is to be sent. (For example, the two message might be something like "DEWEY DEFEATS TRUMAN!" or "TRUMAN DEFEATS

DEWEY!".) If these two messages have different compression ratios, the attacker can determine precisely which message was sent. (For this example, Python's ZLIB compresses "TRUMAN DEFEATS DEWEY!" slightly better than "DEWEY DEFEATS TRUMAN!")

More generally, if the attacker can enumerate the set of possible input messages, and he knows the compression algorithm, he can use the length of the input, plus the compression ratio, as a kind of checksum. This is a very straightforward instance of the side-channel; an attacker is able, by observing compression ratios, to rule out a subset of possible plaintexts.

4.4   Looped Input Streams

Sometimes, an input stream may be "looped," so that after R bytes, the message begins repeating. This is the sort of pattern that encryption should mask, and without compression, using a standard chaining mode will mask it. However, if the compression ratio is visible to an attacker, he will often be able to determine whether or not the message is looping, and may sometimes be able to determine its approximate period.

There are two ways the information can leak. First, if the period of the looping is shorter than the "sliding window" of an LZ77-type compression algorithm, the compression ratio will suddenly become very good. Second, if the period is longer than the sliding window, the compression ratios will start precisely repeating. (Using an LZW-type scheme will leave the compression ratios improving each time through the repeated data, until the dictionary fills up.)

5   String Presence Detection Attacks

The most widely used lossless compression algorithms adapt to the patterns in their input, so that when those patterns are repeated, those repetitions can be encoded very efficiently. This allows a whole class of attacks to learn whether some string S is present within a sequence of compressed and encrypted messages, based on using either known input data (some instances where S is known to have appeared in messages) or chosen input (where S may be appended to some messages before they're compressed and encrypted).

All the attacks in this section require knowledge or control of some part of a set of messages, and generally also some knowledge of the kind of data being sent. They also all require knowledge of either inputs or compressor outputs, or in some cases, compression ratios.

5.1   Detecting a Document or Long String with Partial Chosen Plaintext

The attacker wants to determine whether some long string S appears often in a set of messages $M_0, M_1, ..., M_{N-1}$.

The simplest attack is as follows:

1. The attacker gets the compressed, encrypted versions of all of the M_i. From this he learns their compressed output lengths.
2. The attacker requests the compressed, encrypted versions of $M'_i = M_i, S$, for all M_i. That is, he requests the compressed and encrypted results of appending S to each message.
3. The attacker determines the length of S after compression with the scheme in use.
4. The attacker observes $M'_i - M_i$. If these values average substantially less than the expected length of S after compression, it is very likely that S is present in many of these messages.

5.2   Partial Known Input Attack

A much more demanding and complicated attack may be possible, given only the leakage of some information from each of a set of messages. The attacker can look for correlations between the appearance of substrings of S in the known part of each message, and the compressed length of the message; based on this, he can attempt to determine whether S appears often in those messages. This attack is complicated by the fact that the appearance of substrings of S in the known part of the message may be correlated with the presence of S in the message. (Whether it is correlated or not requires more knowledge about how the messages are being generated, and the specific substrings involved. For example, if S is "global thermonuclear war", the appearance of the substring "thermonuclear" is almost certainly correlated with the appearance of S later in the message.)

A more useful version of an attack like this might be a case where several files are being combined into an archive and compressed, and the attacker knows one of the files. Assuming the other files aren't chosen in some way that correlates their contents with the contents of the known file, the attacker can safely run the attack.

6   String Extraction Attacks

In this section, we consider ways an attacker might use the compression side channel to extract some secret string from the compressor inputs. This kind of attack requires rather special conditions, and so is much less practical than the other attacks considered above. However, in some special situations, these attacks could be made to work. More importantly, these attacks demonstrate a separate path for attacking systems, despite the use of very strong encryption.

The general setting of these attacks is as follows: The system being attacked has some secret string, S, which is of interest to the attacker. The attacker is permitted to build a number of requested plaintexts, each using S, without ever knowing S. For example, the attacker may choose a set of N prefixes, $P_{0,1,...,N-1}$, and request N messages, where the ith message is $P_i||S$.

6.1 An Adaptive Chosen Input Attack

Our first attack is an adaptive chosen input attack. We make a guess about the contents of the first few characters of the secret string, and make a set of queries based on this guess. The output lengths of the results of these queries should be smaller for correct guesses than for incorrect guesses.

We construct our queries in the form

$$\text{Query} = \text{prefix} + \text{guess} + \text{filler} + \text{prefix} + S$$

where

Query is the string which the target of the attack is convinced to compress and encrypt.

prefix is a string that is known not to occur in S.

filler is another string known not to occur in S, and with little in common with *prefix*.

S is the string to be recovered by the attacker.

The idea behind this attack is simple: Suppose the prefix is 8 characters long, and the guess is another 4 characters long. A correct guess guarantees that the query string contains a repeated 12-character substring; a good compression algorithm (and particularly, a compression scheme based on a sliding window, like Zip Deflate) will encode this more efficiently than queries with incorrect guesses, which will contain a string with slightly less redundancy. When we have a good guess, this attack can be iterated to guess another four digits, and so on, until all of S has been guessed.

Experimental Results. We implemented this attack using the Python Zlib package, which provides access to the widely-used Zip Deflate compression algorithm. The search string was a 16-digit PIN, and the guesses were four (and later five) digits each. Our results were mixed: it was possible to find the correct PIN using this attack, but we often would have to manually make the decision to backtrack after a guess. There were several interesting complications that arose in implementing the attack:

1. The compression algorithm is a variant of a sliding-window scheme, in which it is not always guaranteed that the longest match in the window will be used to encode a string. More importantly, this is a two-pass algorithm; the encoding of strings within the sliding window is affected by later strings as well as earlier ones, and this can change the output length enough to change which next four digits appear to be the best match to S[Whi02].

2. Some guesses themselves compress very well. For example, the guess "0000" compresses quite well the first time it occurs.

3. The actual "signal" between two close guesses (e.g., "1234" and "1235") is very close, and is often swamped by the "noise" described above.

4. To make the attack work reasonably well, it is necessary to make each piece of the string guessed pretty long. For our implementation, five digits worked reasonably well.
5. Some backtracking is usually necessary, and the attack doesn't always yield a correct solution.
6. It turns out to also be helpful to add some padding at the end of the string, to keep the processing of the digits uniform.

All of these problems appear to be pretty easy to solve given more queries and more work. However, we dealt with them more directly by developing a different attack—one that requires only chosen-plaintext access, not adaptive chosen plaintext access.

6.2   A Chosen Input Attack

The adaptive chosen input attack seems so restrictive that it is hard to see how it might be extended to a simple chosen or known plaintext attack. However, we can use a related, but different approach, which gives us a straightforward chosen input attack.

The attack works in two phases:

1. Generate a list of all possible subsequences of the string S, and use the compression side-channel to put them in approximate order of likelihood of appearing in S.
2. Piece together the subsequences that fit end-to-end, and use this to reconstruct a set of likely candidate values for S.

The subsequences can be tested in the simplest sense by making queries of the form

$$\text{Query} = \text{Guess} + S$$

However, to avoid interaction between the guess and the first characters of S, it is useful to include some filler between them.

Experimental Results. We were able to implement this attack, with about a 70% success rate for pseudorandomly-generated strings S of 16 digits each, using the Python Zlib. The attack generates a list of the 20 top candidates for S, and we counted it as a success if any of those 20 candidates was S.

There were several tricks we discovered in implementing this attack:

1. In building the queries, it was helpful to generate padding strings before the guessed subsequence, between the guess and the string S, and after the string.
2. It was very helpful to generate several different padding strings, of different lengths, and to sum the lengths of the compressed strings resulting from several queries of the same guess. This tended to average out some of the "noise" of the compression algorithm.
3. There are pathological strings that cause the attack to fail. For example, the string "0000000123000000" will tend to end up with guesses that piece together instances of "00000".

7   Caveats and Countermeasures

The attacks described above make a number of simplifying assumptions. In this section, we will discuss some of those assumptions, and the implications for our attacks when the assumptions turn out to be false. We will also consider some possible countermeasures.

7.1   Obscuring the Compressor Input Size

The precise size of the input may be obscured in some cases. Naturally, some kind of information about relative compression ratios is necessary for the attack to work. However, approximate input information will often be good enough, as when the compression ratio is being used as the side channel. An approximate input size will lead to an approximate compression ratio, but for any reasonably large input, the difference between the approximate and exact compression ratios will be too small to make any difference for the attack.

One natural way for an attacker to learn approximate input size is for the process generating the input to the compressor to have either some known constant rate of generating input, or to have its operations be visible (e.g., because it must wait for the next input, which may be observed, before generating the next output).

7.2   Obscuring the Compressor Output Size

Some encryption modes may automatically pad the compresor output to the next full block for the block cipher being used. Others may append random padding to resist this or other attacks. For example, some standard ways of doing CBC-mode encryption include padding to the next full cipher block, and making the last byte of the padding represent the total number of bytes of padding used. This gives the attacker a function of the compressor output size, $\lceil (len+1)/blocksize \rceil \times blocksize$. These may slightly increase the amount of work done during our attacks, but don't really block any of the attacks.

A more elaborate countermeasure is possible. A system designer may decide to reduce the possible leakage through the compressor to one bit per message, as follows:

1. Decide on a compression ratio that is acceptable for nearly all messages, and is also attainable for nearly all messages.
2. Send the uncompressed version of any messages that don't attain the desired compression ratio.
3. Pad out the compressor output of messages that get at least the desired compression ratio, so that the message effectively gets the desired compression ratio.

This is an effective countermeasure against some of our attacks (for example, it makes it quite hard to determine which file type that compresses reasonably well has been sent), but it does so at the cost of losing some compression performance. Against our chosen-input attacks, this adds a moderate amount of difficulty, but doesn't provide a complete defense.

7.3   Obscuring the Compressor Internal State

It is possible to obscure the internal state of the compressor, in a number of simple ways, including initializing the compressor in a random state, or inserting occasional random blocks of text during the compression operation. In either case, this can cause problems with some of our attacks, because of the lack of precise information about the state of the compressor when an unknown string is being processed. General compression ratios are unlikely to be affected strongly by such countermeasures, however, so the general side channel remains open.

7.4   Preprocessing the Text

The text may be preprocessed in such a way that compression is affected in a somewhat unpredictable way. For example, it is easy to design a very weak stream cipher, which generates a keystream with extremely low Hamming weight. Applying this kind of stream cipher to the input before compression would degrade the compression slightly, in a way not known ahead of time by any attacker. By allowing the Hamming weight of the keystream to be tunable, we could get tunable degradation to the compression.

8   Conclusions

In this paper, we have described a side-channel in widely-used lossless compression algorithms, which permit an attacker to learn some information about the compressor input, based only on the size of the compressor output and whatever additional information about other parts of the input may be available.

We have discussed only a small subset of the available compression algorithms, and only one possible side channel (compression ratio). Some interesting directions for future research include:

1. Timing side-channels for compression algorithms.
2. Attacking other lossless compression algorithms, such as adaptive Huffman encoding, adaptive Markov coders, and Burrows-Wheeler block sorting (with move-to-front and Huffman or Shannon-Fano coding) with this side channel. Adaptive Huffman and Markov coders can be attacked using techniques very similar to the ones described above. Burrows-Wheeler block sorting appears to require rather different techniques, though the same side channels clearly exist and can be exploited.
3. Attacking lossy compression algorithms for image, sound, and other data with this side channel.

4. Attacking lossy image compression by trying to use disclosed parts of a compressed image to learn undisclosed parts of the same image, as might be useful for redacted scanned documents.
5. Reaction attacks against decompressors, such as might be useful when a system cryptographically authenticates plaintext, then compresses and encrypts it. This might lead either to software faults (a change in ciphertext leading to a buffer overrun, for example) or to more general leakage of information about the encryption algorithm or plaintext.

Acknowledgements. The author wishes to thank Paul Crowley, Niels Ferguson, Andrew Fernandes, Pieter Kasselman, Yoshi Kohno, Ulrich Kuehn, Susan Langford, Rene Struik, Ashok Vadekar, David Wagner, Doug Whiting, and the many other people who made helpful comments after seeing these results presented at Certicom, at the Crypto 2001 Rump Session, and at FSE2002. The author also wishes to thank the anonymous referees for several useful suggestions that improved the paper.

References

[BCL02]   Benedetto, Caglioti, and Loreto, *Physical Review Letters*, 28 January 2002.
[CCF01a]   Usenet group comp.compression FAQ file, available at http://www.faqs.org/faqs/compression-faq/, 2001.
[KJY00]   Kocher, Jaffe, Jun, "Differential power analysis: Leaking secrets," in *Advances in Cryptology – CRYPTO'99*, Springer-Verlag, 1999
[Koc96]   Kocher, "Timing Attack on Implementations of Diffie-Hellman, RSA, DSS and other systems," in *Advances in Cryptology – CRYPTO '96*, Springer-Verlag, 1996.
[KSHW00]   Kelsey, Schneier, Wagner, Hall, "Side Channel Cryptanalysis of Product Ciphers," in *Advances in Cryptology–ESORICS 96*, Springer-Verlag, 1996.
[Sal97a]   David Salomon, *Data Compression: The Complete Reference*, Springer-Verlag, 1997.
[Whi02]   Doug Whiting, personal communication, 2002.

Author Index

Lecture Notes in Computer Science

For information about Vols. 1–2322
please contact your bookseller or Springer-Verlag